Trial Designs and Outcomes in Dementia Therapeutic Research

'Dancing Inside' illustrates the type of outcome that means so much to families: the woman had stopped playing the piano some months before any other symptoms became apparent; she began to play again shortly after treatment, as the first indication to her family that she was becoming 'more like her old self'. Such clinically meaningful outcomes can go uncaptured by our current approach to measuring treatment efficacy. (A full description of the experience of artists recording Alzheimer's disease and its treatment can be found in "Lending a helping eye: artists in residence at a memory clinic. Lancet Neurol 2004;3:119–23.)

Trial Designs and Outcomes in Dementia Therapeutic Research

Edited by

Kenneth Rockwood, MD, MPA, FRCPC
Professor of Medicine (Geriatric Medicine & Neurology) &
Kathryn Allen Welden Professor of Alzheimer Research
Centre for Health Care of the Elderly
Dalhousie University
Halifax, Nova Scotia
Canada

Serge Gauthier MD FRCPC
Professor and Director
Alzheimer's Disease Research Unit
McGill Centre for Studies in Aging
Douglas Hospital
Verdun PQ
Canada

Taylor & Francis
Taylor & Francis Group

LONDON AND NEW YORK

© 2006 Taylor & Francis, an imprint of the Taylor & Francis Group

First published in the United Kingdom in 2006
by Taylor & Francis, an imprint of the Taylor & Francis Group, 2 Park Square,
Milton Park, Abingdon, Oxon OX14 4RN

Tel.:   +44 (0)20 7017 6000
Fax.:   +44 (0)20 7017 6699
E-mail: info.medicine@tandf.co.uk
Website: www.tandf.co.uk/medicine

A CIP record for this book is available from the British Library.

Library of Congress Cataloging-in-Publication Data
Data available on application

ISBN 1 84184 321 0
978 1 84184 321 6

Distributed in North and South America by
Taylor & Francis
2000 NW Corporate Blvd
Boca Raton, FL 33431, USA

Within Continental USA
Tel: 800 272 7737; Fax: 800 374 3401
Outside Continental USA
Tel: 561 994 0555; Fax: 561 361 6018
E-mail: orders@crcpress.com

Distributed in the rest of the world by
Thomson Publishing Services
Cheriton House
North Way
Andover, Hampshire SP10 5BE, UK
Tel: +44 (0)1264 332424
E-mail: salesorder.tandf@thomsonpublishingservices.co.uk

Composition by J&L Composition, Filey, North Yorkshire

Printed and bound in Great Britain by CPI Bath.

Contents

List of Contributors

Clive Ballard MMedSci MD MRCPsych
Professor of Age Related Disorders
Institute of Psychiatry
and
The Wolfson Centre for Age-Related Diseases
King's College London
University of London
London
UK

Frederik Barkhof MD
Department of Neurology / Alzheimer Center
Academisch Ziekenhuis
Vrije Universiteit
Amsterdam
The Netherlands

Sandra E Black MD FRCPC
Head, Division of Neurology
Director, Cognitive Neurology Unit and Stroke
Program
Sunnybrook and Women's College Health
Sciences Centre
Professor of Medicine (Neurology)
University of Toronto
Toronto
Canada

David M Blass MD
Department of Psychiatry and Behavioral
Sciences
Division of Geriatric Psychiatry and
Neuropsychiatry
Johns Hopkins University School of Medicine
Baltimore, MD
USA

Michael J Borrie MD FRCPC
Chair, Division of Geriatric Medicine
Department of Medicine
University of Western Ontario
London, ON
Canada

Henry Brodaty AO MBBS MD FRACP FRANZCP
Professor of Psychogeriatrics
School of Psychiatry
University of New South Wales
and
Director, Academic Department for Old Age
Psychiatry
Prince of Wales Hospital
Randwick NSW
Australia

Roger Bullock MD
Kingshill Research Centre
Victoria Hospital
Swindon
Wiltshire
UK

Richard Camicioli MD FRCPC
Department of Medicine
University of Alberta
Glenrose Rehabilitation Hospital
Edmonton, AB
Canada

**Georgina Charlesworth ClinPsyD MPhil
CPsychol AFBPsS**
Department of Psychiatry and Behavioural
Sciences
Royal Free and University College Medical
School
Hampstead
London
UK

John D Fisk PhD
Associate Professor
Departments of Psychiatry and Medicine
Dalhousie University
and
Psychologist, Queen Elizabeth II Health
Sciences Centre
Halifax, NS
Canada

Serge Gauthier MD FRCPC
Professor and Director
Alzheimer's Disease Research Unit
McGill Centre for Studies in Aging
Douglas Hospital
6825 Lazalle Blvd
Verdun, PQ
Canada

Gordon Gubitz MD FRCPC
Assistant Professor of Neurology
Division of Neurology
Department of Medicine
Dalhousie University
and
Queen Elizabeth II Health Sciences Centre
Halifax, NS
Canada

Deborah Gustafson PhD
Sahlgrenska University Hospital
Göteborg
Sweden

David B Hogan MD FACP FRCPC
Division of Geriatric Medicine
University of Calgary
Health Sciences Centre
Calgary, AB
Canada

Christine Joffres DrEd
Geriatric Medicine Research Unit
Queen Elizabeth II Health Sciences Centre
Halifax, NS
Canada

Chris MacKnight MD MSc FRCPC
President of the Canadian Geriatrics Society
and
Assistant Professor of Geriatric Medicine
Department of Medicine
Division of Geriatrics
Dalhousie University
Halifax, NS
Canada

Neelesh Nadkarni MD
Clinical and Research Fellow
Division of Neurology
Department of Medicine
Sunnybrook & Women's College Health Science
Centre
Toronto, ON
Canada

Stanton Newman DPhil
Unit of Health Psychology
Department of Psychiatry and Behavioural
Sciences
Royal Free and University College Medical
School
Hampstead
London
UK

Peter V Rabins MD
Director, Division of Geriatric Psychiatry and
Neuropsychiatry
Department of Psychiatry and Behavioral
Sciences
School of Medicine
Johns Hopkins Hospital
Baltimore, MD
USA

Kenneth Rockwood MD MPA FRCPC
Professor of Medicine
Division of Geriatric Medicine
Department of Medicine
and
Kathryn Allen Weldon Professor of Alzheimer
Research
Dalhousie University
Halifax, NS
Canada

Philip Scheltens MD PhD
Department of Neurology/Alzheimer Center
Academisch Ziekenhuis
Vrije Universiteit
Amsterdam
The Netherlands

Ingmar Skoog MD PhD
Professor
Dept of Psychiatry
Sahlgrenska University Hospital
Göteborg
Sweden

Mathew Smith BSc
Coordinator, Geriatric Clinical Trials Group
Associate Scientist, Lawson Health Research
Institute
Parkwood Hospital
St. Joseph's Health Care
London, ON
Canada
Department of Medicine
University of Western Ontario
London, ON
Canada

David L Streiner PhD CPsych
Baycrest Centre for Geriatric Care
Toronto, ON
Canada

Claire Thompson
Academic Department for Old Age Psychiatry
Euroa Centre
Prince of Wales Hospital
Randwick NSW
Australia

Sarah Voss PhD
Research Psychologist
Kingshill Research Centre
Victoria Hospital
Swindon
Wiltshire
UK

Peter J Whitehouse MD PhD
Director, Integrative Studies
Center Investigator
University Memory and Aging Center
Department of Neurology
Case Western Reserve University
Cleveland, OH
USA

Gordon K Wilcock DM FRCP
Professor of Care of the Elderly
Bristol University
Department of Care of the Elderly
Frenchay Hospital
Bristol
UK

Anders Wimo MD PhD
Department of Clinical Neuroscience and
Family Medicine
Division of Geriatric Medicine
Karolinska Institutet
Stockholm and Department of Family Medicine
Umeå University
Umeå
Sweden

Bengt Winblad MD PhD
Profesor of Geriatric Medicine
Karolinska Institutet
Neurotec
Huddinge University Hospital
Stockholm
Sweden

Preface

From first steps in the early 1960s[1] to a controversial if landmark paper on tacrine[2,3] to the most recent trials, it is clear both that much progress has been made, and that much remains to be done. This book is written to take stock of what is now usefully known, and to speculate on directions for the future. We have invited leading and thoughtful commentators, who have experience with dementia trials, to critically reflect on how we now undertake the testing of medications and other interventions.

A central feature of the current scene is that most observers agree that some patients do remarkably well for extended periods of time on treatment. After this, there is deep controversy. What proportion of patients do well? What do we mean by 'doing well'? How does the perspective on 'doing well' compare between expert and non-expert observers? Does 'doing well' mean that the goals of patients and their carers are met? Does 'doing well' after three months of treatment predict 'doing well' after six, nine or twelve months? If not, how can we differentiate these effects from random fluctuation and regression to the mean? Is stabilisation a treatment effect? Over which time period? Do our current measures capture treatment effects? Are these effects worth paying for?

That successful treatment exists short of a cure is a challenge to how we think of dementia, how we accord a voice to those with dementia, and to those who care for them, and how we design treatments and adjuncts to treatment. This book is meant to bring these challenges to our readers, and to engage them on many levels. We envisage our audience to include not just the academic, industrial and governmental participants in the work-a-day world of dementia trials, but others with an interest in dementia and in clinical trials. We particularly hope that it might be useful to anyone with an interest in trials for chronic conditions in which cure might not be likely, but helpful treatment remains an important goal. An important lesson in the dementia trials is that pride in their accomplishments must be balanced by humility about how much there is to know. Included in what we need to know must be a reappraisal of what we think we know, and a vigilant and rigorous evaluation both of what has worked and what has not.

An important challenge for all our contributors has been to consider whether the information that they have reviewed is likely to be clinically meaningful to treating physicians, patients and their carers. This is a tall order. At present, there are no widely agreed upon guidelines for clinical meaningfulness, although a proposal for how this might be interpreted in the setting of anti-dementia trials is outlined in Chapter 1. In addition, incorporating the perspective of patients and carers can be difficult, as the views can be at odds with each other, in ways that challenge the primacy of the physician-patient relationship.

This is not a book that attempts to account for all the evidence or to summarise the data according to standard guidelines that has been done admirably well in a recent massive textbook.[4] [Rather, our focus is on a critical consideration of what we can take from the trials to date that might inform their interpretation in the context of clinical meaningfulness, and in the expectation that future trials can be made better. Still, we have authors to consider some important methodological issues, including publication bias and drop-out bias. We have also asked them to consider how best to capture the multiplicity of outcomes seen in clinical practice, which remains an important challenge to specifying the model of treatment in dementia. and to address the generalisability.

As physicians who engage in anti-dementia trials, and as co-editors, we are keen to know of ways in which the trials or this book might be improved. We therefore invite comments at our respective e-mail addresses. We hope that this book can contribute to the ongoing international effort to improve the quality of life for persons with dementia; to those people, and to those who take part in the efforts to improve their lives, we dedicate this book.

Kenneth Rockwood
kenneth.rockwood@dal.ca
Serge Gauthier
serge.gauthier@mcgill.ca
May 2005

References

1. Walsh AC. Anticoagulant therapy as a potentially effective method for the prevention of pre-senile dementia: two case reports. J Am Geriatr Soc 1968; 16: 472–81.

2. Summers WK, Majovski LV, Marsh GM, et al. Oral tetrahydroaminoacridine in long-term treatment of senile dementia, Alzheimer type. N Engl J Med. 1986; 315:1241–5.

3. Editor. Tacrine as a treatment for Alzheimer's disease: editor's note. An interim report from the FDA. A response from Summers et al. N Engl J Med 1991; 31; 324:349–52.

4. Qizilbash N, Schneider LS, Chui H, et al. Evidence-based Dementia Practice. Oxford: Blackwell, 2002.

1

Introduction: Expectations of Treatment with the Cholinesterase Inhibition Strategy

Kenneth Rockwood

Introduction

Cholinesterase inhibition is a well-established strategy for the treatment of Alzheimer's disease, and for some other forms of dementia. Across studies, sites, trials and even compounds, the strategy of cholinesterase inhibition produces dose-response effects that favor treatment.[1] These effects are generally large enough to be clinically detectable. At least in economic modelling studies, these impacts appear to be cost-effective,[2–4] and by some accounts might even be cost saving.[5]

Despite these successes, questions persist and there is much progress to be made. As treatment strategies begin to add on to, or even move away from the strategy of cholinesterase inhibition,[6] it is appropriate to take stock about what we have learned. This chapter focuses on issues of particular importance in that regard, and spans many of the other more focused considerations detailed in this volume. Clinical meaningfulness is an important consideration in many of the controversies to date, and is perhaps best understood in the context of the goals of cholinesterase inhibition. After these issues are discussed, we will briefly consider their implications for trial design and measurement, and for the purposes for which clinical trials in dementia are undertaken.

The goal of cholinesterase inhibition in dementia

On the face of it, taking time to spell out the goal of cholinesterase inhibition in dementia might seem banal: surely it is to improve the lives of dementia patients and their carers? But the issue is more complex. If, as seems the usual experience, even successful treatment falls short of cure, what is the model of treatment success? How do we deal with the case of a patient who has better memory only at the expense of worse anxiety? Will longer-term treatment postpone institutionalisation, prolong it, or compress it?

How do we recognise treatment success that falls short of cure?

At this point in the history of cholinesterase inhibition in Alzheimer's disease, it is easy to recognise that some patients enjoy treatment benefits, even though cure is lacking. Treatment benefits are recognised in several ways. Perhaps the least equivocal (and perhaps the least sensitive[7]) are patients identified clinically as having improved in blinded, controlled trials. In general, between one in seven to one in three patients in trials have been identified as showing at least some level of improvement,[8] according to the Clinician's

Interview-Based Impression of Charge, Plus Caregiver Input (CIBIC-Plus).[9] These rates are generally twice as high in patients on active treatment compared with those on placebo, although both groups can inform what we mean by success.

This point has been considered in some detail elsewhere.[10] Briefly, close analyses of how clinicians judge success is revealing. Comparatively few patients show success across the board. Rather, some symptoms improve, others stabilise and others worsen. Within this, the types of symptoms that improve or worsen seem to be particularly informative in clinicians' decision making. For example, we can consider the broad levels of cognition, function and behaviour. Clinicians who are faced with an account of patients in whom function has improved are inclined to rate the patients as having improved, whereas accounts of worse behaviour motivate them to rate patients as being worse.[11,12] By contrast, changes in cognition (either improvement or deterioration) tend to be traded off against other areas.

Those judgements, of course, reflect clinicians' understanding of how patients are better. Patients and their carers have their own views, and they do not always coincide with those of the clinicians. In the Atlantic Canada Alzheimer's Disease Investigation of Expectations (ACADIE) study, patients and carers were asked separately from their treatment physicians about the goals of treatment.[13] In general, patients and carers identified many more goals of treatment than did physicians, but they particularly identified more goals in the areas of social interaction and leisure activities.

The importance of metaphor in defining treatment success

The problem of knowing what constitutes treatment success is an important one (see Table 1.1). The philosopher Evelyn Fox Keller has argued that especially at the cutting edge of inquiry, when what is being sought after is often not readily described *a priori*, the situation is like 'looking for what you do not know is there'.[14] In such situations, metaphor often plays a crucial role in guiding inquiry.

The importance of metaphor in understanding Alzheimer's disease treatment has been considered in detail elsewhere.[15] Briefly, the cholinesterase inhibition strategy appears to have gone through distinct stages. Early on, biological implausibility notwithstanding, there appears to have been some expectation that dramatic improvement might be possible.[16] Although that expectation soon faded, the metaphor as improvement persisted, with improvement largely being understood as the reversal of decline. In other words, things were expected to get better in the reverse order of how they got worse. This metaphor soon gave way to the metaphor of slowing disease decline. Experimental design (such as randomised blinded delayed start, or randomised blinded withdrawal) has been advocated to reveal such slowing.[17]

These metaphors have been criticised as not reflecting the common experience of variable treatment effects (i.e. some symptoms improving, others stabilising, others worsening). Instead of the metaphors of reversal, or stabilisation, it was argued that the metaphor of novel treatment states, not well captured by traditional natural history staging measures, should be considered.[15] This argument was put in the context of identifying the nature of self in dementia. If reversal is the recovery of the old self, and stabilisation is the maintenance of a new, degraded self, then a new

Table 1.1 Challenges in defining treatment success
• Improvement is generally not across the board; treatment success can occur short of cure. • Within domains of treatment, some effects (i.e. improved function, worse behaviour) have more impact than others. • The priorities of clinicians are not necessarily the priorities of patients and their carers.

metaphor might be the reconstruction of a new self. It is in this light that the case of the newly remembering but again anxious patient might be considered, as discussed below. For now, we shall use the achievement of the new self to consider whether such an effect would be clinically important.

What makes a treatment effect clinically important?

There are now more than a dozen pivotal clinical trials of cholinesterase inhibitors in Alzheimer's disease that show statistically significant changes in their primary outcome measures. Still, influential commentators worry about the clinical meaningfulness of these changes.[18–22] Some commentators (and I have joined these calls) argue that employing outcome measures that have more evident clinical meaningfulness than those now in use would help clarify whether the changes observed in standard psychometric test are clinically important.[10,23] In the meantime, however, there is much to learn from the results of existing studies.

A proposal for assessing clinical meaningfulness from published data has been detailed elsewhere.[24] Its essential features are outlined in Table 1.2. These features are modified from the Bradford Hill criteria for determining whether a correlation is causal.[25] The first requirement is *reproducibility*, particularly in the setting of a randomised clinical trial. Reproducibility remains an essential standard in science, and requires little elaboration here. In this context, reproducibility also means that the effects observed in a clinical trial, if judged to be clinically important on other grounds, are almost more likely to be seen in clinical practice.

A second feature likely to support an effect being clinically meaningful is if it shows a *dose response*. The dose response helps grade the clinical sense of what is going on. Larger effect sizes at higher doses, to some upper limit of the therapeutic window (where side-effects supravene) give some sense that clinical effects are detectable.

Table 1.2 Features that suggest a statistically significant difference is clinically important
• Reproducibility
• Dose response
• Sensibility of the measures
• Convergence of the measures
• Effect size >0.20
• Biological plausibility

A third feature of a statistically significant treatment effect to be evaluated is whether the measure in which this effect was demonstrated is clinically *sensible*. 'Sensibility' is a term introduced by Feinstein to aid in the assessment of a clinical measure.[26] It incorporates the classical concept of face validity, by which is meant whether a measure, at face value, appears to be plausible.[27] For example, a thermometer would not, on its own, be a valid measure of Alzheimer's disease treatment effects. Nor would a set of weighing scales be a plausible instrument, even if they were reliable and very sensitive to change. Even though it might be argued that as weight often decreases with dementia progression, so weight gain might be seen as reversing the disease process, most clinicians would argue that weight loss is not a central complaint in dementia, and it would be easy to imagine weight gain without applicable changes in cognition.

Sensibility can be a tricky criterion to apply, however, to a disease or treatment state in which it can be hard to define improvement *a priori*. For example, our group has observed that an important deficit in patients with Alzheimer's disease is that they lose the ability to imagine themselves in the future as being competent agents, i.e. as being people who can effect their own intentions.[28] Thus we are developing a measure that asks patients to project into the future, and imagine various future contingencies. To our group, based on our analyses, these tests appear to make a great deal of sense, but we have witnessed, to be polite about it, some scepticism in our colleagues. Thus notions of what makes sense as a measure will be subjective, and will change over time. In this sense, sensibility is a form of prejudice, but it can still be an important adjunct to clinical judgement, if evaluated carefully, and if stated explicitly at the onset.

Another aspect of sensibility is whether the test is familiar. Thus, for example, the Mini-Mental State Examination (MMSE),[29] though used less often in clinical trials than the Alzheimer's Disease Assessment Scale-Cognitive Subscale (ADAS-Cog),[30] has some potential for an important role in understanding clinical meaningfulness. It can be criticised on many grounds, including, by its lack of sensibility (i.e. it is heavily language dependent, and lacks important items that test executive function). By contrast, the ADAS-Cog is a more comprehensive measure, and is commonly more sensitive to change than is the MMSE.[31] The MMSE, however, is much more widely used than the ADAS-Cog. In consequence, many clinicians, including many with only a secondary interest in dementia, will have a deep experience with the MMSE, whereas many fewer will have any familiarity with the ADAS-Cog. Importantly for our purposes, these clinicians will have learned to calibrate their own clinical judgement in respect of the MMSE, as evidenced in statements such as 'he's an 18/30, but he's a very good 18/30', or 'she's a 24/30, but actually she's quite impaired'. Thus, knowing that actively treated patients had, on average, a two-point less decline over 6 months than placebo-treated patients tells something to most clinicians. It probably tells them something different compared with hearing the algebraically equivalent finding that actively treated patients improved 1.5 points on average, whereas placebo-treated patients declined 0.5 points. (These considerations, by the way, make me sceptical about the analyses of the 'minimal clinically important difference' (MCID) approach to assessing clinical meaningfulness.[32]) In this approach, clinicians are surveyed to ask what difference they would treat as clinically important. Using a variety of techniques, their responses are calibrated to the MCID. This can be helpful, but in my experience is not per-

suasive to clinicians, who in response to the question 'what is the minimal difference in the MMSE that you would judge to be clinically important?' are most likely to answer 'it depends'.

Convergence of the measures refers to more than one measure pointing to a beneficial outcome. Clinicians will appropriately be sceptical if only one measure, even if it is the primary outcome, points to a positive result when all the others are negative. In this sense, convergence of the measures is an aspect of the well-known criterion of convergent construct validity.[27] According to this view, most important constructs are, in fact, latent, and in this sense not provable, so that the best that can be done is to amplify and clarify them, a process best done by comparison with what is known. For example, even anaemia, which has a precise laboratory definition, will have a range of clinical presentations. On an individual basis, varying haemoglobin levels will be associated with disease, as will varying clinical contexts. Thus the problem in a given patient is that of instantiating the general principle of disease in that patient's individual circumstances. This can be clear-cut, making judgement irrelevant or at best banal. In circumstances where judgement is not so straightforward, however, then accumulation of information from varying measures helps clinicians gain a sense of whether the effect is likely to be generally important.

On the other hand, convergence of the measures can be a snare. If measures are highly convergent (say correlations >0.80), then multiple testing is redundant. Perhaps the ideal is somewhat more than nominal correlation (arguably, about 0.30[27]). In such a circumstance, it is evident that, while construct convergent validation is being maintained, the various tests being employed are tapping different aspects of the larger construct of a clinically important treatment effect.

Convergence of the measures can also be a snare even when it is clear that different aspects are being tapped. Consider why different domains are being tested: multiple measures are commonly motivated by the desire to not miss important treatment effects. The result, perversely, can be to dilute the signal of important treatment effects in the noise of multiple measures. Consider, for example, that many patients with Alzheimer's disease have some improvement in their behaviour as a consequence of cholinesterase inhibition. Thus, according to the current means of drug evaluation, it becomes reasonable to add a behavioural measure. Even a short behavioural measure adds about a dozen items to be tested. Of course not every patient will have the same problems; many will have only one or two, and many will have none at all. For a patient whose one behavioural problem is completely improved by a treatment intervention, the effect of multiple domain testing can be simply to dilute this important effect in a sea of 'no change' in items that were not problems to begin with.

Various remedies have been proposed, including methods for summarising multiple test results and individualised outcome measurement.[13,33–35] The latter has been the preferred strategy of our group, and is detailed in Chapter 7. Still, neither method has been widely adopted, so that multiple measurements, presumably in an effort to demonstrate convergence of measures across a range of effects, remains the dominant strategy now employed.

Treatment effects large enough to be clinically detectable

One issue that clouds the enthusiasm for widespread uptake of the ChE inhibitor trials data has been the apparently small numerical size of the treatment effects. Many commentators express scepticism about 'only' a 2–3-point change on, say, the ADAS-Cog, noting that the scale has 70 points. Scepticism about the size of a treatment effect based on the magnitude of the scale is, however, unfounded. For example, a temperature increase to 40°C is only a 3-point increase from normal, on a scale that could measure 100 points or more. Clearly, what is important is not the theoretical range of the measuring device, but rather the experienced range of usual biological variation.

There are quantitative means of assessing whether a given numerical difference is likely to be clinically meaningful, including a calculation known as the effect size. This term has a confusing usage, and is often used simply to refer to the difference in a given measure between the treatment and placebo group at the end of the trial. Again, the problem with the absolute difference is how to contextualise this. Broadly speaking, it is a measure of the 'signal' without a measure of the 'noise'. A more precise method therefore is to calculate what is sometimes referred to as a 'standardised effect size', which incorporates both signal and noise terms. The 'signal' term comes from the differences between the two groups. To be precise, it comes from the difference between the final score and the baseline score in the treatment group compared with the difference between the final score and the baseline score in the placebo group. (Taking the difference between the differences in effect adjusts for small differences between the groups in the scores at baseline.) The 'noise' term is classically taken to be the pooled standard deviation of the test score at baseline. By notation, this measure, also known as Cohen's d is usually represented as:

$$\frac{X_t - X_c}{S_{tc}}$$

where X_t = the difference between the final score and the baseline score in the treatment group, X_c = the difference between the final score and the baseline score in the comparison group, and S_{tc} = the pooled standard deviation of the test score at baseline. This approach has many potential advantages. For example, it allows comparisons across studies, and thus is at the heart of meta-analyses.[36] It also allows comparisons between and across measures.[31] Importantly, it allows for some understanding of the relative magnitude of change, through analogy. For example, as summarised in a comprehensive textbook by Cohen, effect sizes can be grouped as small, medium or large.[37] Considering small effect sizes to be in the range of 0.20–0.40, Cohen points out that this would amount to the average difference in height between 14-year-old girls and 15-year-old girls. By contrast, an effect size of 0.50–0.70 would be moderately large. This amounts to the difference in height between 14-year-old girls and 18-year-old girls.

Several useful points can validly be inferred about the ratio of signal to noise represented by the effect sizes. Small effect sizes are at the edge of clinical detection. They are common in fields in which experimentation is new. They are likely to require either considerable clinical judgement, or special instrumentation, or both. In our example, paediatricians, or for that matter, parents of large families, would be expected to more readily be able to validly assign individual girls to the 14-year-old group or the 15-year-old group. More casual, or less experienced observers would likely require at least small groups of girls before they would be able to assign between the 14-year-old and the 15-year-old girls. Biological measures (e.g. radiographic bone age) would improve the ability of non-experts to assign between groups, but even these would have a considerable degree of judgement in their interpretation in the individual case, especially when rates of change over comparatively short periods of time were to be the essential feature. By contrast, in the case of the moderately large effect sizes, it is easier to imagine that clinical judgements will not be so highly constrained. Of course, from time to time, there will be error: 14-year-old basketball players are more likely to be misassigned, the play of chance can never be eliminated.

There are important variations of the effect size, and potentially subtle differences in interpretation. With respect to the noise term, either the standard deviation of the control group, the pooled standard deviation, or the standard deviation of the change score have been proposed. (The latter has been referred to as a 'standardised response mean' and can also be used as an index of a measure's sensitivity to change.[31,38]) Different circumstances might dictate that one or the other of these terms be used. For example, it has been argued that using baseline change scores can artificially inflate effect sizes in the clinical trials setting. Patients who are enrolled in clinical trials represent a highly selected population, and thus are artificially homogeneous. The result is that the baseline variability is artificially low, so that a given change score is now being divided by a smaller denominator, resulting in a higher ratio. By this line of argument, the standardised response mean is to be preferred, as differential response to treatment will result in higher variability, a larger denominator for any given numerator, and thus a more conservative (i.e. smaller) result. In our experience, however, such differences are more apparent than real, and we have generally found the effect size estimates to be reasonably stable.[31,39]

Implications and unresolved issues

If we accept that there is much more to be understood as we get to grips with clinical meaningfulness in dementia drug trials, there are implications for how future studies might proceed. The first implication is that the question of clinical meaningfulness should be addressed in the course of the study.[40] At present, to the extent that clinical meaningfulness is an object for study, it is felt to emerge from using psychometric tests in a variety of domains. By contrast, clinical meaningfulness studied as such would put more reliance on clinimetric (judgement-based) measures. Today, although it is a primary outcome measure, the CIBIC-Plus does not allow the basis for individual judgements about clinical meaningfulness to be aggregated. This can be accomplished reasonably readily, and elsewhere details of how this might be done have been spelled out.[10] Briefly, we have advocated standardisation of how information is recorded, rather than standardisation of what is recorded, as well as widespread, systematic, qualitative studies. The point here is made that we should specify how information is recorded (e.g. by eliciting patient preferences, by quantifying descriptions, by making clear what behaviours are changes) rather than what is recorded (e.g. two questions about visuo-spatial function, one about orientation) because there is special merit to retaining the value of the CIBIC-Plus as an unspecified measure. In brief, the argument is that, at this point, specification would be premature: we are still not sure what it is that we are looking for. Still, we can take advantage of what is known, and build on it. For example, there is a different rate of decline according to the stage of the disease. Most authorities accept that the declines in conventional testing are steepest in the middle stages[41,42] even though reports continue to extrapolate treatment effects based on linear models of decline.[43] Whether this holds with individualised measures needs specific investigation. In addition, the use of historical controls is not a feasible option for individualised measures, in which the context of treatment is particularly important. In fact, this consideration might also be well obtained with standardised measures. As Thal has pointed out, there is a two-fold difference in the rate of decline in the ADAS-Cog in two, older, one-year studies.[44] The impact of treatment context (and, conceivably) expectations might not be limited to individualised measures, but could extend to supposedly objective ones as well.

In this context, we believe that recording patient preferences needs to be highlighted as a key to understanding clinical meaningfulness. Just as we argue that the goal of the trials is not to show improvement on the ADAS-Cog, neither is it to show improvement on the CIBIC-Plus. If these treatments are to be clinically meaningful, they must be worthwhile to our patients and their families. They are to be trusted in this regard.

Finally, and importantly, we also need to realise that we do not know exactly what to look for because we do not yet understand well enough the role of cholinergic neurotransmission in higher cortical functions. Critically, we cannot rely on animal experimentation to understand how people plan, or the way in which insight is related to emo-

tion. The ChE trials need to be embraced as a means of gaining insight into these most human of issues.

References

1. Rockwood K. Clinical meaningfulness in relation to the size of the treatment effect of cholinesterase inhibition on cognition in Alzheimer's disease. J Neurol Neurosurg Psychiatry 2004; 75:677–85.

2. Hux MJ, O'Brien BJ, Iskedjian M et al. Relation between severity of Alzheimer's disease and costs of caring. CMAJ 1998; 159:457–65.

3. O'Brien BJ, Goeree R, Hux M et al. Economic evaluation of donepezil for the treatment of Alzheimer's disease in Canada. Am Geriatr Soc 1999; 47:570–8.

4. Getsios D, Caro JJ, Caro G et al. Assessment of health economics in Alzheimer's disease (AHEAD): galantamine treatment in Canada. Neurology 2001; 57(6):972–8.

5. Hill W, Futterman S, Duttagupta V et al. Alzheimer's disease and related dementias increase costs of comorbidities in managed medicare. Neurology 2002; 58:62–70.

6. Reisberg B, Doody R, Stoffler A et al. Memantine in moderate-to-severe Alzheimer's disease. N Engl J Med 2003; 348:1333–41.

7. Quinn J, Moore M, Benson DF et al. A videotaped CIBIC for dementia patients: validity and reliability in a simulated clinical trial. Neurology 2002; 58:433–7.

8. Livingston G, Katona C. How useful are cholinesterase inhibitors in the treatment of Alzheimer's disease? A number needed to treat analysis. Int J Geriatr Psychiatry 2000; 15:203–7.

9. Schneider LS, Olin JT, Doody RS et al. Validity and reliability of the Alzheimer's Disease Cooperative Study-Clinical Global Impression of Change. The Alzheimer's Disease Cooperative Study. Alzheimer Dis Assoc Disord 1997; 11(Suppl 2):S22–32.

10. Rockwood K, Joffres C. Improving clinical descriptions to understand the effects of dementia treatment: consensus recommendations. Int J Geriatr Psychiatry 2002; 17:1006–11.

11. Joffres C, Graham J, Rockwood K. A Qualitative Analysis of the Clinician Interview-Based

Impression of Change (Plus): methodological issues and implications for clinical research. Int Psychogeriatr 2000; 12:403–13.

12. Joffres C, Bucks RS, Haworth J et al. Patterns of clinically detectable treatment effects with galantamine: a qualitative analysis. Dement Geriatr Cogn Disord 2003; 15:26–33.

13. Rockwood K, Graham JE, Fay S. Goal setting and attainment in Alzheimer's disease patients treated with donepezil. J Neurol Neurosurg Psychiatry 2002; 73:500–7.

14. Keller EF. Making Sense of Life: Explaining Biological Development with Models, Metaphors, and Machines. Cambridge: Harvard University Press; 2002.

15. Rockwood K, Wallack M, Tallis R. Treating Alzheimer's disease: understanding success short of cure. Lancet Neurol 2003; 2(10):630–3.

16. Summers WK, Majovski LV, March GM et al. Oral tetrahydroaminoacridine in long-term treatment of senile dementia, Alzheimer type. N Engl J Med 1986; 315:1241–5.

17. Leber P. Slowing the progression of Alzheimer disease: methodologic issues. Alzheimer Dis Assoc Disord 1997; 11(Suppl 5):S10–21.

18. Mayeux R, Sano M. Treatment of Alzheimer's disease. N Engl J Med 1999; 341:1670–9.

19. Birks J, Grimley E, Lakovidou V et al. Rivastigmine for Alzheimer's disease (Cochrane Review). In: The Cochrane Library, Issue 2, 2001. Oxford: Update Software.

20. Birks JS, Melzer D, Beppu H. Donepezil for mild and moderate Alzheimer's disease. Cochrane Database Syst Rev 2000; 4:CD001190.

21. Olin J, Schneider L. Galantamine for Alzheimer's disease. Cochrane Database Syst Rev 2002; 3:CD001747.

22. Clegg A, Bryant J, Nicholson T et al. Clinical and cost-effectiveness of donepezil, rivastigmine, and galantamine for Alzheimer's disease. A systematic review. Int J Technol Assess Health Care 2002; 18:497–507.

23. Winblad B, Brodaty H, Gauthier S et al. Pharmacotherapy of Alzheimer's disease: is there a need to redefine treatment success? Int J Geriatr Psychiatry 2001; 16:653–66.

24. Rockwood K, MacKnight C. Assessing the clinical importance of statistically significant improvement in anti-dementia drug trials. Neuroepidemiology 2001; 20:51–6.

25. Hill AB. A Short Textbook of Medical Statistics. London: Hodder and Stoughton; 1984.

26. Feinstein AR. Clinimetrics. New Haven: Yale University Press; 1987.

27. Streiner D, Norman G. Health Measurement Scales (3rd edn). Oxford: Oxford University Press; 2002.

28. Rockwood K, MacKnight C, Wentzel C et al. The diagnosis of 'mixed' dementia in the Consortium for the Investigation of Vascular Impairment of Congition (CIVIC). Ann NY Acad Sci 2000; 903:522–28.

29. Folstein MF, Folstein SE, McHugh PR. 'Mini-mental state'. A practical method for grading the cognitive state of patients for the clinician. J Psychiatr Res 1975; 12:189–98.

30. Rosen WG, Mohs RC, Davis KL. A new rating scale for Alzheimer's disease. Am J Psychiatry 1984; 141:1356–64.

31. Rockwood K, Stolee P. Responsiveness of outcome measures in an anti-dementia drug trial. Alzheimer Dis Assoc Disord 2000; 14(3):182–5.

32. Burback D, Molnar FJ, St John P et al. Key methodological features of randomized controlled trials of Alzheimer's disease therapy. Minimal clinically important difference, sample size and trial duration. Dement Geriatr Cogn Disord 1999; 10:534–40.

33. O'Brien, PC. Procedures for comparing samples with multiple endpoints. Biometrics 1984; 40:1079–87.

34. Rockwood K, Stolee P, Howard K et al. Use of goal attainment scaling to measure treatment effects in an anti-dementia drug trial. Neuroepidemiology 1996; 15:330–8.

35. Bogardus ST, Bradley EH, Williams CS et al. Goals for the care of frail older adults: do caregivers and clinicians agree? Am J Med 2001; 110:97–102.

36. Light RJ, Pillemer DB. Summing Up: the Science of Reviewing Research. Cambridge, MA: Harvard University Press; 1984.

37. Cohen J. Statistical Power Analysis for the Behavioral Sciences (2nd edn). New Jersey: Erlbaum; 1988.

38. Wright JG, Young NL. A comparison of different indices of responsiveness. J Clin Epidemiol 1997; 50:239–46.

39. Rockwood K, Howlett SE, Stadnyk K et al. Responsiveness of goal attainment scaling in a randomized controlled trial of comprehensive geriatric assessment. J Clin Epidemiol 2003; 56:736–43.

40. Burns, A. Meaningful treatment outcomes in Alzheimer's disease. J Neurol Neurosurg Psychiatry 2002; 73:471.

41. Jacobs D, Sano M, Marder K et al. Age at onset of Alzheimer's disease: relation to pattern of cognitive dysfunction and rate of decline. Neurology 1994; 44:1215–20.

42. Mitnitski A, Graham J, Rockwood K. Modeling decline in Alzheimer's disease. Int Psychogeriatr 1999; 11:211–6.

43. Raskind M, Peskind ER, Truyen L et al. The cognitive benefits of galantamine are sustained for at least 36 months: a long-term extension trial. Arch Neurol 2004; 61:252–6.

44. Thal LJ. How to define treatment success using cholinesterase inhibitors. Int J Geriatr Psychiatry 2002; 17:388–90.

2

The History of Therapeutic Trials in Dementia

Peter J Whitehouse

Introduction

The history of therapeutic trials in dementia is not a long one, but it is already full of lessons to guide our current and future efforts. We are still struggling with issues such as 'what is dementia and how should we conceptualise and talk about the possibilities of effective therapy?'. At the turn of the last century, brain psychiatrists, such as Alois Alzheimer, started identifying the biological substrates of progressive cognitive deterioration and testing agents to improve their patients.[1,2] Our nosology of dementia and conceptions of therapeutic trials are continuing to evolve today, but the roots of our work emerged a century ago.[3,4]

The history of therapeutic trials could be seen as a simple chronology of events; starting for example, with the development of treatment for syphilis and ending, for the moment, with the termination of amyloid vaccine trials and the approval by regulatory bodies of a new drug, memantine. However, a simple enumeration of dates and events is not an adequate interpretation of that history. The elaboration of a historical perspective is always guided by present conceptions and future expectations. Much of medical history writing is celebratory. Its purpose is often to note the pioneers and at the same time to gently chide them for their lack of conceptual and methodological sophistication.

This kind of history is often dominated by recounting the development of new technologies and a simplistic conceptual model of scientific progress extending into a limitless future.

Our biological efforts to treat dementias are grounded in the technologies that we have developed to study their pathologies. Brain psychiatrists developed methods of examining autopsy material based on the use of industrial dyes to stain tissues. They developed clinical methods for categorising the diverse set of illnesses that they saw in their clinics and asylums. Models of pathophysiology, such as imbalance in blood flow to different regions of the brain, foreshadowed modern conceptions of disease classification and treatment approaches.[5,6]

A richer historical analysis is necessary, however. As a humanistic discipline, history need not justify itself other than by its fascination with the past activities of men and women. However, it can be useful in understanding the present and creating the future. Just as with anthropology, history can give us a perspective on the range of possible human behaviours and ways of thinking about the world. Frequently one can find the same fundamental conceptual confusions troubling our predecessors in our own contemporary work even though efforts are made to hide the recurring dilemmas by the mask of the inevitability of progress. For example it became politically useful to claim that Alzheimer's disease (AD) is not normal aging.[7] Yet from the beginning of the descriptions of the dementias, fundamental ambiguities about the relationship between aging and degenerative processes were realised. If we are honest, we recognise that these puzzles remain with us today, especially if we broaden our thinking from the biomedical model. In fact, we are likely to hear more 'disease-cure' language as lay organisations and pharmaceutical companies drive to spending more money of these conditions.

History can challenge the heart of what we consider scientific progress. It exposes the often claimed 'value neutrality' of science. Historical analysis can reveal the cultural and political factors that affect what scientists and clinicians consider their purview and how they interpret their data. Much as scientists claim that their data demonstrates fundamental regularities in nature, it does not speak for itself, but requires human interpretation. The history of science, particularly as applied to medicine, is a history of people and ideas often in conflict with each other. Perspectives from the past can enrich our understanding of the ongoing dialogue between science and society. To attempt to neutralise the human element by describing an idea as a hypothesis (for example, the cholinergic or amyloid hypothesis of the pathophysiology of AD) fails to recognise that much of the discussion that follows concerning such ideas relates to academic fashion, personality, power and money. At the fringe of such discourse also lies the powerful issue of scientism. When does our faith in the scientific method and conceptions of progress verge into a religious devotion rather than intellectual activity? How can medicine be energised by the power of science without having its art and soul damaged?

A review of some guiding principles about what disease is and how it is treated begins this chapter. Our attention then turns to the category of dementia and to what we mean by a therapeutic trial. After this general introduction, we examine specific historical episodes in the treatment of dementia including therapy for syphilis, the concept of metabolic enhancers, the approval of tacrine, the development of vitamin E as an intervention to potentially slow progression of disease, and finally the trials and tribulations of the amyloid vaccine. Next, we compare the development of neuropharmacology to treat dementia with psychopharmacology to treat anxiety and depression. Finally, we draw conclusions from this historical analysis relevant to ongoing dilemmas in the development of more effective therapies for dementia. These topics are necessarily rather selective and focused principally on AD. Other interesting histories could be written about the treatment of vascular causes of dementia, for example. Finding coherent language to describe our efforts to develop therapies has been rightfully pointed out by Jason Karlawish as critical. What do we mean by 'modestly' effective drugs with 'minimal' side-effects?[8]

General principles

All diseases are both socially constructed and biologically based. Not all would agree with this simple, provocative statement, as many try to bring the concepts of biology and culture of disease into conflict rather than necessary collaboration.[1] Of course, there are different conceptions of what we mean by social construction. Here we mean that the process of labelling something as disease is a human conceptual task that occurs at the interface between science and healing practices and the rest of society. Biology informs this process but cannot be an ultimate arbitrator. The word illness is appropriately used to characterise the impact of a condition with potential to cause suffering in an individual. Illness represents the personal construction of the explanatory concepts and stories necessary for an individual human being suffering from a disturbance in biological, psychological or sociological disequilibrium.[9]

Calling something a disease has profound implications for how society deals with an issue.[10] The medicalisation of a particular human problem unleashes enormous energies. Although the power of the medical model is great, it can often be distorted by excessive hubris. A disease can be eradicated altogether by simply eliminating it from the list of classified disorders, as psychiatrists did with homosexuality. The quest for biological markers for disease also illustrates the attempt to make a social process more objective and scientific. Applying the label AD is now and will forever remain a social process. The label AD is a socio-marker that occurs when a physician diagnoses a person with this condition. Enormous amounts of money and related exaggerated claims have been made for markers in blood, cerebrospinal fluid and the brain as more powerful indicators for AD. Yet, fundamentally, these markers should remain subordinate to diagnostic process that will remain clinical. The application of a disease label is and will remain the responsibility of a clinician who interprets the entire pattern of disease presentation. Moreover, given the overlap between normal aging and various dementia categories to be discussed later, biological markers are likely to remain of marginal utility in individual clinical practice. In my opinion, and despite numerous exaggerated claims in the past, imaging and other tests based on body fluids are likely to be limited by both cost and predicative value.

All treatment of a disease and illness is and should be recognised as biopsychosocial.[10] When we consider the history of therapeutic trials in dementia we should include both biological and other interventions. Biological interventions are simple when compared to psychosocial and educational interventions. Measuring the effects of a pill compared to placebo is easier than measuring the impact of complex, individualised non-biological interventions. Yet both the pill and psychological processes affect biology and occur in a social context. Biopsychosocial approaches include biology but do not allow it to achieve the dominance that it can in a more limited medical model of therapeutic intervention. Extending such models to include spiritual and educational aspects should also be explored. Even if we find

profound biological fixes, their use clinically will include human relationships and social challenges. Hence even a cure – whatever that would mean – could not drive out the psychosocial aspects of care.

Dementia emerged as a social construct in the late 1800s before systematic neuropathological analysis of brain tissues could be accomplished. Efforts were made to differentiate forms of insanity. The complex negotiations between neurologists and psychiatrists for control of the care of different kinds of patients that continues today began in assigning labels (often eponyms) to various pathological states. The modern conception of psychiatry emerged from the world of alienists, as earlier psychiatrists were called because of their focus on so-called mental alienation as the cause of psychological disturbance. As a part of this process, the concept of dementia emerged as the loss of cognitive functioning in a previously normal individual, in the absence of acute confusional states, i.e. delirium. Mental retardation was differentiated from acquired dementia. With less success, the major psychiatric illnesses such as schizophrenia (also called dementia praecox) were differentiated from so-called organic conditions and from each other. Formally speaking, dementia can be static and the terms degenerative, progressive, or primary dementia reserved for those with decline as an unfortunate part of their course.

Historical examples

Syphilis

Alois Alzheimer contributed to our knowledge of many of the common and not so common dementias of his time. Syphilis was one of the most common causes of dementia in his practice, namely general paresis. He also studied the vascular causes of cognitive impairment and contributed to understanding of pathology of a variety of degenerative dementias including the one which now bears his name. Emile Kraepelin, Alzheimer's senior, was one of the most influential psychiatric nosologists of his day. The fact that he named the condition AD in his 1910 textbook assured at least some recognition, despite the fact that the oral presentation of the first case of AD inspired little interest.[10] This recognition of 'Alzheimer's disease' was prompted in part by Kraepelin's desire to demonstrate a biological basis for behavioural abnormalities, in contrast to the emerging dominance of the Freudian understanding.[11]

As the sciences of clinical description and neuropathology grew, so did the concept of an adequate therapeutic trial. Efforts to develop effective therapies for syphilis illustrate the importance of accurate diagnosis and systematic study of possible interventions. During the time that Alzheimer was practising, the Wasserman test for syphilis and early therapies to treat syphilis with arsenic and other toxic compounds developed. A Nobel Prize was eventually given for the development of fever therapy to treat syphilis, in which artificial means, including deliberate infection with malaria, were used to induce high fevers in individuals to kill spirochetes. The concept of a modern randomised controlled double-blind study evolved over decades.[1] Prior to the time of Alzheimer, the idea of a comparison treatment was introduced, i.e. the beginning of our modern concept of a placebo control arm. The notion of investigator bias and the importance of randomised assignment emerged later.[12] Even today there is much lack of clarity about the design and analysis of trials, particularly those designed to modify the biology of the disease and slow rate of progression.[13,14]

Tensions also existed around the relative importance of biological and psychosocial interventions. Alzheimer's paper on hydrotherapy or the use of baths to treat people with psychiatric problems received more attention than his original case study of AD. The setting for a treatment study was also important as psychiatrists became more associated with inpatient asylums, and neurologists with outpatient practice.[1] Competition among different schools of psychiatric thought also strongly influenced which conditions got labelled as diseases and how their treatments were conceptualised. For example Kraepelin's school of thought

in Munich – to which Alzheimer belonged – was in direct competition with that of Arnold Pick's in Prague – an agon which may have been another factor in accelerating the use of Alzheimer's name to label a disease.[10]

The history of the discovery of the cause of syphilis and the search for a magic bullet to treat it had been well described.[15] The expectations 100 years ago that biological studies of brain diseases would lead to effective therapies were high. The practical challenges of administration and toxicities of earlier therapies made the treatment of syphilis ineffective until the advent of penicillin. The reappearance of syphilis as a public health hazard today should remind us of the importance of humility in the face of nature, and the dangers of overestimating our abilities to deal effectively with diseases – even in those where the causative agent is easy to identify and current therapies relatively effective.

Cerebral metabolic enhancers

The search for biological agents to improve thinking in individuals affected by age-related degenerative dementias is a long one. The early studies of pathology showed the damage to the brain graphically, but provided few clues on how to effectively intervene. Although the identification of the first neurotransmitters, acetylcholine and noradrenaline (norepinephrine), and the emergence of the concept of neurotransmitter receptors occurred when Alzheimer was practising, it took decades for that knowledge to influence the treatment of dementia. The absence of clear-cut conceptions of pathophysiology limited therapeutic developments. Nevertheless, early studies, as well as models of psychiatric disease, suggested a role for blood vessels in the pathogenesis of dementia. Theodore Meynert, one of the early pioneering brain psychiatrists, developed an elaborate theory that attributed psychiatric symptoms such as mania and depression to imbalances in blood flow to various cortical and subcortical structures.[5,6] Attempts to develop vasodilators dominated the early attempts to treat dementia, in which the goal was to increase blood flow to the brain.

Parallel, and in many cases overlapping efforts were directed towards developing so-called cerebral metabolic enhancers.[16] Another widely used term was nootropic, meaning mind growth. Hydergine was the prototypical nootropic or cerebral metabolic enhancer. It was a widely sold drug originally approved for treatment of so-called organic brain syndromes in the elderly, and conferences were organised by its manufacturer including testimonies from some of the leading clinicians of the day. The claimed mechanism of action of this drug changed in part in response to new conceptions of the pathophysiology of dementia. Interestingly, a much later systematic review of published data on hydergine showed that it had some small therapeutic benefit,[17] but the drug quickly faded from the academic literature. Indeed, as early as 1991 a comprehensive review made virtually no mention of hydergine, in favour of approaches based chiefly on the cholinergic hypothesis.[18]

Exaggerated claims were made for a variety of products that were said to improve the health of nerve cells by aiding their metabolism in a variety of ways. Such concepts continue to influence drug usage today, particularly in the area of so-called complementary and alternative medicine, where loose conceptions of metabolic enhancement continue to affect purchases of agents designed to improve thinking.

The cholinergic hypothesis

Although the roots of neurochemistry go back to the late 1800s with figures such as Thudichum,[19] it was not until the 1960s that neurotransmitter systems began to be mapped in the brain. The first major success was the description of the loss of dopamine in Parkinson's disease and the development of effective therapies based on enhancing the action of that neurotransmitter by administration of its precursor, L-dopa. Efforts to mimic the success in this disease led to studies of acetylcholine and bioamines, such as noradrenaline and serotonin in degenerative dementias. In the 1970s, several groups described a loss of markers for cholinergic neurons in the brains of patients with AD. It had been known for some time that drugs that impair cholinergic function can cause confusion and amnesia in intellectually normal people, particularly the elderly. Our work at Hopkins in the 1980s described the anatomical substrate of this loss of cortical cholinergic markers as the so-called substantia innominata or nucleus basalis of Meynert and other associated neurons in what came to be called the cholinergic basal forebrain.[19] Animal studies confirmed the importance of cholinergic mechanisms in cognition. These studies led to the advancement of the so-called cholinergic hypothesis of AD in which dysfunction and death of these cells was claimed to contribute to the cognitive impairment.

The cholinergic hypothesis was often asserted in more political than scientific terms.[20,21] That is to say it was used to position a particular person, institution or viewpoint for power, influence and money. The strongest form of the hypothesis claimed that a loss of these cells explained essentially all cognitive impairment, whereas weaker forms made less extravagant assertions. Nevertheless, this biological knowledge further contributed to an interest in drugs to enhance cholinergic systems in order to treat AD. Early results with drugs such as physostigmine suggested that preventing the breakdown of acetylcholine by inhibiting the enzyme responsible for separating the acetyl and choline groups might improve symptoms in people with AD.[20]

The cholinergic hypothesis under-girded a widely publicised report of four patients in whom drug pumps were implanted neurosurgically, and infusions alternated between bethanecol (a cholinergic agonist) and placebo.[22] Although the study was aimed primarily at feasibility, some efficacy data were gathered from a questionnaire completed by family members. The results were heavily covered in the media. A 1988 report by the United States Congress' Office of Technology Assessment (OTA) strongly criticised the 'bloating of preliminary research data' (in a section titled 'False Hope and

Preliminary Data'.[23] It was also critical of evaluation by family questionnaire, in favour of more standardised methods.

The OTA report evinced less, but still some, scepticism about a then recent report on tacrine. Tacrine as a chemical entity has been known since the 1940s. A rather unexpected and surprisingly positive article published in the *New England Journal of Medicine* claimed dramatic improvements in patients with dementia on this drug.[24] An accompanying editorial celebrated the results as a triumph of the scientific method.[25] However, tacrine was a flawed drug and the paper was a seriously ethically and scientifically flawed article.[8] However, after follow-up studies showed some consistent minimal benefit associated with the drug, it was approved for the treatment of AD. In parallel with the drug's development, the Food and Drugs Administration (FDA) and other regulatory bodies around the world developed guidelines to assist in the development of anti-dementia medications.[14,26] Much of the focus was on appropriate outcome measures. How should society determine that a medication is effective in dementia? Ultimately, the FDA established that a drug should demonstrate its benefit using objective psychometric tests such as mental state examinations and a measure of clinical meaningfulness, such as a global impression of change. The draft guidelines were never published, but the approval of tacrine provided a roadmap for other drugs to follow. Eventually safer and easier to use cholinesterase inhibitors were approved, including donepezil, rivastigmine and galantamine. Each of these agents demonstrated that scores on mental status tests and clinical global scales were better over several months on the active drug than a placebo. Actual improvements in scores were rarely observed. Eventually the medications were also demonstrated to have positive effects on caregiver-rated activities of daily living and have some small effects on behavioural symptoms. Although speculative claims have been made, there is no evidence that these medications actually slow the biological course of the illness.

In 2002 in Europe, and 2003 in the United States, memantine was approved for symptomatic treatment of AD. This drug apparently acts through glutamatergic, not cholinergic mechanisms, and can be co-administered with cholinesterase inhibitors. Its claimed mechanism of action also suggests a possible neuroprotective effect (preventing hypothetical excitatory amino acid damage), but it is only approved for symptomatic use. Nevertheless, opinion leaders and patients alike are confused about disease-modifying language.[8] One could perhaps accurately say that memantine, just like the cholinesterase inhibitors, slows the rate of clinical symptom progression but not disease pathogenesis.

Disease-modifying agents

The ultimate therapeutic goal in the treatment of degenerative dementias is to prevent, cure or at least slow the biological progression of disease. Many claims have been made that such drugs exist but none have been approved by the FDA or other regulatory bodies. Propentofylline was submitted to the Europe Medicines Evaluation Agency for symptomatic as well as disease-modifying indication in AD. It was not approved, due, in part, to the lack of consensus about how one demonstrates an effect on disease progression as well as concerns about small treatment effect sizes.[14] Vitamin E is widely used in the United States but not Europe, with the hope that it might slow progression of the disease. This hope is based on a single study comparing its effects to those of selegeline and placebo.[27] The study played an important historical role not only in terms of some of its design innovations (a survival analysis approach), but also because it was one of the few positive studies conducted by the National Institute on Aging Alzheimer Disease Cooperative Study. The study is controversial, however, and should not be considered conclusive support for the use of vitamin E.[8]

Numerous hypotheses have been put forward to explain why cells die in AD and related conditions. Such theories include lack of growth factors, excess calcium, inflammation, toxic metals and excessive free radicals. These hypotheses are often supported by animal studies, small exploratory trials in humans and large-scale uncontrolled epidemiological studies of human populations, looking for risk and protective factors. Unfortunately randomised controlled studies have in large part not demonstrated any effects of these classes of agents on biological progression.

Amyloid vaccine

We continue to use the approaches developed by Alzheimer himself to diagnose AD in autopsy tissue. The neurofibrillary tangle and amyloid plague remain the cardinal microscopic features. Current thinking emphasises that a better understanding of the formation of these two lesions will likely lead to more effective therapies. Much energy has been placed on understanding plaques, which are composed, in part, of a protein called amyloid that is thought to be toxic to neurons. Many attempts have been made to develop drugs to prevent amyloid formation or enhance its clearance.

A surprising observation was made in mice genetically modified to overproduce amyloid. Vaccinating these mice against amyloid not only prevented the accumulation but prompted the removal of amyloid from their brains.[28,29] In a short period of time, this led to human trials and was associated with excessive expectations concerning the likelihood of success. Although the full results of the study are not published, initial positive reports in a subset of unblinded patients have apparently not been replicated in the full sample, and efficacy remains to be established.[29] Moreover and more importantly, a not unexpected side-effect developed, namely encephalitis which was probably allergic and caused by the immunisation itself. Other attempts at vaccination are likely, but it would be helpful to the field to have the full report of the study to best consider the next steps.

Behavioural and psychological signs of dementia

Auguste D, the first case of AD, illustrated that the behavioural or psychological symptoms in the condition can be a significant cause of suffering.[2] These include agitation, psychosis, depression, apathy, sleep disturbance, wandering and other activity disturbances. A variety of tranquillisers were used to treat such patients at the turn of the last century. Today we are trying to develop more effective medications that treat the behaviour disturbance without causing as many unpleasant side-effects such as sedation. During the history of drug development in AD, there has been considerable uncertainty about the centrality of these so-called secondary behavioural symptoms. Attracting the attention of both the scientific community and regulatory agencies was at times challenging, despite the fact that these symptoms are more important than the cognitive ones in affecting the quality of life of both the patient and the caregiver. An early issue was whether drugs to treat the psychological symptoms should be considered 'pseudo specific' in their effects, i.e. generally sedating in many conditions, rather than targeted on specific symptoms of AD itself. The history of the development of the concept of the BPSD (Behavioral and Psychological Signs of Dementia) is an interesting case study in the social construction of a symptom complex conducted by professional organisations and the pharmaceutical industry.[30]

Caregivers' interventions

Systematic therapeutic trials of psychosocial interventions for patients and caregivers are much less frequently conducted than drug studies, despite a long history of interest in such interventions in dementia.[31] Such trials can be very expensive and for the most part pharmaceutical companies have the resources and interests only in studying biological interventions. Recent caregiver studies have been modelled after drug studies, sometimes as add-ons to existing studies of drugs. The challenges of defining a psychosocial intervention and measuring its impact are greater than for pills. Interestingly, a recent meta-analysis has suggested that the effect size estimated from systematic studies of caregiver interventions is not substantially different from those seen in pharmacological studies.[32]

Historical comparisons between the psychopharmacology of depression and the neuropharmacology of cognition

As suggested above, the role of the pharmaceutical industry in developing the field of therapeutic trials in dementia has been considerable. In fact, it has been said that the development of the field of psychopharmacology was the invention of industry rather than of academics.[1] What interventions are studied and how often has as much to do with the business of drug development as with its science. The pervasive influence of the pharmaceutical industry in the treatment of depression and related conditions has been well documented.[33] A case can be made that in the areas of anxiety, depression, and phobia, diseases were developed for drugs rather than the other way around. Moreover the industry has become sophisticated at influencing individual physicians and entire academic communities in the service of selling drugs. The power of modern molecular science to develop more effective interventions for dementia is real. However, successes at developing truly innovative, new chemical entities have been limited, given the hundreds of millions of dollars invested in the efforts. This has not prevented companies from creating expectations for success and marketing agents with rather modest advantages over existing treatments, in aggressive ways. The pharmaceutical industry, through its marketing and lobbying efforts as well as its scientific successes, creates the very conceptions we use to think about health and disease in our society. The commercialisation of for-profit healthcare which has been rampant – particularly in the United States – over recent decades may not, in the long run, be beneficial for the health of our population. For example, the fascination with the power of molec-ular biology and genetics has left environmental and public health concerns unaddressed.[34]

The implications of history for the future of drug development in Alzheimer's disease

Diagnostic issues

One hundred years ago we were struggling with defining dementia and AD. Especially problematic was the relationship between age-related changes in the brain and the dementias, particularly those of early or presenile onset. A major political victory was won by the Alzheimer's Association and other groups claiming that AD is a disease and not normal aging. However, whether this is a scientific or clinical victory is not clear. There are advantages to labelling causes of human suffering as medical conditions, but there are also disadvantages. Despite the advances in molecular sciences and neuroimaging that have been celebrated especially during the recent 'decade of the brain', one could argue that we are more confused about the nosology of dementia today than we were 20 years ago.[35,36] The clear distinctions between vascular dementia, and degenerative dementias no longer exist. The overlaps in clinical and biological features among Parkinson's disease, Lewy body dementia and AD are all the more confusing. Thus, ironically, greater investment in science and diagnostic characterisation has led to more confusion in classification. This is probably due to Alzheimer's, Parkinson's and related conditions being part of diverse, common and overlapping biological processes that affect the aging brain. Sorting them out into discrete categories will not be easy or even possible or necessary.

Another specific example of the problems of diagnostic labelling relates to changes in brain mechanisms and cognitive capacity as we age normally. The Greeks, and probably those before them, recognised that older people frequently developed problems with short-term memory. In the 1960s, Kral labelled some of these individuals as suffering from benign senile forgetfulness.[37]

More recently, concepts such as aging-associated memory impairment and aging-related cognitive decline have been more precisely operationalised.[38,39] The most popular term in current parlance is mild cognitive impairment (MCI).[40–43] MCI is said to occur when an individual or knowledgeable informant complains of memory difficulty and can be demonstrated to have some degree of cognitive decrement for his age but has no frank dementia and relatively little impairment in his or her functioning in daily life. These labels are best viewed as attempts to categorise people on a continuum of cognitive impairment that occurs with age. MCI was developed to identify those at risk for 'converting' to AD and to include them in therapeutic trials of agents designed to prevent dementia. However, the clinical utility of the term is questionable. Considerable variability exists in how the label is applied by experts, despite numerous consensus conferences. If a person is told that they have MCI, does it mean that they do have AD, do not have AD, may or will get AD or might even get better? The confusion surrounding the term illustrates an active process of social construction still occurring in scientific and lay meetings around the world. It also represents strong attempts by academics to medicalise cognitive aging and apply a drug treatment model. This effort can be viewed as a prototypical example of the attempts to imagine and develop more general anti-aging interventions.[44]

These labelling efforts also intersect with the increasing attention being paid to cognitive enhancement in normal people. Could drugs to treat AD improve cognitive function in people currently functioning intellectually normally?[45] We recently conducted a randomised controlled study of pilots flying in flight simulators and demonstrated that donepezil led to small improvements in abilities to land and handle emergencies compared to placebo.[46] Placing all of us on a continuum of cognitive aging changes has implications not only for therapeutic trials in dementia, but also for how we conceptualise brain aging in our society.

Genetics

Family aggregation of cases of AD has been described since the early decades of the last century. Genuine breakthroughs in our understanding of AD and related disorders occurred in the 1980s through the identification of autosomally inherited genetic abnormalities in rare families with early onset AD. The identification of apolipoprotein E4 as a risk factor in populations at large has served as a prototype for genetic susceptibility testing in medicine in general. Our own work has shown that the application of genetic science in clinical medicine is fraught with many greater difficulties than simplistic models of the development of a personalised molecular medicine would suggest.[47] Developing the appropriate empirical foundation, the necessary mathematical models and the patient education approaches to make genetic information genuinely useful to individuals and societies are much larger challenges than we first thought.

Ethical issues

The history of drug trials in dementia is rich with complex ethical issues. The profit motive has contributed to creating false expectations and dashed hopes on many occasions over the decades. The history of conflict of interest guidelines in dementia trials illustrates the subtleties of the negotiations between clinicians and society, with regard to the independence and trust that society places in professions.[16,48,49]

Other ethical issues in the design of trials are still unresolved. We have not reached a social consensus in this country or others about the appropriate means for obtaining informed consent for persons with cognitive impairment.[50] Abuses have occurred in the past and, although we are more sensitised to ethical issues in research today, problems still exist. Should we be conducting capacity assessments on all those who volunteer for cognitive trials? Should we be employing research advance directives in order to allow individuals who lose cognitive capacity to provide some guidance when cognitively intact about their future wishes? How should we think about risk in trials such as those underway implanting genetically modified cells in the brains of people with mild or moderate dementia? Perhaps the most profound ethical issue that needs further social discussion currently is the ownership of data. Do companies have the right to suppress the results of trials for commercial advantage? I believe that most individuals who agree to participate in a trial believe that they are contributing to public science, i.e. generally available knowledge. In other words, they expect that the data produced through their efforts will become known to other scientists and guide future work. There is a need to renegotiate this aspect of the contract between research participants and commercial sponsors.

Pharmacoeconomics

The history of the study of health economics in dementia trials is a short one. I co-led the development of the first three international conferences on this topic.[51,52] There has been a clear increase in the sophistication of modelling. We are also collecting more economic data in trials to understand the impact of dementia drugs on healthcare costs, but these efforts are rudimentary. The assessment and economic valuing of informal care (offered by families for example) is a major challenge. Many exaggerated claims have been made that currently available drugs save money, especially in studies funded by industry. Efforts have been made to demonstrate that dementia drugs save money by delaying nursing home placement.[53] However, these *post hoc* studies are often affected by a bias in subject recruitment and selectivity in choice of outcomes measures. For example, these analyses often do not include the likely possibility that, even if achieved, a delay in nursing home placement would in fact extend life and therefore thereby increase overall healthcare costs. Ultimately, I believe we should decide what dementia drugs are worth, based on quality-of-life assessment. Does taking any current dementia drug improve the quality-of-life of the patient or the caregiver? Although the study of quality-of-life

in dementia has matured over the last 10 years,[32] the pharmaceutical industry has been slow to incorporate these measures in their trials. Perhaps we can learn from the lessons from studies of anti-depressants. As David Healy has pointed out,[33] much quality-of-life data are unpublished in this field. Perhaps current drugs do not provide sufficient benefit to improve the quality of life of people with either depression or memory problems. The price of drugs is an increasing concern; despite four cholinesterase inhibitors being marketed in the United States and other countries, there is remarkably little price competition, although the pharmaceutical industry is often touted as a model for global capitalism. Thus, we need to learn from the history of pharmacoeconomics in dementia how to better assess the positive and negative impacts of dementia drugs on people, how to consider the opportunity costs to society for paying for drugs of minimal benefit, and how biases from a variety of sources, particularly industry, can affect the design and interpretation of studies.

Conclusion

We stand just a few years from the hundredth anniversary of the description of the first case of AD. It is remarkable to see how far science has come in understanding brain function and dysfunction during this century. The power of genomics, molecular biology, combinatorial chemistry, high throughput screening and other advances in the development of the pharmaceutical sciences are real and are being applied to the challenges of age-related cognitive decline. However, it is scientism not science that allows unrealistic expectations for dramatic therapeutic benefits to be promulgated. Where extrapolations of scientific progress transform from reasonable projections to irrational faith is difficult to discern, however. The chronic diseases of the elderly are intimately related to fundamental biological processes that underlie aging itself. Although the power of science has increased dramatically, the challenges of the chronic diseases that affect contemporary men and women should not be underestimated. Moreover, we are entering a phase of human history in which the very viability of our species is threatened by social injustice and environmental degradation. As we think about the future of therapeutic trials in dementia, we must consider them from a perspective that includes other global priorities. In the United States we have improved the economic and health status of older people considerably. There is more that could be done; there is always more that could be done. However, the lives of individuals of all ages, particularly children, are increasingly threatened by environmental and social causes of disease over which we have more direct influence. As we age as individuals and societies, we must recognise the realities of death and the likelihood that many of us will die with cognitive impairment of varying degrees of severity. We must also live our lives seeing aging as an opportunity for gaining wisdom and perspective. With such wisdom we may recognise the limits of technological solutions to human disease. A study of history can contribute to such collective wisdom.

Acknowledgements

I would like to thank Jason Karlawish, Susie Sami, Jesse Ballenger and Ken Rockwood for comments on this manuscript and the NIA and Takayama Foundation for support of work relevant to this paper.

References

1. Shorter EA. A History of Psychiatry: From the Era of the Asylum to the Age of Prozac. New York: John Wiley & Sons; 1997: 33–68, 190–238.
2. Maurer K, Maurer U. Alzheimer: The Life of a Physician and the Career of a Disease. New York: Columbia University Press; 2003.
3. Katzman R, Bick K. Alzheimer Disease: The Changing View. San Diego: Academic Press; 2000.
4. Whitehouse PJ, Deal WE. Situated beyond modernity: lessons for Alzheimer's disease research J Am Geriatr Soc 1995; 43:1314–15.
5. Whitehouse PJ, Ballenger J. Meynert, Theodor Hermann. In: Encyclopedia of Neurological Sciences. San Diego: Academic Press; 2003: 135–6.
6. Whitehouse PJ. Theodor Meynert: foreshadowing modern concepts of neuropsychiatric pathophysiology. Neurology 1985; 35:389–91.
7. Fox P. From Senility to Alzheimer's Disease: The Rise of the Alzheimer's Disease Movement, Milbank Mem Fund Q 1989; 67(1):58–102.
8. Karlawish JHT. The search for a coherent language: The science and politics of drug testing and approval. In: Kapp MB (ed). Ethics, Law and Aging Review. New York: Springer, New York Publishing Company 2002; 8:39–56.
9. Kleinman A. The Illness Narratives: Suffering, Healing, and the Human Condition. New York: Basic Books; 1989.
10. Whitehouse PJ, Maurer K, Ballenger J (eds). Concepts of Alzheimer Disease: Biological, Clinical, and Cultural Perspectives. Baltimore: John Hopkins Press; 2000.
11. Torack RM. The Pathologic Physiology of Dementia. New York: Springer-Verlag; 1978.
12. Marks HM. The Progress of Experiment: Science and Therapeutic Reform in the United States, 1900–1990 (Cambridge Studies in the History of Medicine), Marks, Rosenberg and Jones (series eds). Cambridge: Cambridge University Press; 1997.
13. The Disease Progression Sub-Group: Bodick N, Forett F, Hadler D et al. Protocols to Demonstrate Slowing of Alzheimer Disease Progression. Alzheimer Dis Assoc Disord 1997; 11(Suppl 3): S50–S53.
14. Whitehouse PJ, Kittner B, Roessner M et al. Clinical trial designs for demonstrating disease-course-altering effects in dementia. Alzheim Dis Assoc Disord 1998; 12(4):281–94.
15. Brandt AM (1987) No Magic Bullet: A Social History of Venereal Disease in the United States Since 1880. New York: Oxford University Press; 1987.
16. Whitehouse PJ, Karlawish J, Ballenger J. Conflicts of interest in dementia drug development. In: Kluwer Series; (In press, 2005).
17. Schneider LS, Olin JT. Overview of clinical trials of hydergine in dementia. Arch Neurol 1994; 51:787–98.
18. Davidson M (ed). Alzheimer's disease. Psychiatr Clin North Am 1991; 14:223–487.

19. Whitehouse PJ. History and the future of Alzheimer disease. In: Whitehouse PJ, Maurer K, Ballenger J (eds). Concepts of Alzheimer Disease: Biological, Clinical, and Cultural Perspectives. Baltimore: John Hopkins Press; 2000; 291–305.

20. Bartus RT, Dean R, Beer B, Lippa A. The cholinergic hypothesis of geriatric memory dysfunction. Science 1982; 217:408–17.

21. Whitehouse PJ, Price DL, Struble RG et al. Alzheimer's disease and senile dementia: loss of neurons in the basal forebrain. Science 1982; 215:1237–9.

22. Harbaugh RE, Robarts DW, Coombs DW et al. Preliminary report: intracranial cholinergic drug infusion in patients with Alzheimer's disease. Neurosurgery 1984; 15:514–8.

23. Office of Technology Assessment. United States Congress. Confronting Alzheimer's Disease and Other Dementias. Washington, DC: Science Information resource Center; 1988.

24. Summers WK, Majovski LV, Marsh GM, Tachiki K, Kling A. Oral tetrhydroaminoacridine in long-term treatment of senile dementia, Alzheimer type. N Engl J Med 1986; 315:1241–5.

25. Davis KL, Mohs RC. Cholinergic drugs in Alzheimer's disease, N Engl J Med 1986; 315:1286–7.

26. Leber P. Guidelines for the Clinical Evaluation of Antidementia Drugs. United States Food Drug Administration, Washington, DC, 1990.

27. Sano M, Ernesto C, Thomas RG et al. A controlled trial of selegiline, alpha-tocopherol, or both as treatment for Alzheimer's disease. N Engl J Med 1997; 336:1216–22.

28. Schenk D, Seubert P, Ciccarelli RB. Immunotherapy with beta-amyloid for Alzheimer's disease: a new frontier. DNA Cell Biol 2001; 20:679–81.

29. Schenk D. Amyloid-β immunotherapy for Alzheimer's disease: the end of the beginning. Nat Rev Neurosci 2002; 3:824–8.

30. International Psychogeriatric Association. Behavioral and Psychological Symptoms of Dementia. Educational Pack. Produced by the International Psychogeriatric Association (IPA) under an unrestricted educational grant provided by Janssen-Cilag and Organon: Belgium: Teknika; 2003.

31. Ballenger JF. Beyond the characteristic plaques and tangles: mid-twentieth century American psychiatry and the fight against senility. In: Whitehouse PJ, Maurer K, Ballenger JF, (eds). Concepts of Alzheimer Disease: Biological, Clinical and Cultural Perspectives. Baltimore: Johns Hopkins University Press; 2000: 83–103.

32a. Brodaty H, Green A, Kochera A. Meta-analysis of psychosocial interventions for caregivers of people with dementia. J Am Geriatr Soc 2003; 51:657–64.

32b. Whitehouse PJ, Patterson M, Sami S. Quality of life in dementia: ten years later. Alzeim Dis Assoc Disord 2003; 17(4)199–200.

33. Healy D. Let them eat Prozac. Toronto: James Lorimer & Company Ltd; 2003.

34. Whitehouse PJ. MCI and AD: Different conditions or diagnostic confusion? Geriatric Times 2003; 4:14–16.

35. Ritchie K, Lovestone S. The dementias. Lancet 2002; 360:1759–66.

36. Whitehouse PJ. 'Classifications of the dementias', commentary on 'The dementias' article by Richie and Lovestone [letter]. Lancet 2003; 360:1759–66, Lancet 361:1227.

37. Kral V. Senescent forgetfulness: Benign and malignant. Can Med Assoc J 1962; 86:257–60.

38. Levy R. Aging-associated cognitive decline. Working Party of the International Psychogeriatric Association: in collaboration with the World Health Organization. International Psychogeriatrics 1994; 6:63–8.

39. Crook T, Bartus RT, Ferris SH et al. Age-associated memory impairment: proposed diagnostic criteria and measures of clinical change – Report of a National Institute of Mental Health Work Group. Dev Neuropsychol 1986; 2:261–76.

40. Whitehouse PJ, Frisoni GB, Post S. Breaking the diagnosis of dementia. Lancet Neurol 2004; 3:124–8.

41. Whitehouse PJ, Juengst ET. Anti-aging medicine and mild cognitive impairment: Practice and policy issues for geriatrics. J Am Geriatr Soc. in press.

42. Petersen RC (ed). Mild Cognitive Impairment: Aging to Alzheimer's Disease Oxford: Oxford University Press; 2003.

43. Petersen RC, Stevens JC, Ganguli M et al. Practice parameter: Early detection of dementia: mild cognitive impairment (an evidence-based review). Report of the Quality Standards Subcommittee of the American Academy of Neurology. Neurology 2001; 56:1133–42.

44. Juengst E, Binstock RH, Mehlman M, Post SG, Whitehouse PJ. Biogerontology, 'Anti-aging medicine', and the challenges of human enhancement. Hastings Center Report 2003; 33:21–30.

45. Whitehouse PJ, Juengst ET, Mehlman M, Murray T. Enhancing cognition in the intellectually intact: possibilities and pitfalls. Hastings Center Report 1997; 3:14–22.

46. Yesavage JA, Mumenthaler MS, Taylor JL et al. Donepezil and flight simulator performance: Effects on retention of complex skills, Neurology 2002; 59:123–5.

47. Barber M, Whitehouse PJ. Susceptibility testing for Alzheimer's disease: race for the future. Lancet 2002; 1:10.

48. Kodish E, Murray T, Whitehouse PJ. Conflict of interest in university-industry research relationships: realities, politics, and values. Acad Med 1996; 71:1287–90.

49. Whitehouse PJ. Interesting conflicts and conflicting interests. J Am Geriatr Soc 1999; 47:1–3.

50. Hougham GW, Sachs GA, Danner D et al. Empirical research on informed consent with the cognitively impaired. IRB: Ethics and Human Research (Suppl) 2003; 25:5:S26–S32.

51. Whitehouse PJ, Winblad B, Shostak D et al. First International Pharmacoeconomic Conference on Alzheimer's Disease: Report and Summary. Alzheim Dis Assoc Disord 1998; 12:266–80.

52. Jonsson L, Jonsson B, Wimo A, Whitehouse PJ, Winblad B. Second International Pharmacoeconomic Conference on Alzheimer's Disease. Alzheim Dis Assoc Disord 2000; 14:137–40.

53. Geldmacher DS. Donepezil is associated with delayed nursing home placement in patients with Alzheimer's disease. J Am Geriatr Soc 2003; 51:937–44.

3

Treatment Hypotheses

Gordon K Wilcock

Introduction

Treatment for dementia has moved into a new era. The advent of cholinesterase inhibitors, and more recently memantine, heralded the end of the somewhat nihilistic therapeutic approach to the needs of people with dementia adopted by many. The availability of these treatments, albeit modest in efficacy, has also resulted in a recognition of the need to properly assess and diagnose people with cognitive impairment, determine the most likely underlying aetiology for their illness, and consider whether or not one of the new medicines is appropriate.

Although they were originally developed for Alzheimer's disease, there is emerging evidence that the cholinergic treatments probably have a wider application, and this may also turn out to be true for some of the other therapeutic hypotheses that are currently being explored.

This chapter will primarily concentrate on existing and emerging treatments for Alzheimer's disease, vascular dementia, and dementia with Lewy bodies, as these constitute the most common causes of dementia.

The hypotheses that are leading to new therapeutic strategies are so numerous that it will not be possible to cover them all in a chapter of this length, which will therefore principally discuss those for which there is already a significant body of evidence.

Treatment strategies for Alzheimer's disease

Alzheimer's disease (AD) is the most prevalent neurodegenerative disorder. After nearly a quarter of a century of scientific endeavour based upon our knowledge of the cholinergic deficit in AD, the cholinesterase inhibitors reached the clinic, firstly with tacrine (now little used because of its adverse event profile), then donepezil, rivastigmine and galantamine. They have been followed by memantine, an NMDA-receptor partial antagonist.

These treatments are predominantly considered to modify symptoms, although there is a possibility that they may also contribute to neuronal survival. The evidence for this is as yet scant, but an awareness of the need for neuroprotection has led to the investigation of a number of other potential strategies over the last ten years or so. These include those based on observational studies of differential rates of development of dementia in cohorts of individuals receiving specific treatment for other non-dementia conditions, who were followed over the years, and other approaches targeted specifically at the molecular pathophysiology of the disease.

Treatment approaches based on the cholinergic hypothesis
Introduction

This is based upon the well-described finding of acetylcholine depletion in the Alzheimer brain, probably secondary to degeneration of basal forebrain nuclei, although there is some uncertainty as to whether this is an early change in the disease, or occurs later than originally believed. A number of therapeutic strategies based upon this knowledge have been explored, including attempts to increase acetylcholine synthesis, the use of postsynaptic receptor agonists, drugs augmenting the release of acetylcholine, and the reduction of acetylcholine degradation at a synaptic level with the cholinesterase inhibitors. The latter is the only strategy that has stood the test of time, and is now available in the form of licensed treatments.

Four cholinesterase inhibitors have been licensed, tacrine, donepezil, galantamine and rivastigmine. Tacrine, as mentioned above, is now rarely prescribed because of the adverse event profile, in particular its potential hepatotoxicity. The other three are variably available in different countries. All these drugs selectively inhibit acetylcholinesterase, and rivastigmine additionally inhibits the action of butyrylcholinesterase, which constitutes about 10% of the total cholinesterase activity in the brain. Galantamine also has an effect on nicotinic acetylcholine receptors, at which it acts as an allosteric ligand and is considered to increase presynaptic acetylcholine release, as well as having a postsynaptic effect.[1]

Clinical experience

The clinical efficacy of each of these drugs is similar, although they differ in pharmacological and pharmacokinetic profiles. In general, however, they are considered to have a modest but worthwhile treatment effect in patients with mild to moderate Alzheimer's disease, although only 50–60% of those for whom they are prescribed achieve significant benefit. Treatment effects range across the spectrum of efficacy variables, e.g. cognition, functional ability and behaviour. Evidence is also emerging that these benefits are sustained over longer treatment periods, e.g. for 12 months or more. There has been only one long-term 'head-to-head' study in this therapeutic class, and this compared donepezil and galantamine over a 12-month period. Both drugs were similarly effective on functional indices, but there was a suggestion of some advantage of galantamine in respect of cognitive outcome measures in a subgroup of patients.[2] However this has to be balanced against the patients' overall needs, e.g. the advantages of once-daily administration with donepezil. In general, there is at present insufficient evidence to routinely recommend one drug over the others.

Important questions that still need to be answered include (a) what is the value of these drugs in the longer term e.g. over several years, (b) is there a place for dual therapy, i.e. combinations of different cholinesterase inhibitors, or a cholinesterase inhibitor and another agent such as memantine,[3,4] and (c) might they in reality be disease modifying, rather than just symptomatic in their effect?

Later modification of the hypothesis

All the early studies on neurochemical changes in the Alzheimer brain used autopsy material that primarily came from end-stage Alzheimer's disease sufferers, in whom the severity of the clinical condition correlated with the degree of pathological change and level of cholinergic markers in the brain.[5,6] The assumption was made, nevertheless, that cholinergic deficits would also occur early in the disease, and were therefore a suitable therapeutic target. However, evidence is now emerging that cholinergic impairment may not arise until later in the disease progression, and that in early Alzheimer's disease it may be cholinergic signal transduction defects, rather than loss of neurotransmitter, that are important.[7–9] Since cholinesterase inhibitors are prescribed for people with early to moderate disease, it is possible that attempts to increase the availability of acetylcholine at a synaptic level may be inadequate as a therapeutic manoeuvre at this stage in the disease process, helping to explain why the results of treatment are modest in those in whom a response is obtained, and possibly even explaining why so many people fail to gain any meaningful benefit. It also raises the possibility that cholinesterase inhibitors could be a realistic treatment for people with more severe disease, and that relative lack of efficacy in this group may reflect the choice of outcome measure, rather than a true lack of efficacy. Trials are currently under way to address this issue.

Another interesting modification of the original hypothesis already alluded to above, is the possibility that cholinergic modulation may have some impact on neuronal survival, and disease progression. A number of studies have suggested a relationship between cholinergic neurotransmission and the formation of beta-amyloid protein in the Alzheimer brain. This derives mainly from *in vitro* experimentation, e.g. evidence supporting the hypothesis that acetylcholinesterase may accelerate the assembly of amyloid-beta peptides into the fibrillary amyloid that is seen in AD,[10] an effect that is not mimicked by butyrylcholinesterase.

There is also evidence that stimulation of protein kinase C-coupled muscarinic acetylcholine receptors increases the non-amyloidogenic secretory pathway of amyloid precursor protein processing, with a consequent potential reduction of the amyloidogenic process. This, and other evidence, suggests that the reduction in cholinergic activity in the Alzheimer brain may favour the amyloidogenic processing of amyloid precursor protein, with increased beta-amyloid production.[11] The interaction between the cholinergic system and amyloid processing is, however, a complex one, and there is also evidence suggesting that β-amyloid protein may, itself, reduce acetylcholine synthesis.[12] This in turn suggests that a self-sustaining cycle of continued degeneration could be present in the AD brain, which cholinergic enhancement might ameliorate. Hypothetically, this then has implications for the treatment of severe Alzheimer's disease with cholinesterase inhibitors. Reducing beta-amyloid production might prolong life at this stage, but unless the clinical benefits are worthwhile, such treatments might not be in the best interests of those with Alzheimer's disease, nor of those who care for, and about, them.

The glutamatergic system

The other neurotransmitter system that has been extensively studied in dementia, and which has resulted in a licensed treatment for Alzheimer's disease, is, of course, the glutamatergic system, where memantine, an NMDA-receptor partial antagonist, is now available for treatment in some countries. There are a number of glutamate receptors, among which the NMDA-receptor is thought to play an important role in the long-term potentiation processes involved in learning and memory function. Memantine is considered to reduce overstimulation of the NMDA-receptor in its resting state, caused by an abnormally high concentration of glutamate. This allows some normalisation of the latter's physiological signalling function, and may also reduce excitotoxicity, which may in turn be neuroprotective by preventing the neuronal calcium overload that has been implicated in neurodegenerative processes.

Treatment of dementia syndromes with memantine goes back to the late 1980s, and early 1990s, but it is the more recent studies that have clarified its role in relation to different types of dementia, and led to the recent successful licensing application. As far as its use in Alzheimer's disease specifically is concerned, a pivotal study was undertaken by Reisberg and colleagues,[13] in which it was evaluated in people with moderate to severe Alzheimer's disease. For an overview of the use of memantine in dementia, the reader is referred to a recent review[14] and to Chapter 17.

In theory at least, there are therefore two complementary treatments for people with Alzheimer's dementia – complementary in the sense that two different neurotransmitter-related strategies are involved, and that between them they offer some hope to people in whom the severity of the disease crosses the spectrum from mild to severe.

Trials of a combination of memantine and a cholinesterase inhibitor have therefore been initiated, one of which has recently reported preliminary findings.[3] This indicated that a memantine–donepezil combination had benefits over a donepezil–placebo combination. However, we require additional information on this approach, and also its role in earlier stages of dementia, where one might expect it to have a more significant effect, at both the symptomatic and possibly neuroprotective level.

Interestingly, it has recently been reported that memantine may have a specific protective effect against neurodegeneration secondary to beta-amyloid. This was a study in rats, in which memantine treatment protected against neurode-generative changes induced by the injection of β-amyloid$_{1-40}$ into the hippocampal fissure.[15] This is, of course, a long way from the treatment of Alzheimer's disease in humans, but nevertheless contributes to the suggestion that memantine may be neuroprotective.

Treatment based upon the findings of 'observational' studies

A number of potential therapeutic strategies have been suggested by findings from predominantly community-based studies, often in relation to other conditions, e.g. the cohorts of subjects treated with anti-inflammatory compounds for arthritis, in whom a lower than expected preva-lence of Alzheimer's disease has been detected. Benefit from oestrogen treatment and statins has also emerged from this approach.

Anti-inflammatory drugs

An association between inflammatory markers in the brain and the pathology of Alzheimer's disease has been known for some time. A recent system-atic review and meta-analysis of observational stud-ies, published between 1966 and 2002, examined the role of non-steroidal anti-inflammatory drugs (NSAIDs) use in the prevention of Alzheimer's disease.[16] This included six cohort and three case-control studies, and determined that the pooled relative risk of AD amongst the users of NSAIDs was 0.72. Furthermore, the magnitude of the ben-efit related to the duration of treatment. There

was also a suggestion that there might be some benefit for those taking aspirin.

This hypothesis has proved disappointing when applied to people who already have Alzheimer's disease, i.e. treating AD sufferers with anti-inflammatory drugs. Placebo-controlled clinical trials of naproxen and the selective cyclo-oxygenase 2 inhibitors celecoxib and rofecoxib have been found ineffective in this context.[17] Similarly negative conclusions have resulted from trials of steroids.[18]

These disappointing findings do not necessarily invalidate the hypothesis. Evidence has emerged that the benefit of the anti-inflammatory approach in the cohort observational studies may not be a class effect, but rather specific to the potential of different anti-inflammatory drugs to modulate the production of beta-amyloid, and that interven-tional studies may not have used the most appro-priate NSAIDs for this effect. There is now evidence indicating that a subset of NSAIDs lowers the production of the amyloid-beta 42 peptide. This was first reported for ibuprofen, indometacin and sulindac in *in vitro* studies.[19,20]

This is now known to be the result of direct modulation of gamma-secretase activity. Gamma-secretase is part of the mechanism that allows the amyloid precursor protein to be cleaved in such a way that a potentially amyloidogenic peptide is produced, as opposed to the pathway involving alpha-secretase, in which the harmful peptide is not created. A further study has shown that meclofenamic acid, racemic flurbiprofen, and its purified enantiomers were the most effective compounds, in terms of reducing amyloid-beta 42 levels.[21] The R-enantiomer of flurbiprofen has fewer side-effects than the racemic mixture, and is currently in clinical trial. We must await the out-come of this, and other clinical trials, to ascertain the effects in those with established disease, and also the potential efficacy of these compounds in delaying disease onset.

The anti-inflammatory story has important les-sons for us all. It is very easy to assume that an asso-ciation between a treatment effect and a particular group of drugs results from the *a priori* hypothesis, in this case that the mechanism is anti-inflammatory in nature. It is, however, important to remember that associations may actually be explained by mechanisms unrelated to the basic hypothesis.

Oestrogens and Alzheimer's disease

There is a considerable body of scientific evidence to support the hypothesis that oestrogen may have important neuroprotective functions in the brain. In addition, a number of observational studies have shown a decreased incidence of Alzheimer's disease amongst women who were long-term users of hormone replacement treatment.[22–24] Early studies of the effects of oestrogen on cognitive impairment in those who already have Alzheimer's disease have been inconsistent, but many of these trials involve small numbers of subjects, evaluated over a short period of time. The Woman's Health Initiative Memory Study (WHIMS) evaluated a treatment consisting of conjugated equine oestrogen combined with medroxyprogesterone against placebo. This study included well over 4000 women, and rather than showing benefit from the treatment, reported that oestrogen plus progestin treatment increased the risk of probable dementia.[25,26] The conclusion was that the risks of oestrogen plus progestin outweighed the benefits. Overall, there is inadequate evidence to justify recommending oestrogen treatment, either for Alzheimer's disease prevention, or treatment of established dementia. This, and the NSAID story, shows how caution is required when developing hypotheses based upon the results of observational studies. These have also been criticised because of the inherent biases that may creep in, especially when the data are examined retrospectively.

Cholesterol-related strategies

The E4 allele of the apolipoprotein E gene is a risk factor for sporadic Alzheimer's disease, and is associated with a small increase in serum cholesterol levels.[27,28] It has also been shown from animal and *in vitro* work that higher levels of cholesterol may increase amyloid production, and conversely cholesterol depletion reduces this. This, and other data, suggests an intricate relationship between cholesterol and amyloid deposition, the mechanism of which is unknown, although it might relate to an interaction between cholesterol within the cell membrane and gamma-secretase, which is also situated here. This potential role has been supported by evidence from two epidemiological cohort studies, again observational in nature, that the use of statins is associated with a decreased risk of developing Alzheimer's disease.[29,30] However, the Prosper Study in which pravastatin was prescribed for individuals at risk of vascular disease failed to show any benefit on cognitive function after three years.[31] Again, it is difficult to understand the intrinsic biases that creep into such studies, but Rockwood et al were able to examine this in a large population in Canada, and although use of lipid-lowering agents occurred more frequently in younger than older subjects in their cohort of over 65 year olds, it was not associated with any other factors indicating a healthier lifestyle, but was associated with a history of smoking and hypertension.[32] In this study, use of these drugs reduced the risk of Alzheimer's disease in those aged under 80 years, an effect that was maintained after adjustment for sex, educational level and self-rated health.

There are a number of hypotheses that have been suggested to explain a potential lipid-related approach. However it is as yet unclear which, if any, are likely to be important in the development of treatments for dementia. If there is any benefit, is this specific for one type of dementia, is it related to a lipid-lowering effect or some other more fundamental relationship with the pathophysiology, e.g. amyloid production, or might it work primarily through a vascular mechanism, or a combination of these and other means?

A number of prospective randomised controlled trials of cholesterol-lowering drugs in people with Alzheimer's disease are under way.

Oxidative stress and Alzheimer's disease

The heterogeneous nature of AD has led to many different hypotheses, and subsequent therapeutic strategies, including the potential involvement of free radicals. The latter are associated with a number of phenomena that have been identified in the brain in AD, including DNA damage, lipid peroxidation, and the presence of advanced glycosylation end-products. Beta-amyloid deposition has been shown to lead to the production of free radicals, and its toxicity is reduced in the presence of free radical scavengers. This and other evidence has been linked to the therapeutic potential of a number of free radical scavengers, including vitamin E, selegiline and ginkgo biloba extracts. Indeed, anti-inflammatory drugs and oestrogens also have antioxidant properties, and have been considered in this context. To date the evidence from studies in humans is conflicting.[33–36] Arguably, the best evidence to date is the study by Sano et al, in which patients with moderately severe impairment caused by Alzheimer's disease demonstrated some slowing of disease progression when treated with vitamin E and/or selegiline.[37] Further studies are under way, and if it can be shown that free radical scavenging, or similar approaches, is effective, this will provide yet more evidence of the heterogeneity of the aetiology of AD. The same may be true for the other dementias.

Anti-amyloid approaches to the treatment of Alzheimer's disease

The central role that has been ascribed to the deposition of beta-amyloid in the pathophysiology of Alzheimer's disease has resulted in many strategies to prevent its formation, reduce its toxicity once deposited, or increase its removal from the brain. One of the main hallmarks of Alzheimer's disease, the neuritic plaque, contains amyloid-beta peptides in beta-pleated sheets, with a halo of surrounding dystrophic neuritic structures, activated microglia and reactive astrocytes. The potential central role of this process in the development of AD, and its implications for therapeutic strategies have been reviewed elsewhere.[38]

The 'building block' of beta-amyloid is cleaved from the amyloid precursor protein (APP) molecule, which is a transmembrane structure, encoded on chromosome 21. It is metabolised by a non-amyloidogenic pathway, alpha-secretase, a membrane-bound protease, which cleaves APP within the potential amyloid beta domain, preventing its formation.

The amyloidogenic process involves cleavage by two different secretases; first, beta-secretase produces a membrane-bound 99-amino-acid residue, which is then further cleaved by gamma-secretase within the transmembrane domain to release both a 40- and a 42-amino-acid peptide. The latter is prone to fibrillary aggregation and is the main component of the amyloid deposits.

There are a number of therapeutic hypotheses based upon this knowledge, of which some will be reviewed here.

Secretase-related strategies

For a detailed consideration of this complex area, the reader is referred to an excellent review by De Wachter and Van Leuven.[39] Briefly, however, the two main targets are the beta- and gamma-secretases. Beta-secretase may be the more logical of the two, as it appears to represent the first step in amyloid beta production. Furthermore, preliminary evidence from beta-secretase knockout mice reveals that they appear to be phenotypically normal, but have reduced beta-amyloid generation,[40] implying that this may be a relatively safe therapeutic option. Gamma-secretase is intricately bound up with the presenilin proteins, which are themselves involved in other systems. Although this originally led to concern that gamma-secretase was a less appropriate therapeutic target, gamma-secretase inhibitors and modulators are presently being evaluated.

Amyloid-beta vaccination

Schenk et al reported a reduction in the deposition of amyloid deposits in the brains of the transgenic PDAPP mice when vaccinated over a period of 11 months with fibrillar human β-amyloid$_{1–42}$.[41] Others have subsequently shown that passive immunisation with both polyclonal and monoclonal antibodies against amyloid beta also reduces amyloid deposition in mouse models, and in addition both active and passive immunisation in transgenic mice AD models have been shown to have positive effects on cognition and behaviour. More recently, similar improvements have also been reported in a small-scale study involving human subjects.[42] The actual mechanism by which vaccination is helpful is as yet unclear, and hypotheses include the possibility that it may interfere with fibrillary aggregation, or facilitate the transport of amyloid out of the brain and into the plasma, possibly by the binding and sequestration of soluble amyloid beta. It may also work via activation of microglia, or indeed a combination of different mechanisms.

The first trial of this approach in sufferers of Alzheimer's disease had to be abandoned in 2002, when a significant number of subjects developed a meningoencephalitis. However, in one subject who died, a substantial reduction of amyloid beta plaques was found in the brain, compared to control AD brains.[43] It is likely that further immunological strategies will be evaluated in the not too distant future.

Other anti-amyloid strategies

There are a number of other therapeutic approaches that may work through an amyloid connection, some of which have already been mentioned in other sections of this chapter. A very interesting concept, however, involves the interrelationship of insulin and amyloid production. Insulin degrading enzyme (IDE) is one of the enzymes responsible for degrading soluble amyloid beta. It has a further potential AD connection, as a region on chromosome 10, close to the IDE gene, has been linked to late-onset AD and also plasma levels of the amyloidogenic peptide. A recent study by Watson revealed that insulin increases CSF β-amyloid 42 levels in normal older adults, indicating that insulin may modulate levels of this potentially toxic peptide in humans.[44,45] This, and other supporting evidence, suggests that hyperinsulinaemia may play a role in amyloid deposition, e.g. by competing for IDE. Further work is

required before these findings can be translated into potential therapeutic strategies, but drugs already exist that reduce insulin resistance and hyperinsulinaemia, and are currently under evaluation.

Other strategies for treating Alzheimer's disease

Hyperphosphorylation of the microtubule-associated protein tau contributes to the formation of the intraneuronal neurofibrillary tangles. Inhibiting the enzymes that facilitate this process constitutes a potential therapeutic strategy, and a number of kinase inhibitors have been considered as potential treatments for evaluation, including lithium and valproate.

Neurotrophic factors, such as nerve growth factor (NGF), may have a role to play, as there is evidence from animal studies that suggests that the intracerebral administration of NGF protects against cholinergic atrophy, and may reverse it in certain animal models. There is, however, no direct evidence that NGF is implicated in the pathogenesis of Alzheimer's disease. Unfortunately, NGF does not cross the blood–brain barrier, and needs to be given intraventricularly or intraparenchymally, unless it can be attached to a carrier molecule that will facilitate its entry into the brain from the blood.

Minor cognitive benefit was found in a Scandinavian study of intraventricular NGF, but there were associated adverse events, and to date this approach has only been evaluated in three individual patients.[46] Research is under way to identify compounds that mimic NGF and which might be administered parenterally, or even orally, and make more realistic therapeutic molecules. However a preliminary trial of fibroblasts genetically modified to produce NGF has suggested this may be helpful in mild AD.[47]

There is also an emerging literature on a relationship between hyperhomocysteinaemia and Alzheimer's disease. An elevated plasma level of homocysteine is well known to be an independent risk factor for stroke, and epidemiological studies have now reported associations between hyperhomocysteinaemia, histologically confirmed Alzheimer's disease, and also the rate of cognitive decline. Whether this works through a vascular mechanism, a more direct causal process, or both, remains to be established. The interrelationship between homocysteine, the B-group vitamins, and cognitive decline is complex, but it has been reported that elevated levels of homocysteine are linked to cognitive impairment, and possibly AD specifically.[48,49] The reader is referred to an excellent review by Morris.[50]

Finally, and more speculatively, is the emerging concept that CSF production and turnover may itself play an important role in the development of Alzheimer's disease. Old age is associated with a trend towards less CSF production, and also a greater resistance to CSF outflow. Part of the function of CSF is to help clear toxic molecules from the interstitial fluid space within the brain, facilitating exchange with the bloodstream. There is a hypothesis that in some people a failure of the CSF circulation leads to a build up of substances such as beta-amyloid in certain vulnerable individuals, which may be a contributing factor to the development of AD.[51] This is an intriguing concept, which is undergoing further exploration.

Treatment strategies for other dementias

As well as treatment for the so-called 'reversible' dementias, for which screening for treatable underlying causes is a routine part of assessment, treatment strategies are now beginning to emerge for vascular dementia and dementia with Lewy bodies.

Attention to risk factors for vascular disease is always going to be an important part of the approach to managing this condition, and also when it exists in combination with other conditions such as AD, as a mixed dementia. More specifically, however, the role of cholinesterase inhibitors has also been evaluated with promising results.[52–54] However, better trials are needed, as those reported to date have relied heavily upon methodology imported from Alzheimer's disease studies, which may not accurately reflect the effect of cholinesterase inhibitors in this context.

Memantine has also been reported to be of some benefit in vascular dementia, but again predominantly using methodology imported from AD trials.[14,55,56] The relationship between cerebrovascular disease and dementia is, however, a complex issue and it is probable that microvascular changes should be separated from other forms of vascular dementia.[57]

Over the years, a number of other therapeutic agents have been evaluated in vascular dementia, with disappointing results, including vasodilators, nootropics, and calcium antagonists. It is, however, arguable that the design of these studies was not optimal, nor was there consideration of different vascular subgroups, and that therefore some of these compounds should be evaluated further, for example propentofylline, for which some evidence of efficacy already exists.

It is also very probable that cholinesterase inhibitors have a role to play in the treatment of dementia associated with Lewy bodies. The cholinergic deficit in this condition is often equal to, or greater than, that found in many cases of Alzheimer's disease, and there is evidence of therapeutic benefit.[58,59] Unfortunately, development of therapeutic strategies for this condition lags behind both the development of treatments for AD and vascular dementia. The reasons for this are complex, and include commercial considerations.

Conclusion

The dramatic increase in our understanding of the pathophysiology of the dementias, especially Alzheimer's disease, over the last ten years, has led to the evolution of a multiplicity of potential therapeutic hypotheses, many of which have been, or are, under evaluation. Others will follow in the future, and hopefully we will soon have disease-modifying treatments to add to the currently available drugs. It is also becoming apparent that some of the treatment approaches for AD may have application in other disorders.

I have attempted in this chapter to summarise some of the more exciting areas in this rapidly developing field, but the choice has, to some extent, had to be a personal one, although it has been influenced by those areas in which research is most active. Readers wishing to explore the emerging treatments for Alzheimer's disease and vascular dementia in more detail are referred to two excellent reviews: Scarpine et al for AD,[60] and O'Brien et al for vascular dementia,[61] the latter also addressing the concept of vascular cognitive impairment in general.

References

1. Scott L, Goa K. Galantamine: a review of its use in Alzheimer's disease. Drugs 2000; 60:1095–122.

2. Wilcock G, Howe I, Coles H et al. A long-term comparison of galantamine and donepezil in the treatment of Alzheimer's disease. Drugs Aging 2003; 20(10):777–89.

3. Farlow MR, Tariot PN, Grossberg GT et al. Memantine/donepezil dual therapy is superior to placebo/donpezil therapy for treatment of moderate to severe Alzheimer's disease. Neurology 2003; 60:A412.

4. Hartmann S, Mobius HJ. Tolerability of memantine in combination with cholinesterase inhibitors in dementia therapy. Int Clin Psychopharmacol 2003; 18(2):81–5.

5. Wilcock GK, Esiri MM, Bowen D, Smith CCT. Alzheimer's Disease: correlation of cortical choline acetyltransferase activity with the severity of dementia and histological abnormalities. J Neurol Sci 1982; 57:407–17.

6. Wilcock GK, Esiri MM. Plaques, tangles and dementia. A quantitative study. J Neurol Sci 1982; 56:343–56.

7. Davis KL, Mohs R, Marin D. Cholinergic markers in elderly patients with early signs of Alzheimer's disease. JAMA 1999; 281:1401–6.

8. Dekosky ST, Ikonomovic MD, Styren SD et al. Upregulation of choline acetyltransferase in hippocampus and frontal cortex of elderly subjects with mild cognitive impairment. Ann Neurol 2002; 51:144–55.

9. Mufson EJ, Ma SY, Cochran EJ et al. Loss of nucleus basalis neurons containing trkA immunoreactivity in individuals with mild cognitive impairment and early Alzheimer's disease. J Comp Neurol 2000; 427(1):19–30.

10. Inestrosa NC, Alvarez A, Perez CA et al. Acetylcholinesterase accelerates assembly of amyloid-beta- peptides into Alzheimer's fibrils: possible role of the peripheral site of the enzyme. Neuron 1996; 16(4):881–91.

11. Rossner S, Ueberham U, Schliebs R, Perez-Polo JR, Bigl V. The regulation of amyloid precursor protein metabolism by cholinergic mechanisms and neurotrophin receptor signaling. [Review]. Prog Neurobiol 1998; 56(5):541–69.

12. Pedersen WA, Kloczewiak MA, Blusztajn JK. Amyloid beta-protein reduces acetylcholine synthesis in a cell line derived from cholinergic neurons of the basal forebrain. Proc Natl Acad Sci U S A 1996; 93(15):8068–71.

13. Reisberg B, Doody R, Stoffler A et al. Memantine in moderate-to-severe Alzheimer's disease. N Engl J Med 2003; 348(14):1333–41.

14. Wilcock GK. Memantine for the treatment of dementia. Lancet Neurol 2003; 2(8):503–5.

15. Miguel-Hidalgo JJ, Alvarez XA, Cacabelos R, Quack G. Neuroprotection by memantine against neurodegeneration induced by beta-amyloid(1–40). Brain Res 2002; 958(1):210–21.

16. Etminan M, Gill S, Samii A. Effect of non-steroidal anti-inflammatory drugs on risk of Alzheimer's disease: systematic review and meta-analysis of observational studies. BMJ 2003; 327(7407):128.

17. Aisen PS, Schafer KA, Grundman M et al. Effects of rofecoxib or naproxen vs placebo on Alzheimer disease progression: a randomized controlled trial. JAMA 2003; 289(21):2819–26.

18. Aisen PS, Davis KL, Berg JD et al. A randomized controlled trial of prednisone in Alzheimer's disease. Alzheimer's Disease Cooperative Study. Neurology 2000; 54(3):588–93.

19. Weggen S, Eriksen JL, Das P et al. A subset of NSAIDs lower amyloidogenic Abeta42 independently of cyclooxygenase activity. Nature 2001; 414(6860):212–16.

20. Weggen S, Eriksen JL, Sagi SA et al. Evidence that nonsteroidal anti-inflammatory drugs decrease amyloid beta 42 production by direct modulation of gamma-secretase activity. J Biol Chem 2003; 278(34):31831–7.

21. Eriksen JL, Sagi SA, Smith TE et al. NSAIDs and enantiomers of flurbiprofen target gamma-secretase and lower Abeta 42 in vivo. J Clin Invest 2003; 112(3):440–9.

22. Kawas C, Resnick S, Morrison A et al. A prospective study of estrogen replacement therapy and the risk of developing Alzheimer's disease: the Baltimore Longitudinal Study of Aging [published erratum appears in Neurology 1998 Aug; 51(2):654]. Neurology 1997; 48(6):1517–21.

23. Slooter AJ, Bronzova J, Witteman JC et al. Estrogen use and early onset Alzheimer's disease: a population- based study. J Neurol Neurosurg Psychiatry 1999; 67(6):779–81.

24. Waring SC, Rocca WA, Petersen RC et al. Post-menopausal estrogen replacement therapy and risk of AD: a population-based study. Neurology 1999; 52(5):965–70.

25. Rapp SR, Espeland MA, Shumaker SA et al. Effect of estrogen plus progestin on global cognitive function in postmenopausal women: the Women's Health Initiative Memory Study: a randomized controlled trial. JAMA 2003; 289(20):2663–72.

26. Shumaker SA, Legault C, Rapp SR et al. Estrogen plus progestin and the incidence of dementia and mild cognitive impairment in postmenopausal women: the Women's Health Initiative Memory Study: a randomized controlled trial. JAMA 2003; 289(20):2651–62.

27. Jarvik GP, Wijsman EM, Kukull WA et al. Interactions of apolipoprotein E genotype, total cholesterol level, age, and sex in prediction of Alzheimer's disease: a case-control study. Neurology 1995; 45(6):1092–6.

28. Singh CF, Davignon J. Role of the apolipoprotein E polymorphism in determining normal plasma lipid and lipoprotein variation. Am J Human Genet 1985; 37:268–85.

29. Jick H, Zornberg GL, Jick SS, Seshadri S, Drachman DA. Statins and the risk of dementia. Lancet 2000; 356(9242):1627–31.

30. Wolozin B, Kellman W, Ruosseau P, Celesia GG, Siegel G. Decreased prevalence of Alzheimer disease associated with 3-hydroxy-3-methyglutaryl coenzyme A reductase inhibitors. Arch Neurol 2000; 57(10):1439–43.

31. Shepherd J, Blauw GJ, Murphy MB, Bollen EL, Buckley BM, Cobbe SM et al. Pravastatin in elderly individuals at risk of vascular disease (PROSPER): a randomised controlled trial. Lancet 2002; 360(9346):1623–30.

32. Rockwood K, Kirkland S, Hogan DB et al. Use of lipid-lowering agents, indication bias, and the risk of dementia in community-dwelling elderly people. Arch Neurol 2002; 59(2):223–7.

33. Luchsinger JA, Tang MX, Shea S, Mayeux R. Antioxidant vitamin intake and risk of Alzheimer disease. Arch Neurol 2003; 60(2):203–208.

34. Repetto MG, Reides CG, Evelson P et al. Peripheral markers of oxidative stress in probable Alzheimer patients. Eur J Clin Invest 1999; 29(7):643–9.

35. Rosler M, Retz W, Thome J, Riederer P. Free radicals in Alzheimer's dementia: currently available therapeutic strategies. [Review]. J Neural Trans Supp 1998; 54:211–19.

36. Wilcock GK, Birks J, Whitehead A, Evans SJ. The effect of selegiline in the treatment of people with Alzheimer's disease: a meta-analysis of published trials. Int J Geriatr Psychiatry 2002; 17(2):175–83.

37. Sano M, Ernesto C, Thomas RG et al. A controlled trial of selegiline, alpha-tocopherol, or both as treatment for Alzheimer's disease. The Alzheimer's Disease Cooperative Study. N Engl J Med 1997; 336(17):1216–22.

38. Hardy J, Selkoe DJ. The amyloid hypothesis of Alzheimer's disease: progress and problems on the road to therapeutics. Science 2002; 297(5580): 353–6.

39. Dewachter I, Van Leuven F. Secretases as targets for the treatment of Alzheimer's disease: the prospects. Lancet Neurol 2002; 1(7):409–16.

40. Luo Y, Bolon B, Kahn S et al. Mice deficient in BACE1, the Alzheimer's beta-secretase, have normal phenotype and abolished beta-amyloid generation. Nat Neurosci 2001; 4(3):231–2.

41. Schenk D, Barbour R, Dunn W et al. Immunization with amyloid-beta attenuates Alzheimer-disease-like pathology in the PDAPP mouse [see comments]. Nature 1999; 400(6740):173–7.

42. Hock C, Konietzko U, Streffer JR et al. Antibodies against beta-amyloid slow cognitive decline in Alzheimer's disease. Neuron 2003; 38(4):547–54.

43. Nicoll JAR, Wilkinson DG, Holmes C et al. Neuro-pathology of human Alzheimer's disease after immunisation with amyloid beta-peptide: a case report. Nat Med 2003; 9:448–52.

44. Watson GS, Craft S. The role of insulin resistance in the pathogenesis of Alzheimer's disease: implications for treatment. [Review]. CNS Drugs 2003; 17(1):27–45.

45. Watson GS, Peskind ER, Asthana S et al. Insulin increases CSF Abeta42 levels in normal older adults. Neurology 2003; 60(12):1899–1903.

46. Eriksdotter Jonhagen M, Nordberg A et al. Intra-cerebroventricular infusion of nerve growth factor in three patients with Alzheimer's disease. Dement Geriatr Cogn Disord 1998; 9(5):246–57.

47. Tuszynski MH, Thal L, Pay M et al. A phase 1 clinical trial of nerve growth factor gene therapy for Alzheimer disease. Nat Med 2005; 11:551–5.

48. Budge MM, de Jager C, Hogervorst E, Smith AD. Total plasma homocysteine, age, systolic blood pressure, and cognitive performance in older people. J Am Geriatr Soc 2002; 50(12):2014–18.

49. Clarke R, Harrison G, Richards S. Effect of vitamins and aspirin on markers of platelet activation, oxidative stress and homocysteine in people at high risk of dementia. J Intern Med 2003; 254(1):67–75.

50. Morris MS. Homocysteine and Alzheimer's disease. Lancet Neurol 2003; 2(7):425–8.

51. Silverberg GD, Mayo M, Saul T, Rubenstein E, McGuire D. Alzheimer's disease, normal-pressure hydrocephalus, and senescent changes in CSF circulatory physiology: a hypothesis. Lancet Neurol 2003; 2(8):506–11.

52. Erkinjuntti T, Kurz A, Gauthier S et al. Efficacy of galantamine in probable vascular dementia and Alzheimer's disease combined with cerebrovascular disease: a randomised trial. Lancet 2002; 359(9314):1283–90.

53. Schneider LS. Cholinesterase inhibitors for vascular dementia? Lancet Neurol 2003; 2(11):658–9.

54. Wilkinson D, Doody R, Helme R et al. Donepezil in vascular dementia: a randomized, placebo-controlled study. Neurology 2003; 61(4):479–86.

55. Orgogozo JM, Rigaud AS, Stoffler A, Mobius HJ, Forette F. Efficacy and safety of memantine in patients with mild to moderate vascular dementia: a randomized, placebo-controlled trial (MMM 300). Stroke 2002; 33(7):1834–9.

56. Wilcock G, Mobius HJ, Stoeffler A, on behalf of the MMM 500 group. A double-blind placebo-controlled multicentre study of memantine in mild to moderate vascular dementia (MMM500). Int Clin Psychopharmacol 2002; 17:297–305.

57. Wilcock G, Stoeffler A, Sahin K, Mobius HJ. Neuro-radiological findings and the magnitude of cognitive benefit by memantine treatment. A subgroup analysis of two placebo-controlled clinical trials in vascular dementia. Eur Neuropsychopharmacol 2000; 10(Suppl 3):S360.

58. Grace J, Daniel S, Stevens T et al. Long-term use of rivastigmine in patients with dementia with Lewy bodies: an open-label trial. Int Psychogeriatr 2001; 13(2):199–205.

59. McKeith I, Del Ser T, Spano P et al. Efficacy of rivastigmine in dementia with Lewy bodies: a randomised, double-blind, placebo-controlled international study. Lancet 2000; 356(9247):2031–6.

60. Scarpini E, Scheltens P, Feldman H. Treatment of Alzheimer's disease: current status and new perspectives. Lancet Neurol 2003; 2(9):539–47.

61. O'Brien JT, Erkinjuntti T, Reisberg B et al. Vascular cognitive impairment. Lancet Neurol 2003; 2(2):89–98.

4

Clinical Trial Designs and Endpoint Selection

Serge Gauthier and Chris MacKnight

Introduction

The pharmacological treatment of Alzheimer's disease (AD) has a relatively short history. Much has been learned since the publication on November 13 1986 of encouraging results from a crossover study using tacrine.[1] This chapter will review the natural history of AD and published experience on trial designs and endpoint selection necessary to establish symptomatic benefit or disease modification. There are still only limited data available on trial designs and endpoint selection in non-AD dementias, although more recently studies in vascular dementia and Parkinson's disease-dementia are being undertaken.

Natural history of Alzheimer's disease

The natural history of AD can be broadly considered as a pre-symptomatic stage, an early symptomatic or prodromal stage with affective and/or cognitive manifestations, and symptomatic mild, moderate and severe stages. Each of these stages could be targeted for therapy, but within the broad constraint of a randomised, controlled model, are likely to require different trial designs and outcomes. For instance, in a healthy elderly sample that is being treated with ginkgo biloba over five years, incident dementia is the primary outcome. Similarly, in persons with the amnestic type of mild cognitive impairment (MCI) being treated with various cholinesterase inhibitors (ChEIs) or vitamin E over three years, diagnosable dementia is a reasonable endpoint. Mild to moderate AD has been studied extensively over 3 to 12 months using parallel groups and dual primary outcomes global impression of change and Alzheimer Disease Assessment Scale – cognitive (ADAS-Cog).[2] Studies are under way in severe stages of AD targeting behaviour as an important outcome.

Disease milestones can be defined in AD (see Table 4.1). Some of these can be a target for treatment, with considerable face validity and impact on care.[3] Delaying loss of autonomy for

---

Table 4.1 Clinical milestones in Alzheimer's disease

- Emergence of cognitive symptoms
- Conversion from amnestic MCI to diagnosable dementia
- Loss of instrumental ADL
- Deterioration to 'worse than expected' performance in individualised outcome measures
- Emergence of BPSD
- Nursing home placement
- Loss of self-care ADL
- Death

self-care and even death in moderate to severe stages of AD are relevant endpoints. That these are well accepted is evidenced by the impact of a study by the Alzheimer Disease Cooperative Study group,[4] so that vitamin E is commonly used in all stages of AD, at least in the US. Similarly, delaying the loss of autonomy for activities of daily living (ADL) or the need for nursing home care would be of great pharmacoeconomic benefit. Delaying emergence of some of the behavioural and psychological symptoms of dementia (BPSD) would reduce caregiver burden and perhaps translate into less need for nursing home placement, although this has been difficult to demonstrate.[5] It might be that nursing placement (and also death) will prove to be poor outcomes in anti-dementia drug trials, because of competing risks and, in the case of placement, biased environmental influences.

A slightly different approach to endpoints is to have patients or caregivers choose their own.[6] In the technique of goal attainment scaling, for example, patients and caregivers define their own goals for treatment.[7] Individualised scales are constructed by defining not just the expected goal (for example, being able to find the way safely to the corner store on most days) but also better and worse than expected outcomes (for example, being completely independent in initiating and carrying out trips to the corner store, or becoming lost or injured in attempting to get there, respectively). The time to achieve a goal, or avoid a worse than expected outcome, can thus become a clinically relevant measure by which treatment effectiveness might be evaluated.[8] Such an approach might help reduce the gap between understanding that a drug produces statistically significant effects, and knowing whether these effects are clinically meaningful.[9]

Even the *a priori* individual choice of milestones can be problematic, however, in understanding their clinical importance. For example, much of our thinking about disease progression in treated AD is dominated by our experience with disease progression in untreated AD. This can be misleading in two ways. One is that reversal of disease

progression might not always be the most apt metaphor for thinking about how treatment works. Similarly, a new 'treated stage' of AD might incorporate elements of mild, moderate and even severe disease.[10] Thus a given symptom might not have the same anticipated effect if it is seen in isolation, compared to the expectation that it would be seen together with other signs of disease progression in the untreated state.

The main symptomatic domains in AD include cognition, ADL and behaviour. There are also motor changes, since most patients with AD will manifest some features of parkinsonism late in the disease. In many cases early changes in mood and anxiety precede the formal diagnosis of AD, with spontaneous improvement as insight about the disease is lost. Cognitive and ADL decline are relatively linear (or perhaps log-linear) over time,[11] whereas BPSD peak midway into the disease course and resolve spontaneously through the severe stage as motor function becomes impaired (see Figure 4.1).[12]

These natural fluctuations in the intensity of individual symptomatic domains through the stages of AD have an impact into trial design and endpoint selection. It should be noted that studies can be of shorter duration and/or of smaller numbers of subjects in moderate compared to mild stages of AD, because of the faster rate of decline in the moderate stage, which may be related in

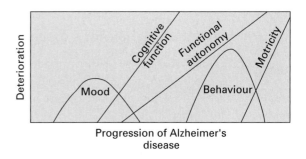

Progression of Alzheimer's disease

Figure 4.1 The intensity of symptoms in various domains throughout the progression of Alzheimer's disease (reproduced with permission from Gauthier et al. 2001[12]).

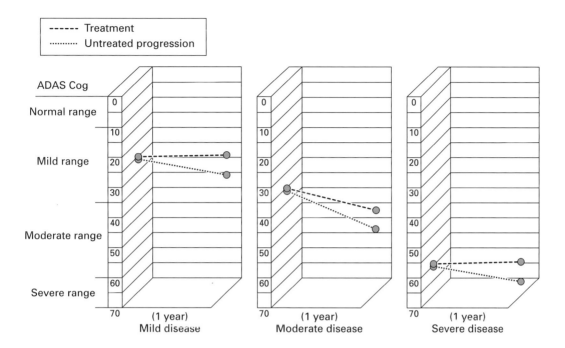

Figure 4.2 Reproduced with permission from Mohr et al. 2004[13]).

part to the sensitivity of measurement scales, or to the natural progression of AD. This is well illustrated in Figure 4.2 which is taken from Mohr et al[13] where the biggest decline over one year in ADAS-Cog is in moderate stages. The floor effect of certain scales such as the Mini-Mental State Examination (MMSE),[14] and the ADAS-Cog in severe stage require change to appropriate scales such as the Severe Impairment Battery (SIB).[15]

Overview of symptomatic trial designs

For the purpose of demonstrating symptomatic benefits in AD, a number of trial designs have been tested, comparing active drugs to placebo, over 3 to 12 months (see Table 4.2).

Since the original study by Summers et al,[1] suggesting the benefit of tacrine using an open-label titration up to a maximally tolerated dose (MTD) up to 200 mg/day followed by a crossover of 3 weeks/3 weeks without washout, the initial

follow-up studies used this design or a variation of it. For example the Canadian Multicentre Study found the MTD up to 100 mg/day and compared tacrine to placebo for two periods of 8 weeks, with washout periods of 4 weeks:[16] a withdrawal effect was detectable on washout after the titration to MTD, whereas a carry-over effect was detected for the ADL scale despite the 4-week washout between randomised treatment periods. In the study of

Table 4.2 Trial designs for symptomatic studies
• Crossover with or without enrichment by responders • Parallel groups • Parallel groups followed by active drug washout • Parallel groups followed by open label active treatment • Parallel groups with survival to a clinical milestone

Davis et al, an enrichment strategy was used, where only patients improving on tacrine up to 80 mg/day were randomised to active drug or placebo in a 6-week parallel double-blind phase;[17] the 'responders' to initial exposure to tacrine were defined as four points or more improvement on the ADAS-Cog, a definition that has been used extensively later in the literature and by regulatory bodies. The crossover design with or without enrichment was felt to be inappropriate for AD considering carry-over effects, changes in baseline between treatment periods, and difficulties in generalisation of study results to the AD population at large.[18]

The parallel group design has been successful in terms of convincing regulators and clinicians of a clinically meaningful symptomatic benefit of four ChEIs. An example of such a study is by Corey-Bloom et al, where two doses of rivastigmine were compared to placebo, demonstrating a clear dose–effect relationship.[19]

An example of the parallel group design followed by a planned but not randomised drug washout is the study by Rogers et al, where patients taken off donepezil lost the symptomatic improvement measured on global impression of change and on cognition within six weeks.[20]

One example of the parallel group design followed by open-label active treatment is the study by Raskind et al, where patients with uninterrupted treatment with galantamine over one year demonstrated no decline in ADL or cognition below their starting point or baseline.[21]

A final example of parallel groups with survival to a clinical milestone is the study of Mohs et al, where patients stayed on the randomly assigned donepezil or placebo group until they reached a predefined decline in function.[22]

Endpoint selection in symptomatic studies

The Food and Drug Administration (FDA) published 'guidelines', which influenced greatly the choice of outcomes for proof of efficacy of drugs improving symptoms in AD:[23] a cognitive performance-based scale such as the ADAS-Cog and an interview-based impression of change became the primary outcomes in mild to moderate AD, defined operationally as scores between 10 and 26 on the MMSE. This has put anti-dementia drug development in a difficult position, as the results of studies (and by extension the measurement of treatment effect in clinical practice) is relatively uninterpretable to patients, their families, frontline healthcare providers, and funders. The development of 'clinimetric' as opposed to 'psychometric' outcome measures may be one way forward. As discussed in Rockwood et al, these measures are clinically meaningful and measurable, yet have been developed with all the rigor of traditional psychometric measures.[6]

The FDA guidelines caution against the 'pseudospecificity' of measurable benefits on neuropsychiatric manifestations in AD. For example, two patients treated with two drugs might both show less aggressive behaviour, but in one case, this might not be a specific treatment effect. Rather it might have come only at the expense of over-sedation, and thus would be a 'pseudo-specific response'. The lack of more precise outcome measures has delayed research in this symptomatic domain. More recent discussions and publications from the FDA and other regulatory agencies have been more open to ADL and behaviour as important outcomes, but important measurement issues remain, as are discussed in depth in the next chapters.

The most difficult domain to study, although very significant clinically, has been behaviour. The availability of general BPSD scales such as the Neuropsychiatric Inventory (NPI),[24] as well as specific scales such as the Cohen-Mansfield Agitation Inventory,[25] has not yet allowed unequivocal demonstration of benefit in severe stages of AD. New methods of analysis of behaviour have been proposed,[26] and will probably be more successful in defining categories of BPSD symptoms most responsive to AChEIs (apathy, anxiety, hallucinations), memantine (agitation) and other drugs.

Memantine as a new therapeutic class has been

found to be effective in a range of studies using parallel groups, in moderate to severe AD.[27] Scales appropriate for this stage of disease, such as the SIB, the Alzheimer Disease Cooperative Study ADL severe scale (ADCS-ADL),[28] and the NPI have been used and accepted by the FDA and other regulatory agencies. Of great importance, the novel design of adding memantine or placebo to a stable dose of an AChEI has been used successfully, paving the way to a number of symptomatic studies where novel drugs or placebo are added to 'standard treatment'.

Disease modification strategies

Although no study has yet led to a successful treatment for disease modification, attempts have been made using parallel groups over one year using ginkgo biloba, I-acetyl-carnitine, prednisone, oestrogens and celecoxib versus placebo in mild to moderate AD.[29–33] Recent refinements of this design include adding the novel drug or a placebo to ChEIs as standard treatment, selection of outcomes which demonstrate relatively linear changes over time (e.g. the Clinical Dementia Rating (CDR) sum of boxes scoring[34]), randomised washout at the end of a year, or volumetric brain measurements using magnetic resonance imaging at the beginning and end of the year of study. The rationale for the 'add-on design' is that one year is the minimum period for meaningful clinical observations in mild to moderate AD considering natural decline (it may need to be longer in mild AD); long duration studies without 'standard treatment' are not possible ethically; there are scales with relatively linear changes over one year, such as the CDR sum of boxes, allowing analysis for slopes of decline; a randomised washout from active treatment at the end of the study could demonstrate a lack of return to placebo; a demonstrable reduction in rate of brain atrophy associated with differences in clinical decline would offer great face validity.

Although this design appears promising, there are uncertainties. For instance although the natural rate of brain atrophy is known in mild to moderate AD, the heterogeneity of disease progression may prevent the demonstration of a reduction in atrophy from a given treatment. Other threats are the (often unknown) duration of carry-over of the treatment effect, the presence of withdrawal effects which could unblind subjects or investigators, and the assumption that drop-outs during the study will not lead to group imbalance at the point of withdrawal. Washout from an active treatment will not be acceptable to all institutional review boards, and a previous attempt at using this design with patients treated with propentofylline has failed to convince regulatory agencies.[35] Delayed start designs have also been proposed for the demonstration of disease modification, but these also rest on assumptions that are difficult to support, such as that response to an agent is independent of disease stage.

One of the difficult issues in disease modification strategies is the decision as to the stage of disease where the proposed drug is most likely to work, based on mechanisms of action. On this 'proof-of-concept' phase II/III efficacy and safety study hinges the entire future of a given drug. For example numerous attempts at treating patients with AD in mild to moderate stages using anti-inflammatory drugs have failed, despite the strong evidence from epidemiological research and the biological plausibility of an inflammatory component to AD pathology. It may be that treatment in the late presymptomatic or in the prodromal stages would be the most appropriate time. On the other hand, studies in these stages of AD would require 3 to 5 years, a very long time for a 'proof-of-concept'. Alternative patient groups could be considered, such as presenilin mutation carriers nearing the onset of symptomatic AD, or amnestic MCI with risk factors for rapid conversion to AD.[36]

Strategies to delay emergence of AD

As hypotheses on the pathophysiology of AD emerge from epidemiological research in human populations, post-mortem and biomarker studies in patients, and animal models, there will be a need to establish if new therapies can delay the onset of symptoms in asymptomatic persons at varying degrees of risk of AD. The prototype of trial design to establish the safety and efficacy of such therapies is the ongoing 5-year survival study comparing ginkgo biloba to placebo in healthy elderly subjects, with incident dementia as the primary endpoint. Variations of this design might also prove to be more practicable, such as enriching the study population with different levels of risk, such as a positive family history of AD and/or selected gene markers,[37] although it should be remembered that any enrichment of a study population will limit the applicability of findings to the population as a whole. Another option is to include careful measurement of cognition and dementia within large-scale preventative studies for other conditions (e.g., the Syst-Eur trial in isolated systolic hypertension).[38]

Design for vaccines and other novel agents

Clinical trial design for the development of vaccines, immunomodulating agents, and other novel therapies is in its infancy.[39] Much of the traditional methodology, through Phases I–III, will need to be modified. For example, maximum tolerated dose is typically studied in Phase I. For immunomodulating agents, the optimal biological dose is a better target (as the maximum tolerated dose may not give the best response).[40] Adaptive designs, where sample sizes and primary outcomes may be adjusted depending on early results, should be considered, both to reduce the costs and duration of drug development.[41]

References

1. Summers WK, Majovski LV, Marsh GM, Tachiki K, Kling A. Oral tetrahydroaminoacridine in long-term treatment of senile dementia, Alzheimer type. N Engl J Med 1986; 315:1241–5.

2. Rosen WG, Mohs RC, Davis KL. A new rating scale for Alzheimer's disease. Am J Psychiatry 1984; 141:1356–64.

3. Galasko D, Edland SD, Morris JC et al. The Consortium to Establish a Registry for Alzheimer's Disease (CERAD). Part IX. Clinical milestones in patients with Alzheimer's disease followed over 3 years. Neurology 1995; 45:1451–5.

4. Sano M, Ernesto C, Thomas RG et al. A controlled trial of selegiline, alpha-tocopherol, or both as treatment for Alzheimer's disease. N Engl J Med 197; 336:1216–22.

5. AD2000 Collaborative Group. Long-term donepezil treatment in 565 patients with Alzheimer's disease (AD2000): randomized double-blind trial. Lancet 2004; 363:2105–15.

6. Rockwood K, Graham JE, Fay S. Goal setting and attainment in Alzheimer's disease patients treated with donepezil. J Neurol Neurosurg Psychiatry 2002; 73:500–507.

7. Kiresuk TJ, Lund SH, Larsen NE. Measurement of goal attainment in clinical and health care programs. Drug Intell Clin Pharm. 1982; 16:145–53.

8. Rockwood K, MacKnight C. Assessing the clinical importance of statistically significant improvement in anti-dementia drug trials. Neuroepidemiology 2001; 20:51–6.

9. Burns A. Meaningful treatment outcomes in Alzheimer's disease. J Neurol Neurosurg Psychiatry 2002; 73:471–2.

10. Rockwood K, Wallack M, Tallis R. Treatment of Alzheimer's disease: understanding success short of cure. Lancet Neurol 2003; 2:630–3.

11. Mitnitski AB, Rockwood K. Modeling decline in Alzheimer's disease. Int Psychogeriatr 1999; 11:211–16.

12. Gauthier S, Thal LJ, Rossor MN. The future diagnosis and management of Alzheimer's disease. In: Gauthier S (ed). Clinical Diagnosis and Management of Alzheimer's Disease. London: Martin Dunitz; 2001: 369–78.

13. Mohr E, Barclay CL, Anderson R, Constant J. Clinical trial design in the age of dementia treatments: challenges and opportunities. In: Gauthier S, Scheltens P, Cummings JL (eds). Alzheimer's Disease and Related Disorders Annual 2004. London: Martin Dunitz; 2004: 97–122.

14. Folstein MF, Folstein SE, McHugh PR. Mini Mental State: a practical method for grading the cognitive state of patients for the clinician. J Psychiatr Res 1975; 12:189–98.

15. Panisset M, Roudier M, Saxton J et al. Severe Impairment Battery: a neuropsychological test for severely demented patients. Arch Neurol 1994; 51:41–5.

16. Gauthier S, Bouchard R, Lamontagne A et al. Tetrahydroaminoacridine–lecithin combination treatment in intermediate stage Alzheimer's disease: results of a Canadian double-blind cross-over multicentre study. N Engl J Med 1990; 322:1272–6.

17. Davis KL, Thal LJ, Gamzu ER et al. A double-blind, placebo-controlled multicentre study of tacrine for Alzheimer's disease. N Engl J Med 1992; 327:1253–9.

18. Gauthier S, Gauthier L. What have we learned from the THA trials to facilitate testing of new AChE inhibitors. In: Cuello AC (ed). Progress in Brain Research. Amsterdam: Elsevier Science Publishers; 1993: 427–9.

19. Corey-Bloom J, Anand R, Veach J, for the ENA 713 B352 Study Group. A randomized trial evaluating the efficacy and safety of ENA 713 (rivastigmine tartrate), a new acetylcholinesterase inhibitor, in patients with mild to moderately severe Alzheimer's disease. Int J Geriatr Psychopharmacol 1998; 1:55–65.

20. Rogers S, Farlow MR, Doody RS, Mohs R, Friedhoff LT and the Donepezil Study Group. A 24-week, double-blind, placebo-controlled trial of donepezil in patients with Alzheimer's disease. Neurology 1998; 50:136–45.

21. Raskind MA, Peskind ER, Wessel T, Yuan W and the Galantamine USA-1 Study Group. Galantamine in AD. A 6-month randomized, placebo-controlled trial with a 6-month extension. Neurology 2000; 54:2261–8.

22. Mohs R, Doody R, Morris J et al. A 1-year, placebo-controlled preservation of function survival study of donepezil in AD patients. Neurology 2001; 57:481–8.

23. Leber P. Guidelines for Clinical Evaluation of Anti-dementia Drugs. Washington, DC: US Food and Drug Administration; 1990.

24. Cummings JL, Mega M, Gray K et al. The Neuropsychiatric Inventory: comprehensive assessment of psychopathology in dementia. Neurology 1994; 44:2308–14.

25. Cohen-Mansfield J, Marx MS, Rosenthal AS. A description of agitation in a nursing home. J Gerontology 1989; 44:M77–M84.

26. Gauthier S, Feldman H, Hecker J et al. Efficacy of donepezil on behavioral symptoms in patients with moderate to severe Alzheimer's disease. Int Psychogeriatrics 2001; 14:389–404.

27. Doody RS, Winblad B, Jelic V. Memantine: a glutamate antagonist for treatment of Alzheimer's disease. In: Gauthier S, Scheltens P, Cummings JL (eds). Alzheimer's Disease and Related Disorders Annual 2004. London: Martin Dunitz; 2004: 137–44.

28. Galasko D, Bennett D, Sano M et al. An inventory to assess activities of daily living for clinical trials in Alzheimer's disease. Alzheimer Dis Assoc Disord 11(Suppl2); 1997:S33–S39.

29. Lebars PL, Katz MM, Berman N et al for the North American EGB Study Group. A placebo-controlled, double-blind, randomized trial of an extract of Ginkgo Biloba for dementia. JAMA 1997; 278:1327–32.

30. Thal LJ, Carta A, Clarke WR et al. A 1-year multicentre placebo-controlled study of acetyl-L-carnitine in patients with Alzheimer's disease. Neurology 1996; 47:705–11.

31. Aisen PS, Davis KL, Berg JD et al for the Alzheimer's Disease Cooperative Study. A randomized controlled trial of prednisone in Alzheimer's disease. Neurology 2000; 54:588–93.

32. Henderson VW, Paganini-Hill A, Miller BL et al. Estrogen for Alzheimer's disease in women. Neurology 2000; 54:295–301.

33. Sainati SM, Ingram DM, Talwalker S, Geis GS. Results of a double-blind, randomized, placebo-controlled study of celecoxib in the treatment of progression of Alzheimer's disease. Presented at the 6th International Stockholm/Springfield Symposium on Advances in Alzheimer Therapy, Stockholm, April 5–8, 2000.

34. Hughes CP, Berg L, Danziger WL, Coben LA, Martin RL. A new clinical scale for the staging of dementia. Br J Psychiatry 1982; 140:566–72.

35. Whitehouse PJ, Kittner B, Roessner M et al. Clinical trial designs for demonstrating disease-course-altering effects in dementia. Alzheimer Dis Assoc Disord 1998; 12:281–94.

36. Gauthier S, Touchon J. Subclassification of mild cognitive impairment in research and clinical practice. In: Gauthier S, Scheltens P, Cummings JL (eds). Alzheimer's Disease and Related Disorders Annual 2004. London: Martin Dunitz; 2004: 61–9.

37. Gauthier S. The benefits of apolipoprotein E e4 screening to research. CMAJ 2004; 171:881.

38. Forette F, Seux ML, Staessen JA et al. The prevention of dementia with antihypertensive treatment: new evidence from the Systolic Hypertension in Europe (Syst-Eur) study. Arch Intern Med 2002; 162:2046–52.

39. Heppner FL, Gandy S, McLaurin J. Current concepts and future prospects for Alzheimer disease vaccines. Alzheimer Dis Assoc Disord. 2004; 18:38–43.

40. Simon RM, Steinberg SM, Hamilton M et al. Clinical trial design for the early clinical development of therapeutic cancer vaccines. J Clin Oncol 2001; 19:1848–54.

41. Fox E, Curt GA, Balis FM. Clinical trial design for target-based therapy. Oncologist 2002; 7:401–9.

5

The Use of Qualitative Research

Christine Joffres and Kenneth Rockwood

This chapter is about qualitative research and its use in anti-dementia drug trials. As these have been generally little used to evaluate anti-dementia drugs, the chapter first introduces readers to general methodological principles of qualitative research, including sampling strategies, data collection, data analyses, different types of qualitative research (e.g. ethnography, grounded theory, phenomenology), mixed studies (i.e. studies that combine qualitative and quantitative methods), and ways to ensure the trustworthiness (or reliability and validity) of qualitative findings. In the second part of the chapter, we illustrated via two of our studies how qualitative data can be collected, analysed, and used in anti-dementia drug trials.[1,2]

Introduction

The purpose of qualitative research is to systematically collect and analyse interviews, observations, and documents to better understand specific phenomena. Qualitative research has gained momentum in the last 10 years, particularly in health research. For example, a PubMed search listed over 2000 manuscripts that included the words 'qualitative research' from 1993 to 2003, compared with only 200 such papers from 1982 to 1992. This increase reflects recognition that qualitative research can allow important research questions to be pursued in ways that are not easily answerable by experimental methods. For example, while drug trials inform practitioners about the efficacy of specific therapeutic agents, qualitative research has been used to better understand clinicians' judgements about the global measures of change commonly used in anti-dementia drug trials,[1] and to build profiles of treatment effects to specific drugs (e.g. galatamine).[2] More generally, researchers agree, that, while qualitative research is not appropriate for every kind of investigation,[3,4] it is particularly useful to:

- explore and develop hypotheses about relatively poorly understood and/or complex phenomena. In her paper about the potential uses of qualitative research in gerontology, Rempusheski explained how a clinical observation of an elderly person with Alzheimer's disease raised her curiosity about the contextual factors associated with specific behavioural manifestations of Alzheimer's disease[3]
- link processes to outcomes. Qualitative studies, based on interviews and observations, have helped health researchers understand how changes brought by a disease were accommodated into the patients' and their families' lives[5]
- study phenomena involving a number of subjects that is insufficient for statistical analysis[6]

• examine situations with ill-defined or uncontrolled-for contextual forces.[3]

Methodologically, qualitative research tends to be hermeneutic ('in depicting individual constructions as accurately as possible'[7]) and dialectic ('comparing and contrasting these existing individual (including the inquirer's) constructions').[7] The qualitative research design is thus necessarily flexible and iterative. Iterations between the selection of respondents, data collection, and data analyses continue until the data yield little or no new information, a technique known as 'sampling to exhaustion'.[8]

Qualitative methods

This section focuses on sampling techniques, data collection, and data analysis strategies specific to qualitative inquiries. It also describes different types of qualitative inquiry (e.g. ethnography, grounded theory, and phenomenology), and the procedures that qualitative researchers used to ensure the internal and external validity (or credibility and transferability) and reliability (or dependability) of qualitative inquiries.

Sampling strategies

Participants in qualitative research are selected purposefully.[9] This sampling strategy (called purposive or purposeful sampling) is based on the selection of information-rich cases, that is, the selection of respondents who know 'a great deal about issues of central importance to the purpose of the research'.[9] While random selection of a representative sample permits generalisation from the sample to a larger population, purposive sampling allows researchers to build propositions, hypotheses, and/or plausible explanations about specific phenomena inductively. There are many strategies for purposively selecting information-rich cases.[9,10] Each has a specific and different purpose. For example, extreme or deviant sampling focuses on cases that are rich in information but unusual or special in some way. Intensity sampling involves the selection of cases that experience or have experienced a specific phenomenon intensely but who are not unusual (e.g. patients with similar conditions who respond well versus patients who respond poorly to the same therapeutic agent). Maximum variation sampling aims at selecting a wide range of responses to a specific experience. In contrast, homogeneous sampling focuses on selecting a small, homogeneous group of a particular subgroup of participants. Other sampling strategies focus on typical (typical case sampling), critical (critical case sampling), confirming and disconfirming (confirming and disconfirming sampling) cases that illuminate the phenomenon under study in ways particularly relevant to the research. Theoretical sampling, specific to grounded theory, aims at picking cases along specific dimensions or conceptual constructs that help researchers generate theory about a specific and, usually, poorly known phenomenon.[11] Still other sampling strategies include snowball or chain sampling, which involves initial participants in identifying other respondents relevant to the research question/s, and opportunistic sampling, whereby researchers decide on the spot to take advantage of new opportunities/cases during data collection. Random sampling can also be employed, depending on the strength of the *a priori* hypothesis and the need for generalisation, as in Brendl et al's study on attitudes about racism.[12] It is not unusual in qualitative research to combine or use sequentially different strategies for selecting participants. A research project can start with the purposeful selection of information-rich respondents, to include later extreme or deviant case sampling or typical case sampling. Studies using mixed methods (i.e. that combine quantitative and qualitative methods) can use random sampling followed by stratified purposeful sampling of above average, average, and below average cases or some kind of purposeful selection strategy or some other combinations of qualitative and quantitative sampling strategies. Table 5.1 includes studies that have used different sampling strategies.

Table 5.1 Examples of sampling strategies in relation to the purpose of the study

Authors	Sampling strategy	Purpose of the study
Joffres et al 2000[1]	Purposive sampling of patients with Alzheimer's disease	To explore clinicians' definitions of change on a clinimetric change scale
Joffres et al 2003[2]	Purposive sampling of patients with Alzheimer's disease	To profile treatment response to galantamine of patients with varying severity of disease
Butcher et al 2001[16]	Maximum variation sampling and random stratified sampling of 30 family caregivers	To examine decision making regarding the placement of a family member with Alzheimer's disease in a special care unit

The exploratory nature of qualitative research generally does not allow researchers to predetermine the size of their sample. Selection of participants continues until the data yield redundant, minimal or no new information, a phenomenon described as data saturation, redundancy or sampling to exhaustion.[13–15]

Data collection

Qualitative data are typically collected via a combination of observations, face-to-face or group interviews, and/or the compilation of written records. Data collection continues until data do not yield new information (data saturation).

Observations

Observations generally involve a detailed description of the phenomenon. Observations can be overt or covert, broad or narrow in focus (e.g. focusing on a few elements of a phenomenon as opposed to an holistic approach), short- or long-term (varying from a single one-hour observation to multiple observations over several months or years), and structured or unstructured.[9,17] The researchers' involvement in the events/phenomena under observation can also vary along a continuum ranging from onlooker/outsider to full participant/insider.[9,17] The degree of structure that guides the collection of observational data can also vary. At one end of the continuum, observations are unstructured, that is, information is

recorded from a holistic perspective (the whole event is observed and not bits or pieces of it). At the other end, structured observations have pre-established observational schedules and pre-specified focus or foci of observation.[17]

Decisions regarding these issues can be complex and are usually a function of the researchers' epistemological beliefs, the nature of the observational data required for the study, and pragmatic concerns (e.g. length of the study, research budget, respondent and researcher availability). For example, marked social, age or physical (e.g. disabilities) differences between researchers and participants may limit the extent to which researchers can participate in the programs/events under observation.

Interviews

As with observations, interviews can be classified according to their degree of predetermined structure. Patton identified three primary types of interviews:[9] (1) the standardised open-ended interview, (2) the general interview guide approach and (3) the informal conversational interview.[9] The purpose of *standardised open-ended interviews* is to collect data in a systematic way while minimising interviewers' effects. Thus questions in such interviews are specified in advance and should be asked exactly as written out in the questionnaire. Questions asking for clarifications and/or elaborations are also written in the questionnaire at appropriate places. Interviewers do not deviate from the questionnaire/survey. They also facilitate data analyses as they set limits to the type of data that can be collected and allow researchers to locate quickly respondents' answers to the same questions.

The *general interview guide* approach is more flexible than the standardised open-ended interview. It includes topics/subject areas that allow the researcher/s to explore similar sets of issues with each respondent, thus ensuring that respondents cover the same material. At the same time, the interviewer is free to probe respondents for more information on the areas of interest within the guide interview. This type of interview facilitates systematic data collection while increasing the comprehensiveness of the data collected.

The *informal conversational interview* is the most open-ended approach. Most of the questions flow directly from the conversations with the respondents. This type of interview allows researchers to be highly responsive to individual differences and to pursue information in whatever direction seemed to be most appropriate for the research questions. The downside of informal interviews is that they are more sensitive to interviewer effects, and depend on the interviewer's conversational skills to a greater extent than more structured interviews. Interviewers must be able to interact easily with respondents, quickly formulate new questions, and guard against asking questions that may lead the respondents' answers. Informal interviews are particularly appropriate for exploratory studies and/or to expand on knowledge collected in the initial phase of a study.

Interviews can be face-to-face or include a small group of people (from 6 to 12). Decisions about conducting face-to-face versus group interviews (also called focus groups) should take into account the following factors: interview topics (sensitive topics may be better discussed in face-to-face interviews); participants (youths may need company to be encouraged to talk); interviewers (focus groups need experienced facilitators as the articulation of group norms may inhibit specific participants with diverging views), research budget (face-to-face interviews tend to be more expensive than focus groups), and time (face-to-face interviews require more time than group interviews).[18–20]

Written documentation

Written documentation (other than the detailed accounts of a researcher's observations) can include any written documents deemed pertinent to the research (e.g. diaries, letters, meeting minutes, organisational policies, patients' files). Our qualitative study of Alzheimer's disease patient treatment responses to galantamine was based on patients' files,[2] as was our study on clinicians' understandings of global change scales.[1]

Written documentation also includes the researcher's field notes. Qualitative researchers are required to write field notes throughout the data collection and data analyses. These consist of the researcher's notes about his/her own feelings, reactions to the phenomenon under study, and memos keeping track of the researchers' insights and emerging hypotheses. Decisions about the sampling and data analysis strategies, and the rationale behind these decisions, are also recorded in the field notes. In a later section, we show how these notes strengthen the credibility of the data.

Table 5.2 includes examples of studies that have used diverse data collection strategies.

Data analyses

Data analyses in qualitative research start at the beginning of and continue throughout data collection. This is congruent with the emergent design of qualitative research. Initial data analyses guide further data gathering (and sampling strategies) in an iterative process. As new data are gathered, they are compared and contrasted with existing data and emerging insights. The new information either validates or challenges (or partly validates and partly challenges) initial findings, highlights gaps in data collection, and increases understanding of the phenomenon

Table 5.2 Examples of data collection strategies in relation to the purpose of a study		
Authors	**Data collection strategy**	**Purpose of the study**
Joffres et al 2000[1]	Patients files including clinicians' notes, patients' tests, and patients' and caregivers' feedback	To explore clinicians' definitions of change on a clinimetric change scale
Morgan et al 2002[21]	Focus groups of decision makers	To obtain input from decision-makers to develop the objectives and design for a study of rural dementia care in Saskatchewan
Smith et al 2001[22]	In-depth interviews of family members of patients with AD	To examine the psychological impact of disclosing a diagnosis of AD on patients and family members
Homan-Rock et al 2001[23]	Two-stage data collection process: experimental observations carried out by 22 physicians in 19 Dutch practices	To develop and validate a short observation of a list of possible early signs of dementia for use in general practice
Hutchinson et al 2000[24]	In-depth interviews with and observations of 21 caregiver/family member with AD dyads	To assess the responses of family caregivers to an activity kit containing 20 therapeutic activities

under investigation. An illustration of the way we analysed our data is provided later in this chapter.

Analytical methods in qualitative research present some common features.[3,25] These include:

- familiarisation with the data by listening to tapes, reading transcripts and other written materials (e.g. field notes)
- breaking the data into meaningful pieces and assigning them codes. Codes are arrived at inductively and are grounded in the data. As data are coded, a coding dictionary with the definitions of the different codes is usually developed
- clustering the data labelled with similar codes into categories. Sorting and sifting through the categories to identify main and subcategories, potential relationships between the categories, as well as the dimensions or properties specific to each category. The linkages of categories are referred to as axial coding in grounded theory because coding occurs around the axis of a cat-

egory, linking this category to 'sub'categories describing its dimensions or properties

- memoing or carefully documenting emerging thoughts, insights, interpretations, or hypotheses with descriptions as 'thick' as possible, given the data available. Memos also include questions to pursue and directions for further data collection[11]
- isolating by comparing and contrasting potential patterns and processes in the data, as well as any deviant or extreme case. Similarities in the data make it easier to recognise patterns and increase the credibility of the data. On the other hand, differences help ensure the richness of the data and subsequent analyses
- verifying emerging patterns, insights and hypotheses against the next wave of data and changing the coding system accordingly
- using new data to develop increasingly thick descriptions of the different aspects of the phenomenon under investigation and to develop small sets of generalisations

- triangulating the data with other sources (see section on trustworthiness of the data, page 53) to ensure that the data are credible.

There are several qualitative software packages available on the market that combine management of textual data with processes for indexing, linking, and searching the data. The most commonly used in health research so far is QSR NUD*IST (Qualitative Solutions and Research Non-Numerical Unstructured Data Indexing Searching and Theory-Building Software, *www.qsrinternational.com*). These packages are very useful to code and manage the data but they are not comprehensive analytical packages.

An illustration of systematic data analyses is provided in the second part of this chapter when we introduce our studies in anti-dementia drug trials.

Types of qualitative research

There are different types of qualitative approaches (also referred to as traditions of inquiry). The most commonly used in healthcare research include ethnography, grounded theory, phenomenology, and fundamental qualitative description.

Ethnography

Ethnography is grounded in culture and its purpose is to describe and interpret a cultural or social group or system.[4,9,10] The researcher immerses him- or herself in the culture under study and examines the group's learned patterns of behaviour. Data are collected through observations, interviews, and the analysis of any other written material deemed helpful in developing 'cultural rules'.[26] Powers described how she immersed herself in observing and recording the details of everyday life in a nursing home.[27] Other examples of ethnographic studies include Henderson's study, which describes the culture of selected special care units in nursing homes,[28] and McAllister and Silverman's study.[29] This study, based on an ethnographic project in a residential Alzheimer's facility and a traditional nursing home, describes the processes of community for-mation and the maintenance of community roles among individuals suffering from dementia in institutional settings.

Grounded theory

The intent of grounded theory is to generate concepts, constructs, or theories about a specific phenomenon that are grounded in the data.[11] Grounded theory is characterised by theoretical sampling (see definition in the sampling strategy section), open, axial, and selective coding (see section on coding and data analyses, page 57 for an explanation and illustration of this coding system), theoretical memos, and constant comparisons. McCarty, Orona and Hurley et al conducted grounded theory studies.[30–32] McCarty explored the process of caregiver stress associated with the care of a parent with Alzheimer's disease.[30] The study findings, which included substantive theory and 13 hypotheses, provide an expanded awareness of the interrelationship between caregiver stress, the contextual aspects of social support, coping, and the nature of the prior filial relationship.[30] Orona walks readers through the abstraction of data bits and the aggregation of verbatim comments to examine the loss of identity attributes in a person with Alzheimer's disease as perceived by a family member.[31] Hurley et al developed an 'achieving consensus' process model about treatment options for patients with AD, using observational and interview data obtained from nurse caregivers and family members of patients with late-stage Alzheimer's disease.[32]

Phenomenology

Phenomenology attempts to describe the 'essential' meaning of lived experiences.[33] It relies primarily on in-depth interviews with a small number of people who share or have shared a common experience, one that is often difficult to measure. Data analyses include horizontalisation of the data (i.e. the researcher finds statements about participants' experiences of the phenomenon under study, clusters them into non-repetitive and non-overlapping statements, and groups these statements into 'meaning units'). The researcher then develops a textural and structural description of the experience (i.e., what happened and how the phenomenon was experienced) and overall description of the 'essence' of the experience. In their study, Butcher et al. (2001) described the experience of caring for a family member with Alzheimer's disease or related disorder (ADRD) living at home.[16] The research involved secondary analysis of in-depth transcribed interview data using vanKaam's 12-step psychophenomenological method. A total of 2115 descriptive expressions were categorised into 38 preliminary structural elements. Eight essential structural elements emerged from an analysis of the preliminary structural elements. The eight elements were then synthesised into a definition of caregiving: caring for a family member living at home with ADRD was experienced as 'being immersed in caregiving; enduring stress and frustration; suffering through the losses; integrating ADRD into our lives and preserving integrity; gathering support; moving with continuous change; and finding meaning and joy'.[16] Vellone et al also explored caregivers' experiences and synthesised their findings under eight main themes: illness, patient, caring, caregiver's life and health, coping, spouse/family, others, and feelings.[34]

Phenomenological research is presently enjoying something of a resurgence in psychiatry.[35] A particularly compelling phenomenological investigation of Alzheimer's disease has drawn attention to ways in which therapeutic interventions can help construct new and adapative relationships between people with AD and those close to them.[36]

Fundamental qualitative description

Fundamental qualitative descriptions are less interpretive than descriptions in other qualitative traditions (e.g. ethnography) as they do not require a conceptual or highly abstract rendering of the data.[37] A qualitative description consists of a comprehensive summary of a case or phenomenon in everyday language. Sandelowski argues that such descriptions have descriptive and interpretive validity in the sense that most participants would agree on the accounting of the phenomenon described and the meanings attributed to this phenomenon.[37] Sampling of participants is purposeful, preferably via maximum variation sampling. Data analyses include qualitative content analysis (data are broken down into meaningful pieces and coded) and may also include quantitative content analyses, which entail counting responses and the number of participants in each category.[38] A few descriptive studies have provided insight into the treatment of patients with AD. For example, Geldmacher detailed experience in six cases in which patients with AD were treated with donepezil to profile treatment responses, that otherwise had largely been based on the MMSE.[39] As well, in a series of 57 patients in whom donepezil was initiatied, Shua-Haim et al. described caregivers' impressions of drug treatment effects using a semi-structured interview.[40] Variable patient responses were noted. Importantly, caregiver satisfaction was not related to standard score results, including the MMSE and standard measures of instrumental activities of daily living (IADL)/activities of daily living (ADL) function and behaviour. Family reports of improvement were notable in several respects: memory (except in one patient); cognitive improvement otherwise was felt to be related largely to improved language and was found not to relate to improvement in function.

Mixed methods

The combination of different methods, either within the same paradigm (e.g. combining observations and focus groups to explore the same phenomenon or variables, a process called within-method triangulation), or across methods (i.e. combining quantitative and qualitative methods within the same study) has been widely advocated by many researchers.[6,8,41–46] Rationales supporting the combination of methods include:

- overcoming the shortcomings and biases of single method studies

- increasing confidence in the results (validation)
- allowing development and validation of instruments used in a research project
- providing a fuller, more accurate picture of the phenomenon under investigation (completeness)
- contributing to theory and knowledge development
- increasing abductive inspiration, which is the process of using one method to generate ideas that are tested via another method.[47]

Quantitative and qualitative approaches provide different kinds of knowledge. While quantitative research provides a broad, general view, or macro perspective of a phenomenon, qualitative data give deeper and more complex insights into the same phenomenon. Hence, the combination of quantitative and qualitative approaches can offset the shortcomings of both strategies and allow researchers to combine knowledge at the macro- and micro-levels. For example, while randomised clinical trials remain the gold standard to investigate the efficacy of a drug in large groups, they rarely tell clinicians if a drug is appropriate for specific subgroups of patients, or how specific subgroups of patients experience this drug. However, medical knowledge generated from large groups cannot be applied indiscriminately to individuals without understanding of the patient specifics and the varied effects of drugs at varying levels of disease severity.[6,48,49] As Lakshman pointed out, clinicians have a 'critical need to understand where generality ends and individuality begins, and that requires merging the two types of knowledge [quantitative and qualitative]'.[6] Qualitative studies have the potential to identify patterns of treatment effects at varying levels of disease severity, as shown by one of our studies on the effects of galantamine on purposefully selected Alzheimer's patients.[2] Importantly too, for our purposes, qualitative methods can draw attention to crucial areas of treatment response that are not captured by standard tests.[50]

Research methods can be combined in three ways: (1) the qualitative study precedes the quantitative study; (2) the quantitative study precedes the qualitative study; and (3) the qualitative and quantitative studies are conducted independently and research findings are compared and/or merged. Where the qualitative study precedes the quantitative one, qualitative evidence typically provides complementary assistance to the subsequent quantitative study.[51] For example, focus groups or face-to-face interviews can be and commonly have been used to develop measurement scales. Informants identify variables or domains that constitute the phenomenon under study (abductive inspiration) and which are integrated in the quantitative instrument. Participants can also provide feedback about the appropriateness, face, and content validity of the resulting scale/instrument.[8] A concrete example of this strategy is provided by Marwit and Meuser.[52] These authors collected 184 statements addressing personal grief reactions from 45 adult child and 42 spouse caregivers in 16 focus groups representing early, middle, late, and post-death stages to develop a psychometrically sound instrument for the assessment of grief in caregivers of persons with Alzheimer's disease.[52]

Focus groups, face-to-face interviews, observations, or a combination of these three strategies have also been used to generate hypotheses about and/or explanations of phenomena that are poorly understood (abductive inspiration). In this case, the quantitative or deductive study is carried out to test the internal validity or credibility of a theory or hypotheses that have emerged from qualitative data (validation). Quantitative findings can also be used to test the generalisability or transferability of previously collected qualitative evidence to a larger population, other populations, or different settings (external validity).

Second, the quantitative study precedes the qualitative study. In this case, the qualitative study provides increased completeness to quantitative findings. For example, when an unexplained quantitative outcome is obtained from a study, participants can be approached to elicit an explanation. The process-approach of qualitative studies is particularly appropriate for this type of exploration (knowledge generation). Similarly quantitative data can reveal outliers or unique experiences,[53,54] which can be explored qualitatively, thereby increasing understanding of the phenomenon under investigation (completeness). The focus of qualitative studies on exploring varying ranges of responses to a phenomenon, as well as deviant cases, helps researchers illuminate a phenomenon by exception,[45] particularly since qualitative researchers are required to develop explanations, theories, or hypotheses that encompass seeming anomalies.

Finally, quantitative data may also be 'qualitised'.[44] This process consists of using scores on instruments to profile participants or develop typologies. These profiles or typologies are verbal descriptions of a group of participants or varying expressions of a phenomenon around their most frequently occurring attributes. Such descriptions yield meaning to and illustrate scoring systems that may not always be interpretable or helpful in clinical practice. Our qualitative studies, which are described in a later section of this chapter, exemplify this process.[1,2]

Third, quantitative and qualitative methods are used separately and findings are compared. In this scenario, the results of each approach are used to cross-validate the study findings. At the same time, the use of two or more instruments to collect data on the same phenomenon (e.g. observations and/or focus groups and structured questionnaires) yields more information than the use of a single method, thus providing a more holistic, comprehensive view of the phenomenon under study by generating macro- and micro-perspectives of the same phenomenon. As well, in many studies, the interpretation and communication of quantitative findings draws upon qualitative methods for helpful illustrations. In other words, qualitative findings are often used to 'spice up' or illustrate quantitative findings.

Trustworthiness of the data

Four criteria are used in qualitative research to increase the trustworthiness of findings. They are (1) credibility (which parallels internal validity in the quantitative approach) (2) transferability (or external validity), (3) dependability (reliability), and (4) conformability (or internal reproducibility).

Lincoln and Guba (1982) defined the issue of credibility as the researcher's ability to represent 'those multiple constructions of reality adequately *from the perspective of the participants*'.[15] The strategies described below strengthen the credibility of the data.

Repeated in-depth interviews

These allow researchers to (1) collect extensive descriptions of the phenomenon under investigation, (2) understand what is pertinent and irrelevant to the experiences under study, and (3) confirm or challenge researchers' emerging hypotheses or insights.

Respondent validation or member checks

These consist of obtaining feedback from the participants about the researchers' descriptions and interpretations of their experiences and incorporating their reactions, comments, and/or challenges into the research findings. Lincoln and Guba regard respondent validation as the strongest available check on the credibility of the findings.[15]

Regular peer debriefing with key informants

Key informants are individuals who are particularly knowledgeable and articulate about the phenomenon under study or the analytical methods. Content experts can validate or challenge the researchers' insights, descriptions and interpretations of the lived experiences, while data analysis experts can validate or challenge the sampling strategies, coding system (i.e. ensure that it is grounded in the data), data analyses, and analytical memos.

Triangulation[47,54–57]

There are five types of triangulation: data sources triangulation, investigator triangulation, data-analysis triangulation, methodologic triangulation, and theoretical triangulation.

- *Data sources triangulation* consists of using multiple data sources (e.g. patients, caregivers, and clinicians), each with a similar focus, to obtain multiple perspectives on the same phenomenon. This type of triangulation also includes time triangulation, which is the collection of data about the same topic/issue, at different times, a process commonly conducted in anti-dementia drug trials. Both types of triangulation increase data completeness and validation.

- *Investigator triangulation* involves using more than one researcher in the data collection and analyses. Confirmation of findings among investigators who collected and coded the data independently increases the credibility of the study findings. Carey et al, Ford et al, Armstrong et al and Bourbon further suggest different ways to strengthen inter-rater reliability when data are coded by several coders via a qualitative data analysis software.[58–61]

- *Data-analysis triangulation* includes constant comparisons, attention to negative cases, and the use of inductive and deductive analytical processes. Systematic comparisons and verifications of the data categorisation ensure that researchers have not overlooked important categories and have properly identified emerging categories and themes. As well, comparisons of what the same participant has said regarding the phenomenon under investigation over time (i.e. intra-respondent comparison), and comparisons between what the overall population of respondents has said about the phenomenon under study, and the diverse elements that constitute this phenomenon reduce researchers' errors or misinterpretations. Search for and attention to negative cases (selected via appropriate sampling strategies such as deviant sampling and maximum variation sampling) allows

researchers to explore alternative explanations and generally shed additional light on elements of the phenomenon that appear contradictory. Similarly, combining inductive and deductive data analysis processes allows researchers to ensure that the categories are grounded in the data and facilitate the emergence of insights and the hypotheses that are tested through subsequent interviews (until data saturation or redundancy).

- *Methodologic triangulation* consists of using two or more collection methods. There are two types of methodological triangulation: within-method triangulation whereby researchers use at least two data collections within the same design approach (e.g. observations and interviews or focus groups) and between- or across-triangulation whereby researchers use qualitative and quantitative collection methods within the same study. These types of triangulation tend to provide a fuller, more accurate, and more holistic picture of the phenomenon under investigation (i.e. completeness).

- *Theoretical triangulation* involves the use of multiple theories or hypotheses when examining a phenomenon. Its purpose is to conduct a study from varied lenses and with multiple questions in mind, to test (i.e. confirm or challenge) the research findings.

The second concern of conducting qualitative research is transferability (paralleling external validity or generalisability). Transferability can be strengthened via a combination of purpose sampling and 'thick descriptions'. The point of purposive sampling is to include respondents across a range of experiences. The inclusion in the sampling of individuals who have or have had different reactions to the same phenomenon, as well as the inclusion of deviant cases, reduces sampling biases and ensures the representation of a wide range of different perspectives.

Thick descriptions of all the factors relevant to the phenomenon under study, as well as of the ways they vary over time and in combination, maximise the amount and range of information about the multiple dimensions of the phenomenon under study, the diverse ways in which a phenomenon can be experienced, and its evolution. Thick descriptions also allow readers to develop a sense of logical consistency and to make adequate decisions regarding the transferability of the phenomenon to different settings and/or different individuals.

The third concern of qualitative inquirers is the dependability (paralleling reliability) of the inquiry. Dependability is the process by which researchers demonstrate that the inquiry processes fall 'within the bounds of good professional practice'.[62] Dependability can be established via regular peer debriefing with key informants knowledgeable in qualitative methods, and an audit trail. First, peer debriefing with qualitative method experts ensures that researchers use adequate sampling and data analysis strategies and that interpretations are valid and accurate. Second, an audit trail includes all the interviews; a log of all the activities conducted during the study, including all the activities of the field contacts (e.g. contacts with key informants and participants), the purpose/s and outcomes of these contacts; and a detailed log of the methodological decisions (sampling and data analysis strategies). The audit trail allows potential auditors to ensure that the methods chosen for the data collection are appropriate to the phenomenon under study and that the techniques of analyses utilised are 'those consonant with the form in which data are collected and assembled'.[62]

Finally, qualitative researchers are concerned by the confirmability (which parallels internal reproducibility) of the findings. Confirmability ensures that the product/s of the research can be substantiated from the data collected. It is also established by keeping a trail of all the data analyses during and after data collection, emerging insights/ hypotheses and their rationale, and the relationships (e.g. between different elements of the phenomenon under study) that have emerged during data analyses. Given that confirmability and

dependability are closely related, the steps taken to ensure the dependability of the data also safeguard their confirmability, as well as other processes such as triangulation and the inclusion of multiple perspectives in the research. Patton, Pope and Mays, Creswell, Giacomini and Cook, and Russell and Gregory provide excellent overviews of the diverse criteria that can be used to evaluate the strength of qualitative research findings.[9,10,13,14,63–65]

Qualitative research in anti-dementia drug trials

In 1990, the Division of Neuropharmacological Drug Products of the US Food and Drug Administration (FDA) recommended that anti-dementia drug trials include clinical global measures of change as primary efficacy outcome. This recommendation was based on the premises that (1) clinically useful drugs must have a clinical effect, not only a cognitive one, and (2) the effectiveness of new treatment should be apparent to experienced clinicians.[66–69] As reviewed elsewhere, the clinical meaningfulness of statistically significant differences in neuropsychological tests has yet to be established for many commentators.[70,71] As a result, the Clinician's Interview Based Impressions of Change (CIBIC) was developed by the Alzheimer's Disease Cooperative Study Group.[72] This instrument is a phenomenological assessment of the patients' symptoms that allows researchers to systematically record clinical observations in a way that can quickly allow for hundreds of observations to be compared. Hence, it is useful in trying to develop detailed profiles of disease treatment effects that have the potential to help researchers and clinicians make more informed decisions about a drug's potential efficacy in specific domains.

The CIBIC has evolved over the years from an unstructured version to a semi-structured version that includes caregivers' information.[73] The Clinician's Interview Based Impressions of Change, plus caregivers' information, or CIBIC-Plus, is now commonly used in anti-dementia trials. It allows researchers and clinicians to record meaningful information about patients' history, general appearance, mental cognitive state, behaviour, and functional ability. It includes seven-point Likert scales recording disease severity and changes during and/or at the end of treatment. These scales have been shown to have face validity and predictive validity.[74,75] Knapp et al and Schneider et al have demonstrated that change scales were sensitive to longitudinal change in 24- and 30-week studies.[74] As well, Schneider and colleagues found that, at 12 months, the change scores of patients' global scales (n = 306) were significantly associated with the change scores of the Clinical Dementia Rating, Global Deterioration Scale, Mini-Mental State Examination, and Functional Assessment Staging (CDR, GDS, MMSE, and Fast).[74] Similarly in two different studies, Morris and colleagues, and Cummings et al found that metrifonate-treated patients exhibited significantly better scores on both the Alzheimer's Disease Assessment Scale-Cognitive (ADAS-Cog) and the CIBIC-Plus than the placebo group, thereby confirming the predictive validity of clinical change scales.[76,77] This profile has held in several other studies in dementia.[78–82] Nonetheless, the criteria that, at least implicitly, underlie the global (numeric) scores of patients' changes have received little formal evaluation, since studies that have used global measures of change have focused on the numeric scores to the exclusion of the textual data. As a result, clear understanding of the criteria that clinicians used to define improvement, decline or 'no change' and eventual patterns of treatment effects were left to one's imagination. To address these shortcomings, our group has focused on analysing the textual data included in the CIBIC-Plus to (a) better understand clinicians' definitions of change and (b) identify eventual patterns of treatment effects.[1,2] The next sections describe the sampling strategy, data collection and analyses, and findings of our studies.

Sampling

Participants in these qualitative studies of the factors that motivate clinicians' judgements of efficacy in anti-dementia trials were selected purposefully and had to be on active treatment. This strategy was consistent with our intent of understanding what patterns might emerge under conditions of treatment, and the clinicians' understanding and evaluations of patient changes. Other selection criteria included relevance (i.e. the information in the file had to be relevant to our studies), richness of information (i.e. we selected files that included as much information as possible about the patients' history, cognition, behaviour–mood, and functional abilities during treatment), diversity (to ensure the inclusion of a wide spectrum of responses), and comprehensiveness (to reach data saturation within each category of change). Our studies included respectively 18 and 42 patients' files. Data came from a 6-month, Phase III, double-blind, randomised, placebo-controlled, multi-centre trial of two cholinesterase inhibitors: metrifonate and galatamine.[1,2]

Data

Data consisted of patients' CIBIC-Plus files. Files were completed by clinicians experienced in treating and/or assessing individuals with AD. Data were collected via semi-structured interviews with patients and their caregivers and clinicians' observations/examinations of patients at baseline, visit 2 (generally within 3 months of baseline), visit 3 (within 6 months of baseline), and, in the metrifonate study, visit 4 (within 9 months of baseline). They included information about the patients' history, general appearance, mental/cognitive state, behaviour and ADL. 'Mental/cognitive state' consisted of information about the patients' recent and remote memory, speech, praxis, orientation (time and space), judgement, concentration, insight and initiative. Behaviour included annotations on the patients' mood, hallucinations, sleep, appetite, and unusual psychomotor activity. ADL contained information about the patients' personal activities of daily living (e.g. dressing, grooming, ambulation, bathing and toilet), instrumental activities of daily living (IADL) (e.g. shopping, food preparation, housekeeping, laundry, ability to handle finances and responsibility for own medication), and social activities–hobbies. Symptoms included in the CIBIC-Plus had to be clinically meaningful.

CIBIC-Plus also included disease change scores. Change scores were rated as follows: 1 = very much improved, 2 = much improved, 3 = minimally improved, 4 = no change, 5 = minimally worse, 6 = much worse and 7 = very much worse. There were no guidelines or descriptors that defined these ratings, which were left to the CIBIC-Plus raters' clinical judgements.

Coding and data analyses

Textual data were broken down into meaningful pieces and assigned a code via QSR NUD*IST (Qualitative Solutions and Research Non-Numerical Unstructured Data Indexing Searching and Theory-Building Software, *www.qsrinternational .com*). This code-based software combines management of textual data with processes for indexing, linking and searching the data. Data coding included open, axial and selective coding.[11] Open coding refers to the identification and labelling of meaningful pieces of information (e.g. the patients' diverse improvements). Axial coding is the process of relating categories. This type of coding is termed 'axial' because coding occurs around the axis of a category, linking a 'main' category (e.g. a specific improvement or decline) to 'sub'categories, each of which describes the dimensions or properties of each identified change. While main categories included information about the nature of the patients' improvements and declines, subcategories included information about the frequency, intensity, scope and duration of the changes, as well as the circumstances under which these changes happened, and the eventual interactions between patients' changes. Selective coding included developing 'core' categories that integrated patterns of change and treatment responses. Initial codes or categories reflected the CIBIC-Plus main domains

(i.e. patients' history, general appearance, mental–cognitive state, behaviour and ADL). Subnodes represented the different areas of each domain (e.g. memory, praxis). Other nodes or subnodes were created as information emerged from the data. Subnodes illuminated the data in ways that were not provided by the already existing nodes.

Coding and the emergence of categories were facilitated by several processes including pattern identification (i.e. the recognition of recurring treatment effects and of their characteristics), clustering of conceptual groupings (i.e. the regrouping of the diverse treatment effects under patterns of treatment effects), constant comparisons and theoretical memos (i.e. memos regarding the development of hypotheses about potential patterns or treatment effects that were subsequently developed via systematic comparisons with the other patients). Patients' changes during treatment were systematically identified, compared and contrasted. Similarities and contrasts between changes within each domain of the CIBIC-Plus and their characteristics were explored both intra- and inter-individually. Improvements, declines, and characteristics of 'no change' were displayed in matrices, as were their characteristics and the time frame at which they had been noted. Textual analyses were carried out blind to the CBIC-Plus change scores and any other change scores . . . , etc). Upon completion of the textual analyses, patterns of change derived from the textual analyses were compared with their corresponding change scores. Transcripts were reviewed several times to ensure that all the relevant data were systematically coded under the appropriate categories.

Trustworthiness (credibility/reliability) of the data

The following strategies strengthened the trustworthiness of the data:

* *purposive sampling* ensured the inclusion of a wide range of responses (comprehensiveness)

- *data triangulation* was ensured via multiple sources of information, including clinicians' observations and tests, caregivers' and patients' input
- *analytic triangulation*. Systematic analyses and inter- and intra-respondent comparisons over time ensured that data were not overlooked and that data were integrated into systematic profiles of treatment effects
- *peer debriefing*. The primary analyst consulted with other qualitative researchers and local geriatricians throughout the data analyses
- *audit trails* of the patients' files, coding procedures, data analyses and emerging profiles were kept
- finally, we presented the data, coding procedures, categories and initial results to two panels of international experts, including qualitative researchers, neurologists, psychiatrists, and geriatricians during two-day meetings.

Findings

Our studies illustrated that clinicians appear to distinguish between a phenomenon being clinically detectable and being clinically meaningful. Clinicians used the CIBIC-Plus notes to render a judgement about meaningfulness and judged detectable changes in that light. They also helped us better understand how clinicians evaluated change and the varied meanings they gave to 'minimally worse', 'minimally improved' and 'no change'. Assessments of change appear to involve a complex process where clinicians weight declines against improvements. However, in general, improved assessments required fewer signs of improvements than declined assessments required signs of deteriorations. This might reflect that improvements tend to be less expected in the course of AD than deterioration. Interestingly, this appears to have been confirmed in a quantitative study, in which CIBIC-Plus interview videotapes were presented to clinician raters in a random order. The raters were more likely to rate deterioration as deterioration, than to recognise improvement.[83]

Both studies brought initial evidence that CIBIC-Plus notes appear to capture aspects of the patients' evolving symptomatology that had not been routinely described. Clinically meaningful improvements and declines were noted in many aspects of cognition, behaviour and function. Clinicians' notes made it also clear that improvements may be accompanied by declines, and that assessments of declines did not necessarily mean that patients did not improve in some areas. Likewise, assessments of improvements did not mean that patients did not decline in some areas. Our analyses further suggested that the clinical correlate of 'no change' on the CIBIC-Plus did not mean that patients had remained stable. In most cases, patients assessed as stable had experienced changes (improvements and declines) in their symptoms during the course of the treatment, but, overall improvements were judged to be offset by declines (e.g. cognitive improvement but functional decline, functional improvement but increased behavioural problems).

Our second study allowed us to develop initial patterns or profiles of treatment responses to galantamine.[2] However, some of the descriptions associated with some of the clinical judgements overlapped across CIBIC-Plus scores. Boundaries between 'very much improved' and 'much improved' and between 'no change' and 'minimally worse' were not always clear from the clinicians' notes. Several patients who appeared identical in the notes were assigned different change scores. These differences suggest that different clinicians might have different models of treatment effects in mind when they rate patients' performance.

Cognitive change appeared to carry less weight in clinicians' assessment of change than did changes in behaviour (especially in the case of worsening) or function (especially in the case of improvement). This was reflected too in the range of CIBIC-Plus scores. In patients in whom the CIBIC-Plus notes identified behavioural worsening, the summary judgement was uniformly scored as 5–7, i.e., worse. For patients in whom functional improvement was identified in the qualitative analysis, CIBIC-Plus scores ranged from 4 ('no change') to 1 ('very much improved'). Patients with notes indicating cognitive improvement (sometimes by the clinicians' own test procedures) could receive scores from 1 to 5 ('minimally worse').

Finally, our studies pointed to inconsistencies in the CIBIC-Plus notes, which limits our findings. Inconsistencies were related to the format and content of the CIBIC-Plus questionnaires. There was marked variability in the length and content of the files. The most complete were 35–40 pages long and included information on several aspects of cognition, behaviour, functional abilities and social activities. Others, which varied from seven to nine pages, contained only a few notes on a limited number of areas of cognition (three or four versus six in the longer files) and one or two aspects of patients' behaviour versus five in the longer files. Lack of specificity was also evident in several areas and impeded assessments of change. Terminology that is too general (e.g. 'irritable at times', 'short-term memory declined', 'spend much time searching') cannot lend itself to interpretation. We also noted seemingly conflicting information between interviewer testing and informant reports (e.g. 'patient praxis unchanged but patient spills and breaks more things').

Recommendations

The following recommendations would strengthen the validity of studies based on CIBIC-Plus questionnaires.[84]

- The file format should be consistent within studies across participating physicians. Questionnaires should be semi-structured, to prompt for topics important to specific drug studies (e.g. cognitive improvements or decline). For individualised measures, semi-structured questionnaires offer the three advantages of being flexible enough that clinicians can decide which information is worth pursuing, but consistent enough that similar domains are found across interviews, and structured enough that they can improve physicians' time-effectiveness.

- Questionnaires should include operational definitions of the domains to be probed (increased consistency). Clinicians' descriptions of patients' symptoms should include specific information about the frequency, duration, extent/scope/severity, and intensity of symptoms at baseline. Data included in the files should be detailed, representative of the patients' unique symptomatology, and contextualised. Any changes in symptoms should be carefully tracked and recorded.

- Disease severity should be scored at baseline and at the end of treatment. As indicated, these scales have been shown to be reliable,[70,79] provided that clinicians are trained in their completion.[76,85] Over time, they should also allow researchers to know whether specific drugs are more effective for patients with moderate, severe or mild AD.

- CIBIC-Plus global change scores need better qualitative descriptors of change. The lack of clear criteria for change scores (e.g. minimally improved, much improved, etc.) makes the change scores minimally more meaningful than a 3- or 4-point change on the MMSE and impairs the CIBIC reliability.[86] Consensus is needed about characteristics of change scores, and clinicians need to be trained to score patients in order to increase the consistency of patients' evaluations across sites.

- Participants' selection should be purposive and focus on participants that are likely to be informative (i.e. to provide personal and specific details; to maintain diaries) and have diverse symptoms and disease severity (to ensure comprehensiveness). Purposive sampling should further extend to participating physicians (i.e. including only those physicians who are willing to record rich and detailed information about patients and have been trained to use the score change scale for a given trial).

Conclusion

In the absence of a well-defined model of disease treatment success, systematic qualitative analyses of large numbers of trial participants, especially when done under double-blind conditions, can help specify the model of successful treatment with the least prejudice. In ongoing studies, qualitative analyses of CIBIC-Plus patients' files holds specific promise as means of helping to understand patterns both of treatment effects and of the clinicians' evaluation of patients' changes during treatment. However, for this potential to be realised, the experiences and observations of patients that motivate clinicians' judgements should be consistently and carefully recorded. They should also reflect aspects of the clinical process that best assure its validity: individualisation and contextualisation of the data. The point of Alzheimer's disease treatment is not to produce 'standard' cognitive function but to help individuals in overcoming their particular deficits. Such individual judgements about treatment effects remain a cornerstone in the evaluation of a drug's efficacy, but better specifications of why individual judgements are critical in understanding both patterns of treatment effects and their clinical meaningfulness. The combination of qualitative data analyses of clinicians' observations of patients, and patients' and carers' information about treatment effects, with quantitative data from clinical drug trials could give researchers and clinicians information about the efficacy of a drug at the population level and a better understanding of typical profiles and the clinical importance of changes seen in the course of dementia treatment within groups of patients with different levels of disease severity.

More broadly, qualitative analyses can also make a critical contribution to the study of treatment effects in dementia by allowing us to see treatment effects that we might not have known to look for.[87] By not requiring premature specification of the model of treatment – a trap that we may have fallen into by using measures that were too narrowly based, or not analysed so as to allow for nat-ural groupings to be revealed – important aspects of the disease treatment experience go unnoticed. Thus, for example, the common report that patients say that they are 'out of the fog' or that 'a veil has lifted',[84] in not being systematically captured, does not motivate early inquiry into the basis for such effects. It is only quite recently, for example, that descriptions of a functional prefrontal compensation circuit in Alzheimer's disease,[88] which appears to be enhanced in the face of treatment with cholinesterase inhibitors,[89] have provided some basis for this phenomenon, which has been witnessed since some of the earliest trails.[68] In consequence, it seems reasonable to us that rigorous qualitative analyses should be employed at the earliest stages in efficacy trials of new compounds.

References

1. Joffres C, Graham J, Rockwood K. Qualitative analysis of the clinician interview-based impression of change (Plus): methodological issues and implications for clinical research. Int Psychogeriatr 2000; 12(3):403–13.

2. Joffres C, Bucks RS, Haworth J, Wilcock GK, Rockwood K. Patterns of clinically detectable treatment effects with galantamine: a qualitative analysis. Dement Geriatr Cogn Disord 2003; 15(1):26–33.

3. Cobb AK, Forbes S. Qualitative research: what does it have to offer to the gerontologist? J Gerontol A Biol Sci Med Sci 2002; 57(4):M197–202.

4. Rempusheski VF. Qualitative research and Alzheimer disease. Alzheimer Dis Assoc Disord. 1999; 13(Suppl 1):S45–9.

5. Green J, Britten N. Qualitative research and evidence-based medicine. BMJ 1998; 316:1230–2.

6. Lakshman M, Sinha L, Biswas M, Charles M, Arora NK. Quantitative vs. qualitative methods. Indian J Pediatr 2000; 67(5):369–77.

7. Guba EG. The Paradigm Dialog. Newbury Park, CA: Sage; 1990.

8. Goering P, Streiner DL. Reconcilable differences: The marriage of qualitative and quantitative methods. Can J Psychiatry 1996; 41:491–7.

9. Patton MQ. Qualitative Evaluation and Research Methods (2nd edn). London: Sage; 1990.

10. Creswell JW. Qualitative inquiry and research design: choosing among five traditions. Thousand Oaks, CA: Sage Publications; 1998.

11. Strauss A, Corbin J. Grounded theory methodology: an overview. In: Denzin NK, Lincoln YS (eds). Handbook of Qualitative Research. Thousand Oaks, CA: Sage Publications; 1998:273–85.

12. Brendl CM, Markman AB, Messner C. How do indirect measures of evaluation work? Evaluating the inference of prejudice in the Implicit Association Test. J Pers Soc Psychol 2001; 81(5):760–73.

13. Russell CK, Gregory DM. Evaluation of qualitative studies. Evid Based Nurs 2003; 6:36–40.

14. Patton MQ. Enhancing the quality and credibility of qualitative research. Health Serv Res 1999; 34(5):1189–1208.

15. Lincoln YS, Guba EG. Establishing dependability and confirmability in naturalistic inquiry through an audit. Paper presented at the Annual Meeting of the American Educational Association, New York, March 19–23 1982. ERIC Document Reproduction Service No. ED 216 019.

16. Butcher HK, Holkup PA, Park M, Maas M. Thematic analysis of the experience of making a decision to place a family member with Alzheimer's disease in a special care unit. Res Nurs Health 2001; 24(6):470–80.

17. Brown C, Lloyd K. Qualitative methods in psychiatric research. Adv Psychiatric Treat 2001; 7:350–6.

18. Lewis M. Focus group interviews in qualitative research: A review of the literature. Action Research Reports, *www.cchs.usyd.edu.au/arow/arer/002.htm*; 2000.

19. Kreuger RA. Focus Groups: a Practical Guide for Applied Research. London: Sage; 1988.

20. Merton RK, Fiske M, Kendall PL. The Focused Interview: A Manual of Problems and Procedures (2nd edn). London: Collier MacMillan; 1990.

21. Morgan DG, Semchuk KM, Stewart NJ, D'Arcy C. Rural families caring for a relative with dementia: barriers to use of formal services. Soc Sci Med 2002; 55:1129–42.

22. Smith AP, Beattie BL. Disclosing a diagnosis of Alzheimer's disease: patient and family experiences. Can J Neurol Sci 2001; 28(Suppl 1):S67–71.

23. Hopman-Rock M, Tak EC, Staats PG. Development and validation of the Observation List for early signs of Dementia (OLD). Int J Geriatr Psychiatry 201; 16:406–14.

24. Hutchinson SA, Marshall M. Responses of family caregivers and family members with Alzheimer's disease to an activity kit: an ethnographic study. J Adv Nurs 2000; 31:44–50.

25. Huberman MA, Miles MB. Data management and analysis methods. In: Denzin NK, Lincoln YS (eds). Handbook of Qualitative Research. Thousand Oaks, CA: Sage Publications; 1994: 428–44.

26. Wolcott BW. Managed care's driving force: demand management. Infocare 1996; 12–15.

27. Powers BA. From the inside out: the world of the institutionalized elderly. In: Henderson JN, Vesperi MD (eds). The Culture of Long-term Care: Nursing Home Ethnography. Westprot, CT: Begin Garvey; 1995: 179–96.

28. Henderson JN. The culture of special care units: an anthropological perspective on ethnographic research in nursing home settings. Alzheimer Dis Assoc Disord 1994; 8(Suppl 1): S410–6.

29. McAllister CL, Silverman MA. Community formation and community roles among persons with Alzheimer's disease: a comparative study of experiences in a residential Alzheimer's facility and a traditional nursing home. Qual Health Res 1999; 9(1):65–85.

30. McCarty EF. Caring for a parent with Alzheimer's disease: process of daughter caregiver stress. J Adv Nurs 1996; 23(4):792–803.

31. Orona CJ. Temporality and identity loss due to Alzheimer's disease.

32. Hurley AC, Volicer L, Rempusheski VF, Fry ST. Reaching consensus: the process of recommending treatment decisions for Alzheimer's patients. ANS Adv Nurs Sci 1995; 18(2):33–43.

33. Van Manen M. Researching lived experience: human science for an action sensitive pedagogy London, Ontario: Althouse Press; 1990.

34. Vellone E, Micci F, Sansoni J, Sinapi N, Cattel C. The lived experience of family member caring for a person affected by Alzheimer's disease: preliminary results. Prof Inferm 2000; 53(3):132–41.

35. Fulford KWM, Morris K, Sadler JZ, Stanghellini G (eds). Nature and Narrative: An Introduction to the New Philosophy of Psychiatry. Oxford: Oxford University Press; 2003.

36. Sabat SR. The Experience of Alzheimer's Disease: Life Through a Tangled Veil. Oxford: Blackwell; 2001.

37. Sandelowski M. Focus on research methods: whatever happened to qualitative description? Res Nurs Health, 2000; 23:334–40.

38. Geldmacher DS. Clinical experience with donepezil hydrochloride: a case study perspective. Adv Ther 1997; 14:305–11.

39. Folstein MF, Folstein SE, McHugh PR. 'Mini-mental state'. A practical method for grading the cognitive state of patients for the clinician. J Psychiatr Res 1975; 12(3):189–98.

40. Shua-Haim JR, Comsti E, Shua-Haim E, Ross JS. Donepezil Aricept (tm): the caregiver voice and clinical impression. Am J Alz Dis 1997; 12: 272–9.

41. Abushaba R, Woefel ML. Qualitative vs. quantitative methods: Two opposites that make a perfect match. J Am Diet Assoc 2003; 103(5):566–9.

42. Foss C, Ellefsen B. Methodological issues in nursing research: The value of combining qualitative and quantitative approaches in nursing research by means of method triangulation. J Adv Nurs 2002; 40:242–52.

43. Burnard P, Hannigan B. Qualitative and quantitative approaches in mental health nursing: Moving above the debate. J Psychiatr Ment Health Nurs 2000; 7:1–6.

44. Sandelowski M. Focus on research methods: combining qualitative and quantitative sampling, data collection, and analysis techniques in mixed-method studies. Res Nurs Health 2000; 23:246–55.

45. Barbour RS. The case for combining qualitative and quantitative approaches in health services research. J Health Serv Res Policy 1999; 4:39–43.

46. Steckler A, McLeroy KR, Goodman RM, Bird ST, McCormick L. Toward integrating qualitative and quantitative methods: an introduction. Health Educ Q 1992; 19:1–8.

47. Risjord MW, Dunbar SB, Moloney MF. A new foundation for methodological triangulation. J Nurs Scholarsh 2002; 34(3):269–75.

48. Madjar I. The role of qualitative research in evidence-based practice. Collegian 2002; 9(4):7–8.

49. Rockwood K, Black SE, Robillard A, Lussier I. Potential treatment effects of donepezil not detected in Alzheimer's disease clinical trials: a physician survey. Int J Geriatr Psychiatry 2004; 19:954–60.

50. Rockwood K, Graham JE, Fay S. ACADIE Investigators. Goal setting and attainment in Alzheimer's disease patients treated with donepezil. J Neurol Neurosurg Psychiatry 2002; 73:500–7.

51. Morgan DL. Practical strategies for combining qualitative and quantitative methods: applications to health research. Qual Health Res 1998; 8:362–76.

52. Marwit SJ, Meuser TM. Development and initial validation of an inventory to assess grief in caregivers of persons with Alzheimer's disease. Gerontologist 2002; 42(6):751–65.

53. Duffy TP. Agamemnon's fate and the medical profession. West New Engl Law Rev 1987; 9(1):21–30.

54. Thurmond VA. The point of triangulation. J Nurs Scholarsh. 2001; 33(3):253–8.

55. Begley CM. Using triangulation in nursing research. J Adv Nurs 1996; 24:122–8.

56. Cutcliffe JR, McKenna HP. Establishing the credibility of qualitative research findings: the plot thickens. J Adv Nurs 1999; 30(2):374–80.

57. Shih FJ. Triangulation in nursing research: Issues of conceptual clarity and purpose. J Adv Nurs 1998; 28(3):631–41.

58. Carey JW, Morgan M, Oxtoby MJ. Intercoder agreement in analysis of response to open-ended interview questions: Examples from tuberculosis research. Cult Anthropol Meth 1996; 8(3):1–5.

59. Ford K, Oberski I, Higgins S. Computer-aided qualitative analysis of interview data: Some recommendations fro collaborative work. The Qualitative Report 2000; (4) (3&4). (*www.nova.edu/ssss/QR/QR4–3/oberski.html*).

60. Armstrong D, Gosling A, Weinmam J, Marteau T. The place of inter-rater reliability in qualitative research: An empirical study. Sociology 1997; 31(3):597–606.

61. Bourbon S. Inter-coder reliability verification using QSRN,UD*IST, Paper presented at Strategies in Qualitative Research: Issues and Results from Analysis Using QSRN,Vivo and QSRN,UD*IST, The Institute of Education, University of London, London, UK; September 30, 2000.

62. Lincoln YS, Guba EG. Effective Evaluation: Improving the Usefulness of Evaluation Results Through Responsive and Naturalistic Approaches. San Francisco: Jossey Bass; 1981.

63. Pope C, Mays N. Qualitative Research in Health Care (2nd edn). London: BMJ Books; 2000.

64. Giacomini MK, Cook DJ. Users' guides to the medical literature. Qualitative research health care: are the results of the study valid? JAMA 2000; 284(3):357–62.

65. Giacomini MK, Cook DJ. Users' guides to the medical literature. Qualitative research health care: what are the results and how do they help me care for my patients? JAMA 2000; 284:478–82.

66. Leber PD. Developing safe and effective antidementia drugs. In: Becker R, Giacobini E (eds). (pp 579–584), Alzheimer Disease: From Molecular Biology to Therapy. Boston, Birkhduser; 1997: 579–84.

67. Reisberg B, Schneider L, Doody R et al. Clinical global measures of dementia: Position paper from the international working group on harmonization of dementia drug guidelines. Alzheimer Dis Assoc Disord 1997; 11:8–18.

68. Rockwood K. Use of global assessment measures in dementia drug trials. J Clin Epidemiol 1994; 47:101–3.

69. Schneider LS, Olin JT. Clinical global impressions in Alzheimer's clinical trials. Int Psychogeriatr 1996; 8:277–88.

70. Rockwood K, MacKnight C. Assessing the clinical importance of statistically significant improvement in anti-dementia drug trials. Neuroepidemiology 2001; 20(2):51–6.

71. Winblad B, Brodaty H, Gauthier S et al. Pharmacotherapy of Alzheimer's disease: Is there a need to redefine treatment success? Int J Geriatr Psychiatry 2001; 16:653–66.

72. Olin JT, Schneider LS, Doody RS et al. Clinical evaluation of global change in Alzheimer's disease: Identifying consensus. J Geriatr Psychiatry Neurol 1996; 9:176–80.

73. Sahadevan S, Rockwood K, Morris JC. Global assessment measures in dementia. In: Gauthier S (ed). Clinical Diagnosis and Management of Alzheimer's Disease (2nd edn revised). London: Martin Dunitz; 2001.

74. Schneider LS, Olin JT, Doody RS et al. Validity and reliability of Alzheimer's disease cooperative study – clinical global impression of change. Alzheimer Dis Assoc Disord 1997; 11(Suppl2):S22–S32.

75. Knapp MJ, Knopman DS, Solomon PR et al. A 30-week randomized controlled trial of high-dose tacrine in patients with Alzheimer's disease. The Tacrine Study Group. JAMA 1994; 271(13):985–91.

76. Morris JC, Cyrus PA, Orazem J et al. Metrifonate benefits cognitive, behavioral, and global function in patients with Alzheimer's disease. Neurology 1998; 48:1508–10.

77. Cummings JL, Cyrus PA, Bieber F et al. Metrifonate treatment of the cognitive deficits of Alzheimer's disease. Neurology 1998; 50:1214–21.

78. Bodick NC, Offen WW, Levey AL et al. Effects of xanomeline, a selective muscarinic receptor agonist, on cognitive function and behavioral symptoms in Alzheimer's disease. Arch Neurol 1997; 54:465–73.

79. Burns A, Rossor M, Hecker J et al. The effects of donepezil in Alzheimer's' disease-Results from a multinational trial. Dement Geriatr Cogn Disord 1999; 10:237–44.

80. Corey-Bloom J, Anand R, Veach J. A randomized trial evaluating the efficacy and safety of ENA 713 (rivastigmine tartrate), a new acetylcholinesterase inhibitor, in patients with mild to moderate severe Alzheimer's disease. Int J Geriatr Psychopharmacol 1998; 1:55–65.

81. Rogers SL, Doody RS, Mohs RC, Friedhoff LT. Donepezil improves cognition and global function in Alzheimer's disease: a 15-week, double-blind, placebo-controlled study. Arch Intern Med 1998; 158:1021–31.

82. Rogers SL, Farlow MR, Doody RS, Mohs RC, Friedhoff LT. A 24-week, double-blind, placebo-controlled trial of donepezil in patients with Alzheimer's disease. Neurology 1998; 50:136–45.

83. Quinn J, Moore M, Benson DF et al. A videotaped CIBIC for dementia patients: validity and reliability in a simulated clinical trial. Neurology 2002; 12(58):433–7.

84. Rockwood K, Joffres C. Improving clinical descriptions to understand the effects of dementia treatment: consensus recommendations. Halifax Consensus Conference on Understanding the Effects of Dementia Treatment. Int J Geriatr Psychiatry 2002; 17(11):1006–11.

85. Rockwood K, Strang D, MacKnight C, Downer R, Morris JC. Inter-rater reliability of the Clinical Dementia Rating in a multicenter trial. J Am Geriatr Soc 2000; 48:558–9.

86. Knopman DS, Knapp MJ, Gracon SI, Davis CS. The Clinician Interview-Based Impression (CIBI): a clinician's global change rating scale in Alzheimer's disease. Neurology 1994; 44(12):2315–21.

87. Rockwood K, Wallackm M, Tallis R. The treatment of Alzheimer's disease: success short of cure. Lancet Neurol 2003; 2:630–3.

88. Grady CL, McIntosh AR, Beig S et al. Evidence from functional neuroimaging of a compensatory prefrontal network in Alzheimer's disease. J Neurosci 2003; 23:986–93.

89. Nobili F, Koulibaly M, Vitali P et al. Brain perfusion follow-up in Alzheimer's patients during treatment with acetylcholinesterase inhibitors. J Nucl Med 2002; 43:983–90.

6

Inclusion and Exclusion Criteria

David B Hogan

Introduction

Study protocols should clearly describe the type of participants on which the intervention is to be tested.[1] Explicit, clear rules should be used in determining which potential subjects will be eligible for the study. Subject selection criteria are designed with three aims in mind: to ensure that those entering a study truly do have the condition the drug is intended to treat; to maximise the likelihood that the study will detect a drug effect; and to minimise the risks for study participants. Knowledge of who is enrolled in a study is important. If the intervention is successful we want to understand what kinds of patients may benefit from the treatment. Is the study population similar to the patients you see in your practice? An effective therapy may seemingly fail in a study if insufficient attention has been placed on recruiting subjects actually suffering from the target condition. Were the selection criteria too loose? Without information on how subjects were selected, planning a confirmatory study would be problematic.

The *study population* is the subset of all those with a condition deemed eligible for the study by the protocol's entry criteria (see Figure 6.1). The *study sample* consists of the members of the study population actually enrolled in the study. A tightrope has to be walked between homogeneity and heterogeneity in the definition of a study population. Too heterogeneous a group may make it hard to show a drug effect because of inadequate targeting and the presence of confounders. It might also be associated with unacceptable risks for participants if medically unstable subjects are allowed to enrol. While homogeneous study populations might provide more interpretable data,

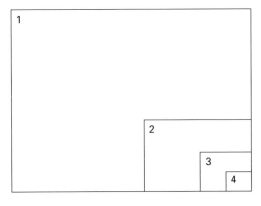

1. Total population
2. Total population with condition
3. Study population (eligible subjects)
4. Study sample (enrolled subjects)

Figure 6.1 Relationship of study sample to the study population, the total population with the condition and the total population.

narrow limits can make subject recruitment challenging and raise concerns about the generalisability of any study results to the broader population of all those with a given condition.

Factors to consider as criteria for subject selection include characteristics of the subject (e.g. sex, age), characteristics of the condition and its treatment (e.g. definition of the disease under study, stage or severity of the study disease), co-morbidities and their treatment (e.g. presence, severity, concomitant medications), protocol issues (e.g. the need to maximise subject adherence, requirement for the presence of a caregiver, subject participation in prior trials), and results of the screening examination (e.g. physical findings, laboratory results).[2]

How to define the presence of the study condition is probably the most important inclusion criterion. Government regulatory agencies approve a drug for marketing based on a determination that it is both effective and safe when used to treat one or more specific conditions (which are called the 'claims' or 'indications' for the drug). Generally, for this to occur it must be possible to detect the condition without undue ambiguity by using widely accepted (and feasible) criteria for diagnosis that are both valid and reliable. Indications for drug therapy can be specific diseases, clinical syndromes, or symptoms such as pain.

Until quite recently drug manufacturers were under no particular obligation to ensure that those enrolled in drug studies were representative of the total population of individuals with the condition of interest.[3] This was justified by holding that these studies were primarily to provide proof that the drug has a beneficial therapeutic effect in at least some patients, in other words that it is efficacious.[4] Regulatory bodies are now encouraging sponsors to test their therapies on the full range of patients who will eventually be using the drug if it is approved for marketing. For example, the Therapeutic Products Programme of Health Canada has issued guidelines for the inclusion of women in clinical trials.[5]

Selection criteria used in dementia studies

In this section we will focus on the phase III randomised controlled trials (RCTs) of pharmaceuticals approved in North America for Alzheimer's disease (AD) – doenepezil, galantamine, memantine and rivastigmine.[4] These agents have also been studied for vascular dementia (VaD) and dementia with Lewy bodies (DLB). While there is a good deal of minor variability between the individual studies, there is relative uniformity in the core inclusion/exclusion criteria (other than in the definition of the condition of interest) used in the studies of these three conditions.

Alzheimer's disease

AD studies have used the DSM-III-R[6] or DSM-IV criteria[7] for the diagnosis of dementia.[8–10] The National Institute of Neurological and Communicative Disorders and Stroke – Alzheimer's Disease and Related Disorders Association (NINCDS-ADRDA) criteria have been employed almost universally for making the diagnosis of AD.[8,9,11–13] Studies typically limit recruitment to those with a diagnosis of probable AD.[11]

An interesting but complex post-randomisation selection criterion was employed in an early dementia study. This trial of tacrine (a cholinesterase inhibitor) used an enrichment/two-phase re-randomisation design.[14] After screening, subjects were randomly assigned to one of three different sequences of therapy with placebo, low-dose tacrine (40 mg), or high-dose tacrine (80 mg). Each subject received, in one of three different sequences, two weeks of therapy with each of the three options (placebo, 40 mg tacrine, 80 mg tacrine). This was done to determine if subjects 'responded' to tacrine (defined as a four points or better response on the Alzheimer's disease Assessment Scale – Cognitive Subscale (ADAS-Cog) with one of the two doses compared to placebo) and, if they responded, what was their best dose. Those who failed to respond were dropped from the study. Responders were then re-randomised under double-blind conditions to either placebo or their best dose of tacrine. It was claimed that this design allowed subjects to be titrated to their best possible dose before starting the study proper, allowing the drug to be compared to placebo under conditions most likely to detect a treatment effect if one existed. Unfortunately this design seemed to fail in what it was intended to do (i.e. enriching the sample with potential responders), and led to concerns about unmasking the treatment blind and confounding by either carryover or withdrawal effects.[14] In addition lessons learned the hard way on the need to slowly titrate cholinesterase inhibitors would speak strongly against this design.

Vascular dementia

VaD studies have used the criteria for probable (and/or possible) VaD of the National Institute of Neurological Disorders and Stroke – Association Internationale pour la Recherche et l'Ensieignement en Neurosciences (NINDS-AIREN) International Workshop,[15–20] DSM-III-R criteria for VaD,[6,18] two modifications of the Hachinski Ischemia Scale (HIS),[17,21–23] and/or a diagnosis of possible AD (using NINCDS-ADRDA criteria) where there is radiological evidence of cerebrovascular disease.[16] The NINDS-AIREN criteria were specifically developed for research studies.

Dementia with Lewy bodies

Consensus clinical criteria for probable DLB[24,25] were used to identify subjects in the one published RCT for this condition.[26]

Other criteria

Subjects recruited into dementia studies have generally been graded as being in a mild to moderate stage of their illness though more severely impaired subjects have been included in some studies.[9,13,23] Mild to moderate severity has been operationalised as: scoring between 10–13 and 20–26 (inclusive) on the Mini-Mental State Examination (MMSE);[27] scoring 12 or more on the Alzheimer's disease Assessment Scale – Cognitive Subscale (ADAS-Cog);[28] and/or, obtaining a

Clinical Dementia Rating (CDR) of one or two.[29] To be eligible, subjects typically had to be residing in the community and have a responsible, consistent caregiver in regular contact with them who would also consent to participate in the study. While this latter requirement was made primarily to obtain informant information for functional, behavioural, and global measures, having such a caregiver for a demented subject probably played an important role in ensuring a reasonable degree of adherence to the study protocol.

Common exclusion criteria were age (e.g. less than 40 or 50 years, over 90 years) and the presence of other (than the condition being studied) neurodegenerative disease(s) that could cause or contribute to dementia, other (than the condition being studied) medical condition(s) that could cause or contribute to dementia, major depression and other psychiatric diagnoses, active alcohol and drug abuse, and the presence of co-morbidities (e.g. obstructive airway disease, cardiac conduction disorders, active peptic ulcers, seizures) that would increase the risk for adverse events and/or make it significantly less likely that the subject would complete the study. Recent (e.g. within 30 days) consumption of investigational drugs, concomitant use of other medications prescribed for dementia (e.g. nootropic agents, cholinomimetic agents, choline), and use of psychotropics have been frequently used to exclude potential subjects.

Critique of selection criteria

Dementia is a clinical diagnosis. Various diagnostic strategies have been proposed for identifying individuals with dementia. Which criteria are used can lead to large differences in the number of subjects classified as having dementia.[30] The most widely used criteria – the DSM-III-R and DSM-IV – are relatively liberal, including more subjects than other diagnostic schemes such as the ICD-9, ICD-10 and CAMDEX.[30] The different ways of conceptualising dementia would lead to very different pools of potential subjects. Both the DSM-III-R and DSM-IV criteria for dementia have been found to have a least moderate inter-rater reliability.[31,32]

With two exceptions, the studies we reviewed required subjects to have one specific aetiology for their dementia. The exceptions were the 9M-Best study of memantine,[23] which included subjects with either presumed AD or VaD, and the galantamine VaD study which, in addition to probable VaD cases, allowed entry of subjects with possible AD and cerebrovascular disease.[16] Most dementia studies are designed on the premise that there is a single cause for the cognitive impairment. This is often not the case. For example a cohort study of patients newly referred to one of eight dementia research clinics in Canada found that 226 (33.7%) of the 670 patients diagnosed with a dementia were felt by the examining clinician to have a mixed aetiology with more than one condition contributing to the dementia.[33] Of the 500 (74.6% of all those with dementia) with a diagnosis of AD, 316 (63.2%) had AD listed as the sole cause (i.e. 36.8% were felt to have AD plus another condition causing the dementia). Among those with VaD/vascular cognitive impairment or DLB, the equivalent figures were 31.7% and 41.9%.

Neuropathological studies also support the contention that many cases of dementia probably have more than one contributing cause. A review of 382 consecutive persons with dementia referred to the State of Florida Brain Bank between 1992 and 2000 showed that 145 (38%) had more than one relevant central nervous system post-mortem diagnosis.[34] Among the 294 with an AD pathological diagnosis, 159 (54%) had 'pure' AD (i.e. they met pathological criteria for AD but failed to meet pathological criteria for any other dementing illness). Mixed dementias were almost as common as 'pure' AD. Among those with VaD or DLB, 'pure' disease was found in only 17% and 30% respectively. A smaller post-mortem study of 80 subjects from the Camberwell Dementia Case Register found that mixed pathologies were present in 27 (33.8%) cases.[35] Forty-four (55%) had 'pure' AD.

Alzheimer's disease

In validation studies using neuropathological confirmation of AD, a clinical diagnosis of probable AD using the NINCDS-ADRDA criteria carries a good sensitivity (approximately 80%) and a somewhat lower specificity (approximately 70%). A diagnosis of possible AD achieves even better sensitivity (90% plus) but at the cost of lower specificity (approximately 50%).[32] There is evidence that diagnostic specificity is getting better over time, presumably reflecting improvements in our ability to diagnose other causes of dementia.[36] The American Academy of Neurology has endorsed the NINCDS-ADRDA criteria for diagnosing probable AD.[32]

Most pharmaceutical studies limit entry to those with probable AD, excluding those with a diagnosis of possible AD. It is argued that probable and possible AD may differ significantly in their underlying pathology (which might influence the likelihood of responding to a given drug) and/or their natural history (which might confound measuring a drug effect). Restricting entry to those with probable AD does lead to a more homogeneous group of subjects but at the cost of generalisability, as many AD patients are excluded. How big a problem this might be is uncertain. The reported ratios between the numbers of probable AD and possible AD cases in a number of populations vary from approximately 80/20 to 60/40.[33,36–38] The neuropathological studies do indicate that subjects with a diagnosis of possible AD are more likely to have non-AD causes for their dementia.[32] On the other hand, a recent study found that select outcomes (i.e. rate of dementia progression as measured by the CDR, nursing home admission, death) of individuals with possible AD were not significantly different from those with probable AD.[38]

How accurately are the NINCDS-ADRDA criteria applied in AD drug studies? There are surprisingly little published data on this question. One studied showed a 22.7% discordance rate between the NINCDS-ADRDA study diagnosis of probable AD and a diagnosis of AD based on an independent CT scan model reportedly able to predict the presence of AD.[39] Of more concern was the finding that the discordance rates across the six sites ranged from zero to 42.9%. This suggested that there might well have been a significant degree of variability between sites on who was being recruited into the study.

Vascular dementia

The diagnosis of VaD is controversial.[40] Different sets of criteria have been advocated for the diagnosis (e.g. HIS, the modifications of Rosen[41] and Loeb[42] of the HIS, DSM-III-R, DSM-IV, ICD-10, and those proposed by the State of California Alzheimer's Disease Diagnostic and Treatment Centers or ADDTC). They are not interchangeable.[31,43–46] Recent autopsy studies have shown that while some vascular pathology might exist in up to 40% of those dying with a dementia, 'pure' vascular dementia is relatively rare, accounting for fewer than 10% of all cases.[32]

Because the NINDS-AIREN criteria have been the ones most commonly used in VaD studies, we will concentrate on them. Moderate inter-rater agreement has been found with them for both possible and probable VaD.[31,47] In studies that compared the clinical diagnosis with neuopathological findings, the NINDS-AIREN clinical criteria for probable VaD were found to have a low sensitivity (20–30%) but high specificity (93–100%).[35,48] Possible VaD as defined by these criteria carried a higher sensitivity (55%) and a lower specificity (84%).[48] Combined possible and probable VaD where most of the cases were possible VaD, had a sensitivity of 58% and a specificity of 80%.[49] While the high specificity is good from the standpoint of restricting study recruitment to subjects who truly do have cerebrovascular disease, the low sensitivity for a condition which is less common than AD does make it more difficult to recruit the required number of subjects in a timely fashion. The low sensitivity would raise concerns about the generalisability of the results obtained.

Notwithstanding the concerns about 'contamination' by AD pathology, the subjects recruited into the RCTs of donepezil for VaD did differ in a number of significant ways from those recruited into similar trials for AD.[50] They were more likely to be male, have clinical and radiological features of cerebrovascular disease, and have a greater burden of co-morbidity. Compared to the Alzheimer trials, those allocated to the placebo arm were more likely not to show declines over the 24 weeks of the trials.

Dementia with Lewy bodies

Neuropathology-confirmed studies of the consensus clinical criteria for DLB have typically shown low sensitivity (60% or less) but high specificity (85% plus).[32] The exception was the study by McKeith and colleagues, which reported a sensitivity of 83% and a specificity of 95% for a clinical diagnosis of probable LBD.[25] A recent report by Lopez and colleagues again reported a low sensitivity (30.7%) and a high specificity (100%).[50] While the relatively high specificity is good from the standpoint of restricting recruitment to subjects who truly do have DLB, it does make it more difficult to recruit the required number of subjects in a timely fashion.

The low sensitivity might be due to the concurrent presence of AD pathology, which occurs frequently with DLB.[24] The presence of AD pathology seems to obscure the diagnosis of DLB by 'Alzheimerising' the clinical manifestations encountered.[51,52]

In most hands, the reliability of the clinical diagnostic criteria has been poor and worse that that seen for other dementias.[53,54] The exception again has been a study published by McKeith and colleagues that showed very acceptable reliability.[55]

Other criteria

As will be seen, the selection criteria used in the studies significantly reduced the pool of potential research subjects. One particular issue is the targeting to those with a particular range of severity. In the late 1980s and 1990s when these studies were being planned, it was felt that subjects with mild to moderate disease severity would be the ones most likely to respond to the investigational agents.[14] While it was acknowledged that this (and other restrictions) would limit the generalisability of the results obtained, it was felt that limiting eligibility in this manner would not undermine the 'internal validity' of the studies.[14] This may have been dealt with in a satisfactory manner by the regulators when they limited the approved indication for marketing to patients whose characteristics matched those in the studies. Unfortunately 'indication drift' has occurred with dementia drugs as it occurs with all medications.[56]

Generalisability of studies

It can be definitively said that the study samples in the AD trials were not truly representative of all those with AD. This has been shown in a number of studies.

Schneider and colleagues found that only 152 (4.4%) and 274 (7.9%) out of a pool of 3470 subjects with possible or probable AD were provisionally eligible for two 'typical' AD trials.[37] Those eligible were younger, better educated, wealthier, and more likely to be white. Women were relatively under-represented in the eligible pool. Common reasons for exclusion were (listed in descending order): a diagnosis of possible AD ($n = 1434$); behavioural and psychological symptoms ($n = 784$); MMSE score outside of desired range or age less than 50 ($n = 720$); co-morbidity ($n = 305$); living alone ($n = 300$); and, an abnormal neurological examination ($n = 284$).

Treves and colleagues reviewed their clinic records of demented patients who were candidates for AD drug trials.[57] They had 279 patients 50 years of age and older who were diagnosed as suffering from probable AD. Only 36 (12.9%) were recruited. Enrolment was more likely in patients with a higher MMSE score, younger age, and more schooling. Common reasons for non-enrolment were (listed in descending order): MMSE score outside of desired range ($n = 88$); behavioural and psychological symptoms ($n = 61$); comorbidity ($n = 30$); non-cooperative patient ($n = 28$); and, refusal of consent ($n = 20$: caregiver 17, patient 3).

Cohen-Mansfield examined participation rates in five dementia studies conducted by her research group.[58] The studies ranged from a descriptive study of agitation in a nursing home to a double-blind drug study of behavioural and cognitive problems in nursing home residents. For the intervention studies, 9.4–30.7% of the initial population met screening criteria. Of those meeting criteria for an intervention study, 40–92.3% consented to it. Of those starting an interventional study, 29–60.3% completed it. The low recruitment rate for participants in the drug trial (1% of the population felt to have a dementia) was highlighted. Common reasons for excluding potential subjects in the drug study were (listed in descending order): MMSE outside of desired range ($n = 228$; total initial population 457); refused/unable to obtain MMSE ($n = 90$); co-morbidity ($n = 69$); no MMSE available ($n = 37$); consent refused ($n = 26$); and, medical/behavioural reasons or lack of cooperation ($n = 12$). Cohen-Mansfield asked '. . . if only <2% of persons suffering from dementia are recruited into a trial, do the results represent what is likely to happen to others suffering from dementia?'.

In contrast to the previously reviewed studies which started with a potential pool then looked at study eligibility, Gill and colleagues sought to determine what proportion of those receiving donepezil in Ontario, Canada from September 2001 to March 2002 would have been ineligible for the RCTs of this agent that led to its approval.[56] Between 51% and 78% of those prescribed donepezil would have been ineligible for the RCTs.

A unique consideration in dementia studies is the role of the caregiver. The RCTs reviewed for this chapter all required the presence of an involved, willing caregiver. It is more than just their presence that is important to consider. They often play a key role in determining whether an eligible subject enrols in a study. The characteristics of caregivers might well constitute a 'hidden' inclusion/exclusion criterion.

One small descriptive study examined the factors that influence a caregiver's decision to consent to participate in a clinical trial.[59] Among participants, hope for improvement in the subject's condition was frequently (63%) endorsed by the caregiver as a factor that influenced decision making. Concern about side-effects was the most frequently (60%) cited reason for declining by caregivers of non-participants. Non-participating subjects were more likely to have a child as their caregiver and to have had prior experience in a clinical trial. Caregivers in both groups felt that they had the right and the responsibility to make decision about the subjects' therapy. While most consulted with other family members about participation in the study, only a fifth involved the subject in the decision making.

Discussion

Clearly there is room for improving our ability to diagnose the various conditions that can lead to dementia. Having an exquisitely sensitive and specific biological marker for AD would enhance our abilities to not only diagnose AD but also other causes of dementia such as VaD or DLB. For studies we would be better able to wean out those who have concurrent AD, leaving us with cases of 'pure' VaD or DLB. At a recent conference I attended on vascular cognitive impairment I heard a speaker say that the most useful advance for the diagnosis of VaD would be a good test for AD.

But what should we do now? There are advocates for continuing to try to differentiate the various potential causes of dementia.[40] They would argue that we should even look at developing and refining clinical subgroups within AD. Grouping together dementia patients could hide differing responses to therapy. Apparent disease homogeneity can be a transient phenomenon. As our knowledge advances, we may be able to differentiate further. AD in the future may be restructured into a variety of distinct entities. This, though, flies in the face of the growing appreciation that multiple disease entities might be operant in a given individual suffering from dementia. Should we intensify our efforts to classify, subclassify and reclassify? Or, should we consider targeting as a symptom cognitive impairment? There are cogent arguments in the camps of both the 'splitters' and 'lumpers'.

For a medication to receive marketing approval for an indication, government regulatory agencies hold that it must be possible to diagnose the condition in a practical and reasonably precise manner. While arguably we can do that for AD (which is the most common cause – either alone or in combination with another contributing disease – for dementia) it is suspect that we can diagnose less common dementias like 'pure' VaD or LBD precisely enough to satisfy the regulatory agencies.

Most would concur that dementia can arise from cerebrovascular disease and that there is an entity called VaD. Things start falling apart when we talk about the criteria we will use to diagnose VaD. It can be further argued that an unknown proportion of subjects who fulfil any criteria for a diagnosis of VaD (e.g. the NINDS-AIREN criteria for possible or probable VaD) do not have 'pure' VaD. Many will also have the pathological hallmarks of another dementing illness, in particular AD. Any benefit seen with treatment might be from treating their concurrent (often AD) pathology, not their VaD. A similar argument can be made about DLB.

Trying to focus on 'pure' cases can be difficult. Diagnostic criteria are useful in so far as they are able to identify the presence (positive predictive value, PPV) or absence (negative predictive value) of a condition in a patient. In addition to sensitivity and specificity, this is dependent on the prevalence of the condition. For conditions with relatively low prevalences, even criteria with high specificities yield only moderate PPVs. For example, in the Camberwell Dementia Case Register study, a clinical diagnosis of probable VaD carried a PPV of 43% in predicting 'pure' VaD cases confirmed by neuropathology – in spite of a specificity for the criteria of 95%.[35] In other words if we recruited for a study using the NINDS-AIREN criteria for probable VaD from a pool of patients with a similar prevalence of VaD (7/80 or 8.75% had cerebrovascular findings only), over half of those eligible for the study would not have 'pure' VaD. In the Camberwell study, a clinical diagnosis of probable AD, because of the higher background prevalence of the disease (44/80 or 55% had neuritic plaques only), had a higher PPV (76%) even though the specificity of the NINCDS-ADRDA criteria was lower at 75%.[35] Barker and colleagues, using their frequencies and reported sensitivities/specificities for the NINCDS-ADRDA and NINDS-AIREN criteria, estimated that the PPV for a diagnosis of 'pure' AD and 'pure' VaD were 62% and 8% respectively.[34]

With our current state of knowledge is it worthwhile trying to mount studies of 'pure' DLB or VaD? Would they ever be able to convince the doubters? Would we be better off acknowledging the high likelihood of multiple pathologies and study mixed AD/VaD and/or AD/DLB?

One particular issue is the representativeness of the subjects who have been enrolled in the critical phase III trials. Drugs usually behave in a similar manner across the various subsets of patients with a condition, but important differences may be present. Formerly it was acceptable to show that an intervention worked in the theoretical optimal state under ideal conditions. The rules of the game are changing. Most would now hold that enrolment should be more inclusive. Up to 95.6% of the total population with AD were judged ineligible for at least certain studies.[37] One is reminded of the biblical quotation, 'For many are called, but few are chosen' (Matthew, King James Bible, 1611; 21:8–9). This desire to open up eligibility and improve the representativeness of subjects enrolled in studies will probably persist and strengthen.

Unfortunately for many of the dilemmas facing those planning studies, the most we can do is highlight the issues that must be considered, the trade-offs that might have to be made. For all making these difficult choices, we sincerely hope that on a consistent basis, 'of two evils I have chose the least' (Matthew Prior: Ode in Imitation of Horace, 1692).

References

1. Friedman LM, Furberg CD, DeMets DL. Study population. In: Fundamentals of Clinical Trials (3rd edn). St Louis: Mosby; 1996: 30–40.

2. Spilker B. Establishing criteria for patient inclusion. In: Guide to Clinical Trials. New York: Raven Press; 1991: 147–58.

3. Leber PD, Davis CS. Threats to the validity of clinical trials employing enrichment strategies for sample selection. Control Clin Trials 1998; 19:178–87.

4. Day S. Dictionary for Clinical Trials. Chichester: John Wiley and Sons, Ltd; 1999.

5. Health Canada. Therapeutic Products Programme Guidelines: Inclusion of Women in Clinical Trials. Ottawa: Health Canada; April 17 1997.

6. American Psychiatric Association. Diagnostic and Statistical Manual of Mental Disorders (3rd edn). Washington: American Psychiatric Association; 1987.

7. American Psychiatric Association. Diagnostic and Statistical Manual of Mental Disorders (4th edn). Washington: American Psychiatric Association; 1994.

8. Birks J, Grimley Evans J, Iakovidou V, Tsollaki M. Rivastigmine for Alzheimer's disease. Cochrane Database Syst Rev 2000; (4):CD001191.

9. Birks JS, Harvey R. Donepezil for dementia due Alzheimer's disease. Cochrane Database Syst Rev 2003; (3):CD001190.

10. Olin J, Scneider L. Galantamine for Alzheimer's disease. Cochrane Database Syst Rev 2002; (3):CD001747.

11. McKhann G, Drachman D, Folstein M et al. Clinical diagnosis of Alzheimer's disease: report of the NINCDS/ADRDA Work Group under the auspices of the Department of Health and Human Services Task Force on Alzheimer's Disease. Neurology 1984; 34:939–44.

12. Aisen PS, Schafer KA, Grundman M et al for the Alzheimer's Disease Cooperative Study. Effects of rofecoxib or naproxen vs placebo on Alzheimer disease progression – A randomized controlled trial. JAMA 2003; 289:2819–26.

13. Reisberg B, Doody R, Stoffler A et al. Memantine in moderate-to-severe Alzheimer's disease. N Engl J Med 2003; 348:1333–41.

14. Leber PD, Davis CS. Threats to the validity of clinical trials employing enrichment strategies for sample selection. Control Clin Trials 1998; 19:178–87.

15. Roman GC, Tatemichi TC, Erkinjuntti et al. Vascular dementia: diagnostic criteria for research studies – report of the NINDS-AIREN International Workshop. Neurology 1993; 43:250–60.

16. Erkinjuntti T, Kurz A, Gauthier S et al. Efficacy of galantamine in probable vascular dementia and Alzheimer's disease combined with cerebrovascular disease: a randomized trial. Lancet 2002; 359:1283–90.

17. Orgogozo J-M, Rigaud A-S, Stoffler A, Mobius H-J, Forette F. Efficacy and safety of memantine in patients with mild to moderate vascular dementia: a randomized, placebo-controlled trial (MMM 300). Stroke 2002; 33:1834–9.

18. Wilcock G, Mobius HJ, Stoffler A. A double-blind, placebo-controlled multicentre study of memantine in mild to moderate vascular dementia (MMM500). Int Clin Psychoparmacol 2002; 17:297–305.

19. Black S, Roman GC, Geldmacher DS et al. Efficacy and tolerability of donepezil in vascular dementia. Stroke 2003; 34:2323–32.

20. Wilkinson D, Doody R, Helme R et al. Donepezil in vascular dementia: a randomized, placebo-controlled study. Neurology 2003; 61:479–86.

21. Rosen WG, Terry RD, Fuid PA, Katzmann R, Peck A. Pathological verification of ischaemic score in differentiation of dementias. Ann Neurol 1980; 7:486–88.

22. Loeb C, Gandolfo C. Diagnostic evaluation of degenerative and vascular dementia. Stroke 1983; 14:399–401.

23. Winblad B, Portis N. Memantine in severe dementia: results of the 9M-Best Study (Benefit and Efficay in Severely Demented Patients During Treatment with Memantine). Int J Geriatr Psychiatry 1999; 14:135–46.

24. McKeith IG, Galasko D, Kosaka K et al. Consensus guidelines for the clinical and pathological diagnosis of dementia with Lewy bodies: report of the consortium on dementia with Lewy bodies international workshop. Neurology 1996; 47:1113–24.

25. McKeith IG, Ballard CG, Perry RH et al. Prospective validation of Consensus criteria for the diagnosis of dementia with Lewy bodies. Neurology 2000; 54:1050–58.

26. McKeith IG, Del Ser T, Spano P et al. Efficacy of rivastigmine in dementia with Lewy bodies: a randomised, double-blind, placebo-controlled international study. Lancet 2000; 356:2031–36.

27. Folstein MF, Folstein SE, McHugh PR. 'Mini-Mental State': a practical method for grading the cognitive state of patients for the clinician. J Psychiatr Res 1975; 12:189–98.

28. Rosen W, Mohs RC, Davis KL. A new rating scale for Alzheimer's disease. Am J Psychiatry 1984; 141:1356–64.

29. Morris JC. The Clinical Dementia Rating (CDR): current version and scoring rules. Neurology 1993; 43:2412–14.

30. Erkinjuntti T, Ostbye T, Steenhuis R, Hachinski V. The effect of different diagnostic criteria on the prevalence of dementia. N Engl J Med 1997; 33:1667–74.

31. Chui HC, Mack W, Jackson JE et al. Clinical criteria for the diagnosis of vascular dementia: a multicenter study of comparability and interrater reliability. Arch Neurol 2000; 57:191–6.

32. Knopman DS, DeKosky ST, Cummings JL et al. Practice parameter: diagnosis of dementia (an evidence-based review): Report of the Quality Standards Subcommittee of the American Academy of Neurology. Neurology 2001; 56:1143–53.

33. Feldman H, Levy AR, Hsiung GY et al. A Canadian Cohort Study of Cognitive Impairment and Related Dementias (ACCORD): study methods and baseline results. Neuroepidemiology 2003; 22:265–74.

34. Barker WW, Luis CA, Kashuba A et al. Relative frequencies of Alzheimer Disease, Lewy body, vascular and frontotemporal dementia, and hippocampal sclerosis in the State of Florida Brain Bank. Alzheimer Dis Assoc Disord 2002; 16:203–12.

35. Holmes C, Cairns N, Lantos P, Mann A. Validity of current clinical criteria for Alzheimer's disease, vascular dementia, and dementia with Lewy bodies. Br J Psychiatry 1999; 174:45–50.

36. Lopez OL, Becker JT, Klunk W et al. Research evaluation and diagnosis of probable Alzheimer's disease over the last two decades: I. Neurology 2000; 55:1854–62.

37. Schneider LS, Olin JT, Lyness SA, Chui HC. Eligibility of Alzheimer's disease clinic patients for clinical trials. J Am Geriatr Soc 1997; 45:923–8.

38. Villareal DT, Grant E, Miller JP et al. Clinical outcomes of possible versus probable Alzheimer's disease. Neurology 2003; 61:661–67.

39. Willmer J, Mohr E. Evaluation of selection criteria used in Alzheimer's disease clinical trials. Can J Neurol Sci 1998; 25:39–43.

40. Ritchie K, Lovestone S. The dementias. Lancet 2002; 360:1759–66.

41. Rosen WG, Terry RD, Fuld PA, Katzman R, Peck A. Pathological verification of ischemic score in differentiation of dementia. Ann Neurol 1980; 7:486–88.

42. Loeb C, Gandolfo C. Diagnostic evaluation of degenerative and vascular dementia. Stroke 1983; 14:399–401.

43. Verhey FR, Lodder J, Rozendaal N, Jolles J. Comparison of seven sets of criteria used for the diagnosis of vascular dementia. Neuroepidemiology 1996; 15:166–72.

44. Wetterling T, Kanitz RD, Borgis KJ. Comparison of different diagnostic criteria for vascular dementia (ADDTC, DSM-IV, ICD-10, NINDS-AIREN). Stroke 1996; 27:30–36.

45. Pohjasvaara T, Mantyla R, Ylikoski R, Kaste M, Erkinjuntti T. Comparison of different clinical criteria (DSM-III, ADDTC, ICD-10, NINDS-AIREN, DSM-IV) for the diagnosis of vascular dementia. Stroke 2000; 31:2952–7.

46. Rockwood K, Davis H, MacKnight C et al. The Consortium to Investigate Vascular Impairment of Cognition: methods and first findings. Can J Neurol Sci 2003; 30:237–43.

47. Lopez OL, Larumbe MR, Becker JT et al. Reliability of NINDS-AIREN clinical criteria for the diagnosis of vascular dementia. Neurology 1994; 44:1240–5.

48. Gold G, Bouras C, Canuto A et al. Clinicopathological validation study of four sets of clinical criteria for vascular dementia. Am J Psychiatry 2002; 159:82–7.

49. Gold G, Giannakopoulos P, Montes-Paixao C et al. Sensitivity and specificity of newly proposed clinical criteria for possible vascular dementia. Neurology 1997; 49:690–4.

50. Pratt RD. Patient populations in clinical trials of the efficacy and tolerability of donepezil in patients with vascular dementia. J Neurol Sci 2002; 203–204:57–65.

51. Lopez OL, Becker JT, Kaufer DI et al. Research evaluation and prospective diagnosis of dementia with Lewy bodies. Arch Neurol 2002; 59:43–6.

52. Merdes AR, Hansen LA, Jeste DV et al. Influence of Alzheimer pathology on clinical diagnostic accuracy in dementia with Lewy bodies. Neurology 2003; 60:1586–90.

53. Litvan I, MacIntyre A, Goetz CG et al. Accuracy of the clinical diagnoses of Lewy body disease, Parkinson disease, and dementia with lewy bodies. Arch Neurol 1998; 55:969–78.

54. Lopez OL, Litvan I, Catt KE et al. Accuracy of four clinical diagnostic criteria for the diagnosis of neurodegenerative dementias. Neurology 1999; 53:1292–9.

55. McKeith IG, Fairbairn AF, Bothwell RA et al. An evaluation of the predictive validity and inter-rater reliability of clinical diagnostic criteria for senile dementia of the Lewy body type. Neurology 1994; 44:872–7.

56. Gill SS, Bronskill SE, Sykora K, Li P, Rochon PA. Representation and eligibility of real-world subjects with dementia in clinical trails of donepezil. Geriatrics Today. J Can Geriatr Soc 2003; 6:67 [abstract].

57. Treves TA, Verchovsky R, Klimovitsky S, Korczyn AD. Recruitment rate to drug trials for dementia of the Alzheimer Type. Alzheimer Dis Assoc Disord 2000; 14:209–11.

58. Cohen-Mansfield J. Recruitment rates in gerontological research: the situation for drug trials in dementia may be worse than previously reported. Alzheimer Dis Assoc Disord 2002; 16:279–82.

59. Elad P, Treves TA, Drory M et al. Demented patients' participation in a clinical trial: factors affecting the caregivers' decision. Int J Geriatr Psychiatry 2000; 15:325–30.

7

Global Assessment Measures

Kenneth Rockwood

Global assessment measures have played an important role in dementia drug trials, where they serve two purposes. They allow judgements to be made about whether patients have changed. Secondly, they allow a summary score of these changes to be calculated. This chapter reviews the use of global measures in dementia drug trials. While some summary measures such as the SF-36 have been used as global measures in clinical trials in dementia,[1,2] here we focus on global measures that require clinical judgement in their scoring. We also chiefly examine trials of pharmacological interventions. Global measures are placed first in an historical and conceptual context. Next, some empirical studies are reviewed and suggestions made for how global measures might best be used in future trials. The chapter argues that the essential feature of global measures is that they rely on clinical judgements about multi-faceted phenomena. In consequence, a measurement scheme suited to judgement-based instruments needs to be employed if their full potential is to be realised. This is a worthy goal, as the careful evaluation of new phenomena by expert observers is an essential means of advancing understanding about new entities, such as treated Alzheimer's disease. As argued in detail elsewhere,[3] 'treated Alzheimer's disease' is best understood as a new entity, and not simply the reversal of untreated AD. In short, 'disease progression in reverse' is a poor model for detecting beneficial effects, and looking for that pattern, to the exclusion of looking for other items, inhibits our understanding of benefit.

Global measures, clinical judgement and clinimetrics

Global measures of dementia severity are commonly used as criteria for inclusion and exclusion in dementia drug trials, and are little considered in this chapter, where the emphasis is on their use as outcome measures (see Chapter 6 for the consideration of inclusion and exclusion criteria.) The use of global measures as outcomes is longstanding. Indeed, from the beginning of the modern era of dementia drug trials, there has been a regulatory requirement that a global measure be used to evaluate new compounds. The original rationale for the use of a global measure was that it could provide a means of ensuring the clinical meaningfulness of any changes observed on standard neuropsychological testing. In 1990, guidelines promulgated by the United States Food and Drug Administration stipulated that a global outcome measure should be a primary outcome in registration trials, as the 'ultimate test of the clinical utility of a drug's anti-dementia effects'.[4] After initial variability in the use and construction of global change scores, the standard has evolved to the Clinician's Interview-Based Impression of Change *with caregiver input* (CIBIC-Plus).

Despite enjoying widespread use, the CIBIC-Plus has often been criticised. Its provenance is suspect, in that its ubiquity springs from regulatory fiat rather than from the scientific assessment of its measurement properties, a traditional means by which some measures supplant others. On its face, the CIBIC-Plus also seems less likely to be valid. In contrast to the highly specified way in which psychometric tests are employed, an early version, (the Global Clinical Impression of Change – CGIC) required only an unspecified clinical interview, after which a single judgement was made using a Likert scale.[5] The latter was anchored at 4 ('no change') and scaled from 1 (the greatest degree of improvement) to 7 (the maximal degree of worsening). Exactly what constituted 'better' or 'worse' was not specified. In consequence, there has been a move, as detailed below, to standardise the clinical interview on which the global impression of change is made.[6] This has been seen as an appropriate alternative to abandoning global measures. A strong constituency has existed, however, for eliminating global measures in favour of 'appropriate combinations' of 'assessments such as well-designed activities of daily living, cognition, and behaviour measures'.[7]

While standardisation of some aspects of the clinical interview can be helpful, many of the calls for standardisation stem from an attempt to impose a particular measurement ethos on a subject matter (expert clinical judgement) to which it may well be unsuited.[3,8,9] All global measures share an essential feature, which is that they incorporate clinical judgements. The class of measures that are based on clinical judgement has been referred to as 'clinimetric', and a theory of clinimetric measurement has been elaborated in a way that is particularly suited to studying new – and decidedly clinical – phenomena such as treated Alzheimer's disease.[10] Clinimetric measures can be contrasted with 'psychometric' measures in which the opportunity for judgement is minimised. Despite having many strong proponents,[11] psychometric measures also have their detractors, who argue that while the distinction might once have been necessary, it overstates the case, and that advanced psychometric techniques have now caught up with whatever insights might have been afforded by clinimetric analyses.[12] However that might be, much of the current discussion of methodology in dementia lags far behind the more advanced psychometric analyses, so that a consideration of the aims and insights from clinimetrics remains of value.

As envisaged by Feinstein,[10] clinimetric measures (e.g. rating scales of symptoms, signs or treatment responses, indexes) share several essential features. They consist of items that should be selected by clinical judgement (rather than using a statistical criterion). The items should use patients' reports of what is troubling them, rather than coming from a theoretical scheme. Importantly, too, weighting of items should reflect patient preferences. Items should be heterogeneous, to capture all symptoms that are important to the phenomenon being investigated (rather than giving priority to homogeneity); similarly, all items should be included, rather than statistical criteria being used to eliminate items. They should be easy to use, and easy to score.

Apart from the last, the criteria for clinimetric measures derive from Feinstein's concern about how the measurement of clinical phenomena is conceptualised. In psychometric theory, a central idea is that a single latent construct gives rise to a number of phenomena. These several phenomena can be measured individually, but theoretical constructs or statistical rules are used to combine items into a parsimonious measure. Cognitive impairment in Alzheimer's disease is one example, where items of worsening in memory, language, visuo-spatial construction and attention and concentration, for example, can be linked in a single measure (such as the Global Deterioration Scale).[13] The latent construct (disease progression) causes the individual items to behave as they do, even though the items that are affected by disease progression are not all the same.

By contrast, dementia, although based in cognitive impairment, is not just cognitive impairment. Rather, it is the range of cognitive, behavioural and functional features that give rise to individually defined disease expression. Until treatments were developed, there was unity enough in dementia expression (even in a 'single' disease such as AD) that individual differences could be seen as anything from curiosities to subtypes. For example, the stage at which dementia presents has been related to the social supports available to the patient prior to presentation.[14] Thus a stereotypical professor of mathematics who lived alone with his wife in an environment of heavy cognitive demands (lectures, journal clubs, attendance at the opera) might present far earlier than an illiterate, older person who has a rich social support mechanism, but on whom very few cognitive or functional demands are placed. Given the relative homogeneity of the nature of disease progression, such discrepancies could be readily accommodated, especially in longitudinal studies. The presence of cholinergic treatment, however, has changed the nature of disease progression in ways that undermine the central psychometric assumption of a single latent construct. As yet, however, much of the easily recognised variability in treatment response (as exemplified by scores across the Likert range on the CIBIC-Plus) has not been studied carefully enough to allow patterns in variability to be readily identified. Indeed, it can be argued that it is the failure of the assumption of homogeneity, as much as anything, that has given rise to calls in the lay press that the clinical meaningfulness of treatments be made more clear if the cost of drug treatment is to be justified.[15] In other words, the failure to see an evident link between what the tests measure and what patients experience can be seen as an appreciation that 'improvement' is not a unidimensional construct. In consequence, there is new potential for global measures that fully exploit clinimetric principles, rather than aping psychometric ones.

Types of global clinimetric measures in anti-dementia drug trials

As reviewed elsewhere, there are two types of global, judgement-based measures: specified and unspecified.[16] Specified measures, which typically are made up of a series of generalisable descriptors, include instruments such as the Clinical Dementia Rating (CDR),[17] and the Global Deterioration Scale.[18] In such measures, clinicians make a judgement about which descriptor from a specified list best describes the patient under consideration. For the most part, these measures were developed prior to the advent of anti-dementia drug trials. Thus, their specification generally includes a hierarchical ordering of the profiles usually associated with untreated decline. For example, in the Functional Assessment Staging Tool (FAST),[19] the prompts for functional disease progression, roughly specify a staging of dependence in instrumental activities of daily living (IADL) followed by a stage in which there is prompting for personal activities of daily living (ADL), followed by a stage in which there is dependence in personal ADL. The judgement about the level of dependence is, however, based on inquiries by the clinician. In short, specified clinical assessment measures often follow the format of a semi-structured interview, and provides a means for clinicians to investigate whether the patient conforms to a known disease profile.

In contrast to the specified clinimetric measures, the early versions of the CGIC and the CIBIC-Plus can be considered as 'unspecified' measures. Such measures often provide no structure for the interview that underlies the clinical judgement to be made. In this context, somewhat more structured, but still unspecified global measures include individualised measures such as Goal Attainment Scaling (GAS).[20]

Specified global measures
As noted, many early measures in dementia drug trials were based on staging measures for dementia, including the classic Dementia Rating Scale.[21]

The most commonly used specified global measure in dementia drug trials, however, has been the CDR.[17,22] The CDR was developed at the Washington University School of Medicine and has been widely use since its first publication in 1982. Originally it was used as a staging measure, to aid in clinical discrimination from 0 (no cognitive impairment), 0.5 (mild cognitive impairment that does not meet criteria for dementia), 1 (mild dementia) to more severe stages. The CDR has been subjected to several studies on its measurement properties, including autopsy confirmation of Alzheimer's disease. A number of inter-rater reliability studies have been carried out, including studies in dementia drug trials.[23,24] The CDR also benefits from having been used in a number of longitudinal studies, which refines the possibility for modelling the expected course of long-term decline. Although it too was originally intended as a staging tool, and the CDR can be scored to give global measure, suitable for the calculation of change scores. The so-called 'sum-of-boxes' scoring measure allows the scores to range from 0–27. It has been used in studies of both cholinergic and non-cholinergic compounds.[25]

The CDR has been adapted in a variety of ways, including a version for use in a long-term care setting,[26] and the extension of its domains in the Functional Rating Scale.[27,28] Most recently in the drug trials context, the CDR has been proposed as a means of screening patents by distinguishing between mild cognitive impairment and Alzheimer's disease.[29]

A contemporary to the CDR, the Gottfries–Brane–Steen scale consists of subscales that measure intellectual (12 items), emotional (3 items) and items of self-care and other ADL (6 items); as well as six items of behavioural and psychological symptoms of dementia.[30,31] As reviewed by Brane et al, the GBS has been used in clinical trials, including non-cholinergic compounds such as propentofylline.[31,32] Its responsiveness makes it an attractive global scale.

An interesting recent development in global dementia scales that might have important applicability in dementia drug trials is the Dementia Deficits Scale (DDS).[33] The scale measures self-awareness of cognitive, emotional and functional deficits in dementia and can be completed independently by the patient, clinician and informant, to yield two measures of deficit awareness. The discrepancy between clinician and patient assessments of the patient's deficits, and the discrepancy between informant and patient assessments provide a hitherto largely unexplored means of evaluating important treatment effects.

The need for greater reliability has been emphasised, and standardisation of the assessment on which the clinical judgement was made was seen as an important remedy. Standardisation, it was also argued, would help ensure that important treatment effects were not missed.[34,35] This concern ultimately resulted in the Alzheimer's Disease Cooperative Study Clinician's Global Impression of Change (ADCS-CGIC) in which 15 domains which should be assessed in the clinical interview are described, together with prompts for items to be covered.[6] Within each of the domains, change is again rated on a seven-point scale. Specific patient preferences are not elicited. In essence, using this approach, the clinical interview goes from unstructured to structured. Given the individual nature of the deficits at baseline, however, exactly what constitutes a level of improvement for any given item remains at best incompletely specified.

Unspecified clinimetric global measures

As noted, the CIBIC-Plus in its unstructured form remains the most common unspecified global measure. Its individualisation and its essential reliance on clinical judgement also qualify it as a clinimetric measure, and although it can reflect patient preferences, there is no requirement for it to do so. While it is unspecified, reproducible patterns of judgements about treatment effects have demonstrated in formal qualitative analyses of the notes that clinicians have used in the CIBIC-Plus interviews.[36,37] These studies suggest that clinicians tend to privilege caregiver reports of function and behaviour over their own estimates of cognition, especially when function has been reported to improve, or behaviour disturbances have been reported to decline. Although clinicians' narratives potentially contain important insights into the response to enhanced cholinergic neurotransmission, this resource has been little exploited.[38] The CIBIC-Plus has also recently been used with non-cholinergic agents such as memantine in moderate to severe stages of Alzheimer's disease, and even in that setting, where there is comparatively little treatment experience on which to base clinical judgements at those stages, it has been shown to be responsive.[39,40]

Although the CIBIC-Plus has been the dominant clinimetric measure in dementia, it is but one of several potentially relevant individualised outcome measures available.[41] To the extent that it is used at all, perhaps the next most widely used clinimetric measure in dementia trials is GAS.[42,43] GAS is a technique for both measurement and management. At baseline, patients and their family members are interviewed and asked, of the problems that they have with dementia, which they would most like to see resolve with treatment. For each problem, the current state is described, and a treatment goal is set. In addition, a range of expectations is scaled, so that other levels (from 'very much worse [than the present state]' to 'very much better [than the present state]' are also described. At subsequent interviews, the description that best matches the patient's present state is scored. A standardised formula is used to produce a summary score that reflects the overall degree of goal attainment, taking into account the number of goals and how they have been weighted. Clearly, this exercise calls for clinical skills to elicit the description, and clinical judgement about how to describe the levels. The technique has been used as the primary outcome measure in one trial and is being used in another due to be complete in 2005. It offers the potential to understand patient preferences, to know whether trials produce clinically meaningful results and to gain insights into cholinergic neurotransmission. So far, it has been a sensitive measure of detecting clinically important change, and more responsive to change than the Alzheimer's Disease Assessment Scale – Cognitive Subscale (ADAS-Cog).[44] Used in conjunction with the CIBIC-Plus, it provides a means of enhancing the latter's sensitivity, without constraining its individualised nature.

Challenges to clinimetric global measures

Despite their potential, considerable scepticism exists about global scales and especially about the CIBIC-Plus. This scepticism about clinimetric global measures goes beyond methodological concerns, although these concerns are real. A review in 2000 concluded that there has been a dearth of published information about the reliability and validity of global scales in dementia drug trials, and there has been little new information since then.[45] One important challenge to the CIBIC since that review is an intriguing inter-rater reliability study. Quinn et al had a videotaped version of the CIBIC evaluated by raters who were deceived for study purposes.[46] The raters were told that the CIBIC interview videos that they were evaluating had been taken at baseline and at 6 to 12 months later, when in fact half of the interviews were shown in reverse order. The authors compared ratings on 'true order' interviews with those on 'reverse order' interviews and found that, while absolute agreement was poor (kappa = 0.18), the agreement on whether patients were better, worse or the same was better (kappa = 0.51). Of some interest, reliability was better in the 'true order' group (in which more patients had deteriorated) than in the 'reverse order' group, in which more patients had improved. This suggests that raters are sceptical about improvement, even when it conforms to a 'reversal of progression' model. Nevertheless, in a recent meta-analysis, the impression of benefit conferred with a global measure was similar, in the aggregate, to that conferred by the ADAS-Cog.[47] This suggests that, whatever individual patient assessment difficulties, group effects are detectable. The problem – and this holds for the ADAS-Cog[48] – is that whatever is being detected is not readily translated into clinical practice.

It is also interesting to note that in some early trials of anti-dementia compounds, unspecified clinimetric global measures are less sensitive to change than the ADAS-Cog. This was the case with the agent linopirdine.[49,50] It also is the pattern seen with the gingko biloba extract EGb 761, where an early report showed a difference in the ADAS-Cog at a level that was large enough to be clinically detectable, but without any difference detected in the global clinical measure employed in the trial.[51] A more recent study, however, showed changes in both the ADAS-Cog and the global clinical measure,[52] although this was not replicated in another later study that employed both a standard psychometric measure and a global clinical one,[53] nor in a study for symptomatic memory enhancement.[54] Still, the possibility that 'knowing what to look for' in early dementia might yield replicable results has not yet entirely been disproved.

Lessons from the use of global measures in dementia drug trials

This chapter has argued that global clinical measures have played a useful role in detecting treatment effects in dementia. Important potential exists for these measures to further that role, especially in advancing how we view treatment in a way that does not imply simple disease reversal. But if disease progression in reverse is a poor model for detecting beneficial effects, how might global measures be helpful? Two approaches hold promise. With respect to specified measures, analysing not just the changes in scores, but the changes in patterns of scores can yield insights about new treatment effects. For example, multi-domain staging measures, by their nature, are designed so that there is congruity between domains. In other words, the scoring is arranged so that the descriptor consistent with mild dementia in one domain

(e.g. memory) is consistent with the description of impairment with mild dementia in each of the other domains (e.g. language, praxis, attention and concentration). Specified global measures can therefore yield insights about treatment effects by alerting physicians to look not for uniform, across-all-domains responses, but for common variations (e.g. improved attention and concentration versus less effect on recent memory). To date, there has been little attention paid to these types of analyses.

Clearly, one limitation of specified measures is that they do not contain enough specificity in the domains that they cover. For example, none of the measures here reviewed would allow the following pattern in language to be detected: more spontaneous language, less repetition, and longer and more complete sentences. This is the case even though, in the author's experience, this is a common pattern of treatment response. The paucity of such clinically relevant measurement means that we are left with information about language effects, other than the sorts of anecdotes here illustrated, that does not go much beyond reports of individual test items from psychometric scales.

Another limitation of specified global measures is that they do not measure items that are of great clinical importance to caregivers. 'Recovery of initiative', improved 'alertness', being 'more present and more in tune' were symptoms that have been identified in a survey of experienced physicians as the most readily detectable symptomatic benefits from donepezil treatment, yet they are not captured by any of the specified measures (psychometric or clinimetric) now in use.[55] To this end, and given that they are common and detectable, such treatment effects are now readily measured only by unspecified global measures. In short, until we have enough experience about dementia treatment to allow a model of treatment profiles to be specified, an unspecified measure seems like a reasonable way to capture clinical observations, as long as some caveats are followed. The observations need to be documented. It is not enough to say to what extent change has occurred. The factors that have given rise to the judgement of change need to be spelled out. The information thus gathered needs to be analysed systematically. Prospective qualitative analyses can achieve this, and timely analyses can also allow important insights about the performance of other measures. For example, if greater 'alertness' or 'initiative' is cited, some consideration needs to be given to whether these effects are being measured in other ways (for example, by measures of behaviour or of function). If not, the reasons for this need to be pursued.

Finally, two innovations in how we think about clinical trials are needed. The first is that they can be used to test our instrumentation. When a clinical trial employs more than one measure, the measures themselves – and especially their sensitivity to change – can be analysed. In effect, the trial can be viewed as one in which patients and treatments are held constant, but the measures are varied. Secondly, the trials need to be understood as a means of studying human cholinergic neurotransmission. For both of these purposes, the supplementation of standardised psychometric tests by individualised clinimetric ones offers important opportunities to advance the science not just of treatment but of dementia pathophysiology.

References

1. McHorney CA, Ware JE. Construction and validation of an alternate form general mental health scale for the Medical Outcomes Study short-form 36-item health survey. Med Care 1995; 33:15–28.

2. Teri L, Gibbons LE, McCurry SM et al. Exercise plus behavioral management in patients with Alzheimer disease: a randomized controlled trial. JAMA 2003; 290:2015–22.

3. Rockwood K, Wallack M, Tallis R. Treating dementia: understanding success short of cure. Lancet Neurol 2003; 2:630–3.

4. Leber PD. Developing safe and effective anti-dementia drugs. In: Becker R, Giacobini E (eds). Alzheimer Disease: From Molecular Biology to Therapy. Boston: Birkhauser; 1997.

5. Knopman DS, Knapp MJ, Gracon SI, Davis CS. The Clinician Interview-Based Impression (CIBI): A

clinician's global change rating scale in Alzheimer's disease. Neurology 1994; 44:2315–21.

6. Schneider LS, Olin JT, Doody R et al. Validity and reliability of the Alzheimer's Disease Cooperative Study-Clinical Global Impression of Change. The Alzheimer's Disease Cooperative Study. Alzheimer Dis Assoc Disord 1997; 11(Suppl 2):S22–32.

7. Reisberg B, Schneider L, Doody R et al. Clinical global measures of dementia. Position paper from the International Working Group on Harmonization of Dementia Drug Guidelines. Alzheimer Dis Assoc Disord 1997; 11(Suppl 3):8–18.

8. Rockwood K. Use of global assessment measures in dementia drug trials. J Clin Epidemiol 1994; 47:101–3.

9. Rockwood K, MacKnight C. Assessing the clinical importance of statistically significant improvement in anti-dementia drug trials. Neuroepidemiology 2001; 20:51–6.

10. Feinstein AR. Clinimetrics. New Haven: Yale University Press; 1987.

11. De Vet HC, Terwee, CB, Bouter LM. Current challenges in clinimetrics. J Clin Epidemiol 2003; 56:1137–41.

12. Streiner D. Clinimetrics vs. psychometrics: an unnecessary distinction. J Clin Epidemiol 2003; 56:1142–5.

13. Reisberg B, Ferris SH, de Leon MJ, Crook T. The Global Deterioration Scale for assessment of primary degenerative dementia. Am J Psychiatry 1982; 139:1136–9.

14. Sibley A, MacKnight C, Rockwood K et al. The effect of the living situation on the severity of dementia at diagnosis. Dement Geriatr Cogn Disord 2002; 13:40–5.

15. Grady D. Minimal benefit is seen in drugs for Alzheimer's disease. New York Times; April 7, 2004, Wednesday Section A: page 1, column 2.

16. Rockwood K, Morris JC. Global staging measures. In: Gauthier S (ed). Alzheimer's Disease: Diagnosis and Treatment. London: Martin Dunitz; 1996.

17. Morris JC. The Clinical Dementia Rating (CDR): current version and scoring rules. Neurology 1993; 43:2412–4.

18. Auer S, Reisberg B. The GDS/FAST staging system. Int Psychogeriatr 1997; 9(Suppl):167–71.

19. Reisberg B. Functional assessment staging (FAST). Psychopharmacol Bull 1988; 24:653–9.

20. Kiresuk TJ, Smith RE, Cardillo JE. Goal Attainment Scaling: Applications Theory and Measurement. Hillside, NJ; Lawrence Erlbaum Associates; 1994.

21. Blessed G, Tomlinson BE, Roth M. Blessed-Roth Dementia Scale (DS). Psychopharmacol Bull 1988; 24:705–8.

22. Hughes CP, Berg L, Danziger WL, Coben LA, Martin RL. A new clinical scale for the staging of dementia. Br J Psychiatry 1982; 140:566–72.

23. Morris JC, Ernesto C, Schafer K et al. Clinical dementia rating training and reliability in multi-center studies: the Alzheimer's Disease Cooperative Study experience. Neurology 1997; 48(6):1508–10.

24. Rockwood K, Strang D, MacKnight C, Downer R, Morris JC. Interrater reliability of the Clinical Dementia Rating in a multicenter trial. J Am Geriatr Soc 2000; 48:558–9.

25. Akhondzadeh S, Noroozian M, Mohammadi M et al. *Melissa officinalis* extract in the treatment of patients with mild to moderate Alzheimer's disease: a double blind, randomised, placebo controlled trial. J Neurol Neurosurg Psychiatry 2003; 74:863–6.

26. Marin DB, Flynn S, Mare M et al. Reliability and validity of a chronic care facility adaptation of the Clinical Dementia Rating scale. Int J Geriatr Psychiatry 2001; 16:745–50.

27. Feldman H, Schulzer M, Wang S et al. The Functional Rating Scale in Alzheimer's disease assessment: a longitudinal study. In: Iqbal K, Mortimer JA, Winblad B, Wisniewski HM (eds). Research Advances in Alzheimer's Disease and Related Disorders. Chichester: Wiley; 1995: 235–41.

28. Feldman H, Gauthier S, Hecker J et al. A 24-week, randomized, double-blind study of donepezil in moderate to severe Alzheimer's disease. Neurology 2001; 57:613–20.

29. Grundman M, Petersen RC, Ferris SH et al. Mild cognitive impairment can be distinguished from Alzheimer disease and normal aging for clinical trials. Arch Neurol 2004; 61:59–66.

30. Gottfries CG, Brane G, Gullberg B, Steen G. A new rating scale for dementia syndromes. Arch Gerontol Geriatr 1982; 1:311–30.

31. Brane G, Gottfries CG, Winblad B. The Gottfries–Brane–Steen scale: validity, reliability and application in anti-dementia drug trials. Dement Geriatr Cogn Disord 2001; 12(1):1–14.

32. Mielke R, Moller HJ, Erkinjuntti T et al. Propento-fylline in the treatment of vascular dementia and Alzheimer-type dementia: overview of phase I and phase II clinical trials. Alzheimer Dis Assoc Disord 1998; 12(Suppl 2):S29–35.

33. Snow AL, Norris MP, Doody R et al. Dementia Deficits Scale. Rating self-awareness of deficits. Alzheimer Dis Assoc Disord 2004; 18:22–31.

34. Schneider LS, Olin JT. Clinical global impressions in Alzheimer's clinical trials. Int Psychogeriatr 1996; 8:277–88.

35. Knopman DS. Global change assessments in anti-Alzheimer clinical drug trials. Dement Geriatr Cogn Disord 1998; 9(Suppl 3):8–15.

36. Joffres C, Graham J, Rockwood K. A qualitative analysis of the clinician interview-based impression of change (Plus): Methodological issues and implications for clinical research. Int Psychogeriatr 2000; 12:403–13.

37. Joffres C, Bucks RS, Haworth J, Wilcock GK, Rockwood K. Patterns of clinically detectable treatment effects with galantamine: a qualitative analysis. Dement Geriatr Cogn Disord 2003; 15:26–33.

38. Rockwood K, Joffres C. Improving clinical descriptions to understand the effects of dementia treatment: consensus recommendations. Int J Geriatr Psychiatry 2002; 17:1006–11.

39. Reisberg B, Doody R, Stoffler A et al. Memantine in moderate-to-severe Alzheimer's disease. N Engl J Med 2003; 348:1333–41.

40. Tariot PN, Farlow MR, Grossberg GT et al. Memantine treatment in patients with moderate to severe Alzheimer disease already receiving donepezil: a randomized controlled trial. JAMA 2004; 291:317–24.

41. Donnelly C, Carswell A. Individualized outcome measures: a review of the literature. Can J Occup Ther 2002; 69:84–94.

42. Rockwood K, Stolee P, Howard K, Mallery L. Use of goal attainment scaling to measure treatment effects in an anti-dementia drug trial. Neuroepidemiology 1996; 15:330–8.

43. Rockwood K, Graham J, Fay S. Goal setting and attainment I Alzheimer's disease patients treated with donepezil. J Neurol Neurosurg Psychiatry 2002; 73:500–7.

44. Rosen WG, Mohs RC, Davis KL. A new rating scale for Alzheimer's disease. Am J Psychiatry 1984; 141:1356–64.

45. Oremus M, Perrault A, Demers L, Wolfson C. Review of outcome measurement instruments in Alzheimer's disease drug trials: psychometric properties of global scales. J Geriatr Psychiatry Neurol 2000; 13:197–205.

46. Quinn J, Moore M, Benson DF et al. A videotaped CIBIC for dementia patients: validity and reliability in a simulated clinical trial. Neurology 2002; 58:433–7.

47. Whitehead A, Perdomo C, Pratt RD et al. Donepezil for the symptomatic treatment of patients with mild to moderate Alzheimer's disease: a meta-analysis of individual patient data from randomised controlled trials. Int J Geriatr Psychiatry 2004; 19:624–33.

48. Rockwood K. Size of the treatment effect on cognition of cholincstcrasc inhibition in Alzheimer's disease. J Neurol Neurosurg Psychiatry 2004; 75:677–85.

49. Rockwood K, Beattie BL, Eastwood MR et al. A randomized, controlled trial of linopirdine in the treatment of Alzheimer's disease. Can J Neurol Sci 1997; 24:140–5.

50. Rockwood K, Stolee P. Responsiveness of outcome measures used in an antidementia drug trial. Alzheimer Dis Assoc Disord 2000; 14:182–5.

51. Le Bars PL, Velasco FM, Ferguson JM et al. A placebo-controlled, double-blind, randomized trial of an extract of Ginkgo biloba for dementia. North American EGb Study Group. JAMA 1997; 278(16):1327–32.

52. Kanowski S, Hoerr R. Ginkgo biloba extract EGb 761 in dementia: intent-to-treat analyses of a 24-week, multi-center, double-blind, placebo-controlled, randomized trial. Pharmacopsychiatry 2003; 36:297–303.

53. van Dongen M, van Rossum E, Kessels A, Sielhorst H, Knipschild P. Ginkgo for elderly people with dementia and age-associated memory impairment: a randomized clinical trial. J Clin Epidemiol 2003; 56:367–76.

54. Solomon PR, Adams F, Silver A, Zimmer J, DeVeaux R. Ginkgo for memory enhancement: a randomized controlled trial. JAMA 2002; 288:835–40.

55. Rockwood K, Black S, Robilliard A, Lussier I. Potential treatment effects of donepezil not detected in Alzheimer's disease clinical trials: a physician survey. Int J Geriatric Psychiatry 2004; 19:954–50.

8

Cognitive Outcomes

Neelesh K Nadkarni and Sandra E Black

Introduction

An important challenge facing therapeutic research aimed at restoring or enhancing cognition is the selection of appropriate and robust outcome measures capable of documenting meaningful change in cognition over time. These cognitive outcome measures have to be sensitive enough to capture the expected change, broad enough to be applicable to a representative population within a particular syndrome and evidence-based. At the same time, particularly in a demented population, they need to be comprehensive but easy to administer and well tolerated. In dementia clinical trials, the cognitive measures selected have often reflected national preference and once successfully used, they have become the industry standard, leading sometimes to difficulty in comparing results. Standardisation, and issues of validity of translation to different languages, remain important challenges to efficacy evaluation.

Current diagnostic criteria for Alzheimer's disease (AD) strongly emphasise cognitive impairment, specifically memory loss, which is a core symptom of AD. Neuropsychological tools have been developed to evaluate specific cognitive domains in a standardised, quantitative manner and to monitor their change over time. Measures of cognitive abilities have therefore been commonly used as primary and secondary outcome measures in anti-dementia drug trials. For example, the Alzheimer's Disease Assessment Scale – cognitive subscale (ADAS-Cog) has been widely used to illustrate the beneficial effects in clinical trials of cholinesterase inhibitors (ChEI) in mild to moderate AD. In this chapter we will critically analyse the cognitive outcome measures that have become the industry standard. We will discuss the rationale for using cognitive assessment as an outcome measure, the expected temporal course of cognitive decline in AD, common cognitive measures used in clinical trials and treatment effect seen. We will also highlight cognitive domains not well tested by current popular outcome measures, and comment on outcome measures used in patients in the early and more severe stages of AD. From this review, we suggest possible augmentations in designing future trials in dementia.

Cognitive impairment in AD and other dementias

Both the National Institute of Neurological and Communicative Disorders and Stroke/ Alzheimer's Disease and Related Disorders Association criteria and the Diagnostic and Statistical Manual (4th edition) (DSM-IV) criteria stress the need for documenting cognitive loss in dementia.[1,2] AD is clinically defined by early memory impairment with concurrent or subsequent deficits in attention, language, visuoperceptual and constructional abilities.[1] Other domains usually affected are praxis and executive function, such as planning, abstraction, decision-making capacity and speed of information processing. In comparison with AD, executive deficits may be more common in patients with vascular dementia (VaD), particularly when associated with subcortical vascular disease, which is often referred to as subcortical ischaemic vascular dementia (SIVD).[3–6] Within the VaD spectrum itself there is considerable heterogeneity as far as cognitive and clinical presentations are concerned,[7] with executive deficits predominating in lacunar state dementia and SIVD.[8,9] For example, phonemic fluency, regarded as a frontal lobe task, and recognition memory, regarded as a hippocampal function, were found to reliably differentiate SIVD from AD.[9] The cognitive profile of Lewy body dementia (LBD) shows relatively more executive and visuospatial deficits and relatively less episodic memory loss, in comparison to AD.[10] Cognitive changes in fronto-temporal dementia (FTD) are characterised by deficits in attention, language and executive function with relative preservation of visuo-spatial function and recognition memory even in the late stages.[11–16] In contrast to AD, short-term episodic memory loss is less prominent in early FTD though memory impairment worsens as the disease progresses.[17–19] Additionally, impaired social cognition and other behavioural manifestations differentiate FTD from AD and VaD.[20]

Rationale for cognitive assessment as an outcome measure

As detailed in Chapter 2, when guidelines for testing 'anti-dementia' drugs for AD were formulated in the late 1980s, their repertoire was pragmatically limited to cognitive, behavioural and functional assessments. In 1990, the Food and Drug Administration (FDA) proposed guidelines for assaying treatment-specific effects in AD and other dementias.[21] These guidelines emphasised the importance of cognitive measures not only for staging dementia, but also for demonstrating efficacy. Improving cognition or retarding its decline was set as one of the required primary outcome measures, the other one being a global-functioning assessment. Canadian guidelines (1995) required that symptomatic treatment effect should include at least one individual cognitive domain.[22] The European regulatory agencies added the requirement for a functional scale as a primary outcome measure and a responder analysis[23,24] (see Whitehouse[25] for a more detailed review).

Behavioural impairment, functional decline in daily activities and caregiver stress correlate to some extent with cognitive decline and the cost of dementia care.[26] Assessing cognitive outcomes, therefore, remains key for determining efficacy of dementia-specific treatment. Hence psychometric testing is a central outcome measure in clinical trials of dementia.[27]

Although cognitive decline correlates with loss of functional independence, the specific cognitive drivers of functional decline are still being investigated. For example, episodic memory deficits may have less impact on self-care than visuo-spatial dysfunction.[28] Deficits in executive function, rather than memory, language and spatial skills, account for the majority of variance in instrumental activities of daily living.[29–31] Memory and visuo-spatial deficits may influence the development of paranoid delusions, misidentification and hallucinations,[32,33] whereas executive dysfunction may influence other behaviours, as discussed later. Since certain cognitive deficits appear specifically to influence functional or behavioural symptoms, it is important to document separate cognitive domains, not just composite outcomes, to better understand therapeutic effects.

Cognitive outcome measures in dementia trials (Table 8.1)

Mini-Mental Status Examination (MMSE)

Though the Folstein MMSE was initially intended to differentiate organic from functional brain disorders in elderly psychiatry patients, it is ubiquitous as a screening instrument for dementia and AD, and a secondary outcome measure of efficacy in many dementia trials.[34] It includes questions on orientation, memory, attention, language and visuo-constructive abilities, and scores range from 0 to 30, with a higher score indicating better cognitive status.

Factor analysis of the MMSE has revealed that item relationships depend on whether scores are assessed cross-sectionally or longitudinally.[35] With a single test administration a two-factor solution emerged, a *general cognitive functioning factor* and a *language comprehension factor*, accounting for 62% of variance.[35,36] When used as a dynamic tool, measuring change over time, a five-factor structure emerged, accounting for 75% of variance, pertaining to orientation, following commands, repetition, language expression and recall.[35] In correlating the Weschler Adult Intelligence Scale, to these five longitudinal factors,[37] Brooks et al found that picture arrangement and object assembly (which draw on temporal ordering, sequencing and psychomotor performance[38]), comprehension and digit symbol subtests best

Table 8.1 Cognitive outcome measures used in clinical trials of treatment in dementia

Scale	Range of scores (worst to best performance)	Annual decline in points	Stage of disease best suited	Domains tested
ADAS-Cog	70 to 0	9–11	Mild–moderate	Orientation, memory, language, praxis
MMSE	0 to 30	2–5	Mild–moderate	Orientation, memory praxis
SIB	0 to 100	18–19	Severe	Social interaction, attention, naming, orientation, memory, language, praxis
CAMCOG	0 to 107	12–14	Early–mild	Attention/calculation, executive function, orientation, memory, language, praxis
SKT	27 to 0	2–4	Mild–moderate	Memory, attention, language, praxis
EXIT25	50 to 0	0.89 (data only in elderly without dementia)	Early–mild	Executive function
DRS	0 to 144	11–13	Mild–severe	Conceptualisation, initiation, attention, memory, construction

predicted longitudinal decline on the following commands, language repetition and language expression factors of the MMSE.[35] In another study, three-word recall on the MMSE correlated with verbal memory on the California Verbal Learning Test (Pearson's $r = 0.52$, $P < 0.05$), but it explained only 10% of the variance.[39] Attention and memory subscores on the MMSE correlated with verbal ability, and copying and naming were poor indices of the purported cognitive function as well, in another study.[40] Given this loading for verbal ability, a low MMSE score can be misleading in patients with isolated language impairment, as can a high score in patients with predominant executive dysfunction or visuo-spatial dysfunction. This is supported by reports that approximately 50% of individuals with a normal MMSE score perform poorly on a clock-drawing test, which entails visuo-spatial and executive function components.[41,42]

Longitudinal studies have demonstrated a 2- to 5-point annual decline in the MMSE score in patients with AD depending on the stage.[43,44] However, decline over one year may not be predictive of future decline,[45] and the variance in measurement may equal the highly variable average annual score change.[46] Hence, the use of MMSE as the sole outcome measure for periods less than three years is dubious.[46] This limited degree of decline, sometimes within the range of variability, is another reason for questioning its utility as an outcome measure in dementia trials lasting only six months or one year. For example, beneficial cognitive effects of treatments based on improvements in the MMSE score by 1.3 points over 15 months,[47] or 1.23 points within 3 months,[48] or for that matter, even a 0.24-point decline over a 6-month period in the placebo-group,[49] is unlikely to be clinically significant, given that only a change greater than three points would be considered clinically meaningful.[46]

The MMSE also displays a stage-dependent sensitivity to decline. When change in longitudinal MMSE scores are plotted along baseline MMSE scores, a curvilinear relationship is seen, indicating that scores at the extremes of scale show less decline over time.[50] Sensitivity to decline diminishes particularly in later stages of the disease when language function is severely compromised.[51] The MMSE displays floor and ceiling effects as performance on the MMSE plateaus out in early and late stages of the disease, hindering interpretation of any 'meaningful' change.[52,53]

The MMSE has a high test–retest reliability (range 0.8. to 0.95),[34,44] but even so, test–retest and inter-rater reliability may be inadequate for detecting small changes in cognitive function.[54] A 'standardised' version of the MMSE (sMMSE) that provides clear guidelines for administration, scoring and timing can improve reliability, and has been used as a secondary outcome measure in a moderate to severe AD study.[55,56] The treated group showed a significantly higher improvement of 1.79 points on the sMMSE compared to the placebo, but what constitutes a meaningful change on the sMMSE has not been clearly defined.

Age and, to a greater extent, education, affect performance on the MMSE, with advancing age and lower educational level being associated with poorer performance on the MMSE.[44] Studies from the US, UK, Italy and Brazil have attempted to ameliorate this by correcting scores based on normative distributions of the MMSE in the general population.[57–60] However, these corrections derived from the general population do not necessarily apply during the course of the dementia. As was reported in one study, educational level correlated with MMSE scores in normal as well as the AD population, but age correlated only in the normal elderly population.[61] Many of these population norms have been criticised as age and cohort could affect cross-sectional data on which the results were based.[62] Therefore, population-norms using longitudinal data were reported as percentile distributions by age, sex and educational level and corrected for loss due to drop-outs during the 9-year longitudinal study.[62]

Many translations of the MMSE have appeared and have been validated in different populations around the world,[63–68] though not without some concerns regarding the translations used, for example, the Spanish translation of the MMSE.[69] Validation of such scales may be limited by inter-cultural, educational, socio-economic and demographic diversity and must be taken into account in considering its psychometric properties.[70]

Alzheimer's Disease Assessment Scale-Cognitive Subscale (ADAS-Cog)

The ADAS-Cog is by far the most widely used primary outcome measure in clinical trials.[71,72] This scale emerged in the 1980s as a potential research tool for assessment of cognitive dysfunction for natural history and psychopharmacological intervention studies. The scale emphasised the cognitive characteristics of AD as then understood.[71,72] Its domains were selected carefully, from clinical charactcristics of autopsy-confirmed or clinically diagnosed cases that were further classified into two broad categories: cognitive and non-cognitive. The former included components of memory, language and praxis (ADAS-Cog) and the latter included mood and behavioural changes (ADAS-noncog). In the original paper, 27 AD participants and 28 normal controls were rated. Fifteen participants from both groups were age-, sex- and education-matched to detect group differences between scores on various ADAS-Cog items. Additionally, longitudinal data were obtained at 12 months from 10 participants from each group. This scale was found to have a good inter- and intra-rater reliability and validity, meeting key criteria for an outcome rating scale.[73,74] Most importantly, unlike scales then in common use (Blessed Dementia Scale, Pfeiffer's Short Portable Mental Status Questionnaire and the Face-Hand Test),[75,76] only the ADAS-Cog provided a reasonable estimate of cognitive decline for longitudinal studies.[50,77,78]

The ADAS-Cog is an 11-item scale that provides a measure of orientation, verbal episodic memory (immediate recall and recognition of a word list), language and praxis, four primary cognitive impairments that differentiate AD from normal elderly controls.[72] Items of memory and orienta-

tion are weighted heavily compared to those of language and praxis. The ADAS-Cog takes about 30 to 40 minutes to administer, and is scored from 0 to 70 with a higher score indicating greater dysfunction.

The initial data for evaluating longitudinal change on the ADAS-Cog were obtained by following 111 patients with AD and 72 healthy elderly controls enrolled in a follow-up protocol at the Mount Sinai Medical Center and VA Medical Center in New York. The study began in 1979, involved serial administration of a battery of tests at regular intervals, including the ADAS-Cog, and estimated the average annual change score to be a 9.55 ($\pm$ 8.21) point decline in AD and an improvement of 0.23 ($\pm$ 1.98) points per year in healthy controls, consistent with other studies.[77,78] However, in the placebo-treated arm of clinical trials, AD patients decline annually by approximately 5–6 points.[79] This slower rate of decline could reflect a placebo effect or more likely, selection criteria that yielded a healthier AD sample, and timely management of concomitant illnesses. Based on the longitudinal studies, the FDA stipulated that a treatment that reverscd the natural history of cognitive decline by at least 6 months, would be considered clinically significant.[80] This was equated to a 4-point or greater difference on the ADAS-Cog scale.[77] In a progressive degenerative condition, maintenance of the score or slowing the decline over time would be beneficial for patients and their caregivers.[81,82] As with the MMSE, the decline in ADAS-Cog is stage dependent; patients with mild- or severe-stage dementia display slower deterioration over one year than those in the moderate stage.[78,83] For example, the annual change in ADAS-Cog scores is affected by the baseline score, with an inverted-U relationship between the axes.[78] However, in comparison to other scales available at that time, such as the Blessed Memory Test[84] the ADAS-Cog was still more sensitive to change in the early and late stages of the disease.[52,85] Nevertheless, as discussed later, the ADAS-Cog appears to be best suited for mild to moderate stages of dementia. Still, at present, stratification into groups based on base-line ADAS-Cog scores is usually *post hoc*. Given the stage dependency of ADAS-Cog change scores, it would be advisable to consider a stratified approach in future trials, so that the effect size on mild and moderate patients could be determined by design, rather than afterthought.

The cognitive domains evaluated by the ADAS-Cog include spoken language, comprehension, word-finding in spontaneous speech, following commands, naming, constructional and ideational praxis, orientation and word-list recall, test-instruction recall and word-list recognition. A factor analysis, however, revealed a pattern of intercorrelations among these 11 cognitive sub-tests that yielded three prominent factors inter-preted as overall mental status, verbal fluency and praxis.[86] Another application of factor analyses to the ADAS-Cog from two large clinical trials yielded three factors, interpreted as memory, language and praxis.[87] This may indicate that the samples in clinical trials may be different from general popu-lations, or that the scale itself may lack robustness in its underlying factors. Both these studies also identified the presence of a composite cognitive dysfunction factor.

Executive function is an important domain that plays an essential role in daily activities and psychi-atric symptoms, but it is missing in the original ver-sion of the ADAS-Cog.[88–90] Impairment in these so-called 'frontal lobe' functions is important not just in FTD or VaD, but it plays an important role in most dementias, including AD. Other domains that are not captured by the ADAS-Cog are visuo-spatial attention, working memory, and other memory components such as autobiographical and remote memory,[91] which are differentially affected in AD and other dementias. In consequence, the ADAS-Cog might underestimate the benefit of treatment seen in clinical practice, as it measures only three strongly intercorrelated domains and under-samples executive and visuo-spatial function as well as speed of information processing.

In assessing cognitive deficits and beneficial effects on cognition, it is imperative to attend not only to statistically significant differences, but also to the clinical meaningfulness of these differ-ences.[92] Statistical significance depends on sample size, which must be taken into account when com-paring outcomes across multiple studies. Effect size calculation provides an estimate of overlap between groups, as well as a measure of the pro-portion of differences between groups.[93] The effect size calculations (Cohen's d) for the ADAS-Cog original study data approximate to 1.24 and 1.53 for the baseline score and 12-month follow-up score respectively, corresponding to an overlap of 37.8% and 29.3% between groups at the respective time-points.[72] The effect size at the 12-month interval for the Stern et al study was 1.08,[78] which corresponds to an overlap of about 44.6%. Ideally a diagnostic marker should be capable of differen-tiating 100% of healthy from diseased individuals. In terms of effect size, a value of 3.0 corresponds to about 7% overlap between two comparison groups. A large meta-analytic study compiled cog-nitive test results from 190 studies conducted between 1984 and 1997, and compared neuropsy-chological test performance in 7156 patients with AD and 8772 normal healthy controls.[94] Effect sizes were calculated for all neuropsychological tests and listed by domains in order of their mag-nitude. It was found that delayed recall on several standardised measures (e.g. Buschke Selective Reminding Test, California Verbal Learning Test) provided the least overlap (<7.2%) with an effect size of 3.0.[94]

Calculation of effect sizes from published clini-cal trials is only possible if a clear representation of the outcome data is provided in a tabular form. While most studies provide the data in terms of change in mean scores (standard error/standard deviation from which calculations for effect sizes can be made, not all studies (e.g. rivastigmine[95,96]) provide the data needed to derive the effect sizes. Rockwood measured effect sizes of the ADAS-Cog, in terms of Cohen's d and standardised-response-means, from various ChEI outcome data, and found that the effect sizes ranged from 0.1 to 0.4, depending on dosage and the type of analysis reported (intention to treat/last observation carried forward or observed case analyses).[97] Among the reported trials, larger effect-sizes were seen with higher dosage of ChEI, standardised-response-means methodology, and observed case analyses. An effect size of 0.1 to 0.5 denotes a small to medium effect as defined by Cohen.[93] The ADAS-Cog is heavily biased towards memory and language functions, and this small to medium effect size may not reflect the benefits seen in clinical practice, as improvements in executive functioning and attention, not captured by the ADAS-Cog, may drive the clinical effect.

In the original paper, subtest scores on the ADAS-Cog were not discussed,[72] but subtest scores in addition to the total scores were reported in some clinical trials using ADAS-Cog as an outcome measure, for example in studies of metrifonate and tacrine.[83] This led researchers to study subtests. One subtest analysis revealed a differential sensitivity of the ADAS-Cog to varying stages of AD.[83] Naming and constructional praxis were most sensitive to progression from moderate to severe stage of AD. A subgroup of these patients, however, had deficits in naming and praxis on the Boston Naming Test and Rey Osterrieth Figure Test,[98] but not on the naming and praxis subtests of the ADAS-Cog. Some subtests show good sensitivity to transition between different stages in AD. For example, word-list recall was sensitive to change from normal- to mild-stage AD. Memory and spontaneous language were more sensitive to the early-stage AD, while naming, commands and praxis were more sensitive in the moderate- to severe-stage transition. In short, detection of improvement in the specific cognitive functions using the ADAS-Cog subtests may only be possible for patients in certain stages of dementia, and this should be considered when interpreting efficacy studies in AD clinical trials.[83]

The stage-dependent sensitivity of ADAS-Cog makes it subject to both floor and ceiling effects. Both consistently low scores across 'low-functioning' patients in the severe stages (floor effect),[99] and very high scores in the 'high-functioning' patients with either mild cognitive impairment or early stages of disease (ceiling effect) are reported.[86] However, Ihl et al found no such effect on the ADAS-Cog.[52] Mattes attempted to identify floor and ceiling effects in individual subtests and reported that cognitive ability could be more accurately evaluated by reducing the number of words on the word-recall task and the number of trials on the word recognition task.[100] These modifications, these authors argued, would increase the likelihood of identifying medication effects. However, these results have not yet been replicated. Due to the floor effects of the ADAS-Cog, residual cognitive capacities that are preserved even in advanced disease, or are amenable to improvement with therapy, may be overlooked. In spite of these limitations, the ADAS-Cog has been used in clinical trials involving patients with moderate to severe AD.[101,102] However, tests such as the Severe Impairment Battery (SIB) may be better suited to this stage of disease, as discussed below.[103,104]

The guidelines for administering the ADAS-Cog have been criticised as being too permissive, thereby compromising reliability.[105] A standardised ADAS (SADAS) was therefore devised and compared to the regular ADAS scale.[105] While the inter- and intra-rater reliability of the total score and non-cognitive subscale of the SADAS substantially improved, the reliability of the cognitive subscales of the SADAS was similar to that of the original ADAS-Cog. Hence the advantage of the SADAS is unclear.

The conduct of clinical trials in different countries has led to multi-lingual versions of the ADAS-Cog, which have been validated and found to be applicable in patient populations in different cultures.[106–114] However, during the process of translation and validation, several changes from the original have been introduced.[115] Other limitations arise from translations unless the translated versions are culturally adapted, and on re-translation to the original language by a different translator, are found to be similar to the original.[70] To overcome these limitations, the EUROpean HAR-monisation Project for Instruments in

Dementia (EURO-HARPID) compared translated versions of the ADAS-Cog used in eight European countries. From this, the harmonised EURO-HARPID versions were developed and validated against other cognitive scales.[116] This project demonstrated the potential for harmonisation of outcome measures, which would be useful in the design of large multi-national trials and in inter-continental comparison of cognitive outcomes. The lack of normative and longitudinal data on these scales, however, potentially hinders utility in clinical trials.

The effects of age, education and intellectual ability on the ADAS-Cog are inconsistent in studies to date. While earlier studies showed no effect of education and a small effect of age,[83,117] Doraiswamy et al reported that age and education do play a role in the performance of the ADAS-Cog.[118,119] Stern et al reported that age does not influence the rate of decline on the ADAS-Cog,[78] but others suggest that age may affect baseline performance, and thereby the rate of decline.[78,83,115]

To summarise, the ADAS-Cog measures several aspects of cognitive performance with reasonable reliability and validity in AD. It remains the most common outcome measure, but is limited in the number of cognitive domains it captures, consistency of administration and scoring and variable sensitivity to change at different dementia stages. Also, the baseline score on the ADAS-Cog and the specific version used, need to be accounted for whenever cognitive outcomes are reported. Supplementation with executive function tests, key amongst these limitations, should be considered, especially in the non-Alzheimer's dementias.

Attention and executive deficits in dementia

As indicated, executive function (EF) refers to the ability to conceptualise, abstract, organise, initiate and regulate behaviour.[120] These abilities reflect higher-level coordinating functions such as planning, attention, working memory, selective and sustained attention and self-monitoring and self-

control. Impairments in attention and executive functions often occur early in the course of dementia irrespective of its underlying aetiology; some diseases affect attention and EF earlier than other domains.[15] For instance, EF deficits frequently predominate in VaD.[6] Inclusion of EF measures has been suggested for clinical trials in VaD and AD with cerebrovascular disease (CVD).[88,121] Thus additional tasks probing these 'frontal lobe functions' have been added to an enhanced battery called the VaDAS-Cog or VaD-specific ADAS-Cog.[88,89] Following preliminary studies indicating efficacy of cholinesterase inhibitors on ADAS-Cog in VaD,[122,123] clinical trials are underway to evaluate their efficacy in other cognitive domains in VaD, specifically EF. Impairments in attention and EF, in concert with memory, language and visuo-spatial deficits, however, have been consistently documented even in the early stage of AD.[3,124–129] Specifically, deficits in cognitive flexibility or set shifting and concurrent manipulation of information are seen early.[129,130] Deficits in attention are detectable early in AD even when all other cognitive domains, except memory, show no significant deficits on neuropsychological testing.[130,131] Furthermore, baseline performance on EF tests is found to predict the progression of Mild Cognitive Impairment (MCI) to AD.[132–134] When neuropsychological measures are repeated over long duration such as five years, tests of EF, such as verbal fluency and other non-memory tests were superior to memory tests in detecting cognitive change over time.[135]

Even though EF deficits are common in VaD and AD, the extent and nature of specific dysfunction can differ.[5,6,136–140] For example, patients with VaD exhibit more perseveration on the Modified Card Sorting Test, while patients with AD exhibit more perseveration on a category fluency testing, even though short-term memory, comprehension, semantic fluency, conceptualisation, problem-solving and concurrent manipulation of information are equally affected in both groups.[136] Given that EF comprises different cognitive capacities, the selection of EF tasks may help differentiate dementia groups. Another component of EF is working memory, which involves online maintenance of information and its concurrent manipulation for a goal-directed utilisation, can be affected early in AD.[125,141,142] Furthermore, working memory deficits may distinguish AD from SIVD; whereas AD show deficits secondary to impaired attentional shifting, SIVD may manifest working memory deficits due to impaired inhibitory control and inability to manipulate complex information.[140]

If the characterisation of AD as a pure amnestic disorder is misleading, we must address executive functioning in clinical trials in AD. Only a few studies in AD and LBD have used EF tasks as outcome measures, with statistically significant improvement being reported in attention and working memory in the galatamine- and rivastigmine-treated groups compared to placebo.[143–145]

In a Canadian postal survey of physicians experienced in the treatment of AD, potential benefits of ChEI, classified into various clinically recognisable cognitive domains, were sought.[146] Items most often rated as being improved were related to initiation and attention, suggesting that EF, which has been largely undocumented by use of the ADAS-Cog and the MMSE in clinical trials, may show the best response pragmatically to ChEI therapy.[146]

Executive functioning in dementia and other neurological disorders is associated with functional disabilities, behavioural symptoms and progression of dementia. While visuo-spatial deficits in AD can adversely affect the ability to perform some instrumental activities of daily living (IADL),[28] the progression in disability from compromised IADL to decline in the self-care activities of daily living is most closely linked to EF.[22] EF abilities are essential to performing IADL such as managing a home, dealing with finances, using transport services, using a telephone, etc. Some popular functional scales such as the Disability Assessment in Dementia, rate function in relation to initiation, planning and action, potentially capturing EF components.[22] A recent study

even reported EF as a predictor of falls.[147] Neuropsychological studies have shown that functional impairment is associated with executive dysfunction in normal aging,[30,148,149] and both neurodegenerative and vascular dementias.[150,151] Common neuropsychological tests of EF have been used to predict functional status and assess competence and driving ability.[152–155]

The relationship and predictive value of EF is not limited to function and extends to behavioural and psychological symptoms.[156] Specifically, EF correlates with agitation, anxiety/depression and apathy.[150,156–158]

The cognitive profile in very old AD patients (>80 years) is also different from that of young-old AD patients (mean age of 70 years) in that the older group outperforms the younger group on executive function tasks.[45,159–161] Age-dependent greater degree of executive dysfunction was particularly noted in one of these studies in the early stages of the AD.[161] As the patients enrolled in clinical trials in AD are often in the young-old age group (mean age ranging from 73 to 75), it is quite important to assess EF deficits in this age group.

The limitation of the ADAS-Cog in assessing components of EF has been addressed by the Alzheimer's Disease Cooperative Study in a revised ADAS-Cog that incorporates EF tests, by including cancellation tasks and maze completion tasks.[90] For clinical trials in mild-stage AD patients (MMSE >20), inclusion of word list learning and delayed spontaneous recall and maze tests, and for trials targeting AD patients with MMSE ≤15, adding tests of praxis (drawn from the Boston Diagnostic Aphasia Examination[162]), were also suggested.[90] The harmonisation guidelines echoed concerns about the limitations of the ADAS-Cog, and stressed the need for EF and attention/concentration tasks as cognitive outcome measures.[163] An expanded version of the ADAS-Cog that included subtests for assessing concentration/distractibility and delayed recall was used in measuring the effect of galantamine in patients with AD, and improvements in scores on both this scale and the regular ADAS-Cog were found.[164,165] In other studies, working memory functions improved after treatment with rivastigmine in LBD and AD, and attention improved after treatment with galantamine in AD.[144,145,166] These studies suggest that ChEI may improve executive functioning. Although the concentration/distractibility subtest of the ADAS-Cog measures attention to a certain extent and was utilised in pivotal studies on rivastigmine, these subtest results are generally not reported, partly because of sample size limitations.[95,96]

In addition to the VADAS-Cog, which supplements the ADAS-Cog with frontal lobe tasks, additional EF measures such as Executive Interview (EXIT25)[167] and the Executive Clock-drawing Task (CLOX)[168] have been advocated for use in clinical trials of VaD.[167,168]

Executive Interview (EXIT25)

The EXIT25[167] is a 25-item bedside scale that takes 10–15 minutes to administer. The scores range from 0 to 50. High scores signify impairment and the cut-off for discriminating healthy elderly from dementia is 15/50. The EXIT25 scale has been found to have good reliability and validity and performance on the EXIT25 correlates well with standard neuropsychometric EF and functional measures.[149] While longitudinal data are now emerging on elderly who have otherwise normal scores on general cognitive measures,[169] there is limited information on its longitudinal sensitivity in patients with dementia.

Clock-drawing tasks

The Executive Clock-drawing task (CLOX) taps executive and visuo-spatial functions, as well as other general cognitive resources needed to draw a clock.[168,170] In the two-part task, CLOX-1 assesses EF, where subjects are asked to draw a clock from memory. The CLOX-2 task assesses the subject's ability to copy a clock picture, tapping more into visuo-constructional praxis. Freedman et al devised another method for scoring clock drawings based on 15 critical elements involving the drawing, the contour, entering the numbers, cen-

tring and setting the hands.[171] This method of scoring which has been used successfully in differentiating AD from healthy elderly and patients with depression,[42] can also be used in evaluation and longitudinal assessment in clinical trials. Other methods of scoring a clock-drawing task have been described by Sunderland et al, Mendez et al, Watson et al and Shulman.[172–175]

Mattis Dementia Rating Scale (MDRS)

The Mattis Dementia Rating Scale (MDRS) is another general cognitive battery used in the US, which samples the EF domain more than the ADAS-Cog.[176] It is well validated and normed.[177–183] Its sensitivity for longitudinal decline is superior to that of the MMSE, especially in the advanced stages.[45] The annual decline ranges from 11–13 points.[43,45,53,184] The scores range from 0 to 144, higher scores indicating better performance. It takes 30–45 minutes to administer. Except for its limited use in demonstrating cognitive benefits in an antioxidant study,[185] it has been little used in randomised controlled trials.

Frontal Assessment Battery

The Frontal Assessment Battery (FAB) is another recent addition to the EF assessment armamentarium.[186] The battery consists of six subtests that assess conceptualisation, mental flexibility, motor programming, interference, response inhibition and environmental autonomy. It can be administered in 10 minutes and scores range from 0 to 18, high scores indicating better performance. However, there are very little data on its standardisation, normative values and longitudinal sensitivity to change.

The Cambridge Neuropsychological Test Automated Battery (CANTAB) and Cognitive Drug Research (CDR) computerised assessment system

Computerised assessment batteries can comprehensively tap into the executive components of cognition. The Cambridge Neuropsychological Test Automated Battery (CANTAB) offers sensitive and specific cognitive assessment that is administered via a touch-sensitive screen (*www.camcog.com*).[187] It provides assessment of EF, visual memory and attention and provides consistency of administration across testers and test centres. It has been widely used in Europe to assess dementia. Preliminary studies have reported small yet statistically significant improvement in attention with ChEI treatment, using tools such as the CANTAB.[187,188] Other computerised methods such as the Cognitive Drug Research (CDR) computerised assessment system, validated in patients with dementia, have been developed to ascertain treatment effects in attention and executive functions.[145,189–191] This battery includes tests of working memory, processing speed, interference, response inhibition and psychomotor processing speed.

Cambridge Cognitive Examination (CAMCOG)

The CAMCOG has been reported to be a useful tool in preclinical detection of AD.[192] CAMCOG's memory component outperformed measures of medial temporal atrophy on MRI as predictors of progression to AD.[193] CAMCOG is the cognitive subscale of the Cambridge Mental Disorders of the Elderly Examination (CAMDEX), which takes 25 minutes to administer and has gained widespread popularity in the UK. It has been used in many other countries with the exception of North America, and equivalent translations have been validated in many different languages.[194–197] It is designed to cover seven areas of cognition: orientation, language, memory, attention/calculation, praxis, abstract thinking and perception. It provides a subscale score for each of these categories, as well as a total CAMCOG score. Scores range from 0 to 107, higher scores indicating better cognitive function; scores over 100 are rarely achieved. It can differentiate between individuals when used in patients with a premorbid 'high functional status', thus overcoming ceiling effects seen with other scales used in early stages.[198,199] The rate of decline in dementia (either AD, VaD or DLB) is 12–14 points per year or 28 points in

two-years.[200,201] This compares with an annual decline rate of 1.6 points in normal age-matched controls.[202] Age-, sex- and education-normed values have been developed in a large population of individuals aged over 65 years.[203]

One drawback of the CAMCOG is its relative lack of EF assessment, limited to a measure of abstraction (similarities). Recently, a shorter form of the scale has been introduced, the Rotterdam-CAMCOG (R-CAMCOG).[204] Though this reduced the administration time by more than half, executive functioning is still poorly sampled. The R-CAMCOG should not be mistaken for the revised version of the CAMCOG, the CAMCOG-R, which includes two additional tests of EF, namely ideational fluency and a visual reasoning task.[205] The sum of performance on these two new tasks provides a separate 'EF score' (maximum of 14) and does not contribute to the total CAMCOG score. The sum of this 'EF score' in combination with the score of the abstract thinking and attention EF tests on the original CAMCOG provides the 'total EF score' (maximum of 28). However, the inclusion of these additional EF measures correlates only moderately with the standard EF measures, providing little benefit for the additional time taken to administer the CAMCOG-R.[206] Some authors question the sensitivity of the CAMCOG to memory decline in the early stages, reporting that the paired associate learning and the visual recognition tasks of the CANTAB had superior sensitivity in early AD, especially when CAMCOG and MMSE scores were essentially within the normal range.[207]

Syndrome Kurtz Test (SKT)

The SKT is the oldest available cognitive measure and continues to be popular in Europe.[208] Developed in Germany in the 1970s at the University of Erlangen, it has since been translated into several languages. The nine subtests assess naming of objects and numbers, sequencing and rearrangement of numbers, counting symbols, immediate, delayed and cued recall and cognitive 'rigidity' (interference). Each task is allotted 60 seconds and the entire test takes 10–15 minutes to complete. Raw scores are transformed to age- and intelligence-appropriate normative values, the cut-off for dementia being a score greater than 5. Normative values have been studied in German and Spanish populations, but these normative values may not be applicable to other populations.[208,209] The SKT is best suited for mild to moderate severity of dementia. Factor-analysis of the SKT revealed two factors, described as memory and attention (i.e. speed of information processing), and this factorial stability has been reported across cultural diversity within Europe.[210,211] The factor structure in demented groups is reported to show a three-factor structure, interpreted as memory, attention and language. The SKT provides a reliable measure of attention that is lacking in other outcome measures used so far. A limitation of the SKT, namely its floor effects, is acknowledged in the SKT manual. Also it is reported to overestimate deficits in patients in the mild stages and may even mis-classify up to 10.2% of normal subjects as impaired.[52]

Others

The Modified-Mini Mental Status Examination (3MS),[212] which was developed to improve the reliability and validity of the MMSE with the addition of more items and graded scoring system (0–100), has gained popularity in epidemiological studies such as the Canadian Study of Health and Aging.[213] Data on its normative values and reference ranges are now emerging from such studies.[214–216] Batteries such as the Behavioural Neurology Assessment and Addenbrook Cognitive Examination are gaining popularity in dementia clinics as comprehensive bedside measures of cognitive functioning.[217,218] These measures, including the 3MS, tap into executive functions besides other domains and are relatively easy and quick to administer.

Outcome measure in early stages of dementia

Individuals with subjective memory complaints and objective memory impairment (relative to age- and education-matched normal controls) who are otherwise intact on activities of daily living and performing well within normal limits in other cognitive domains, do not meet criteria for dementia and have been described as *amnestic* mild cognitive impairment (MCI).[219] This term, used in context of the continuum from normal aging to probable AD to denote prodromal AD,[220] has grown in popularity in the last few years, though other terms are still used, such as cognitively impaired not demented (CIND),[221] and criteria for what constitutes cognitive impairment have varied. The progression rate of MCI to AD has been reported to be 10–15% per year,[219,222] and great effort has focused on finding predictors for decline. Since this group is considered at high risk for developing AD, therapeutic trials are currently in progress targeting prevention of decline, using psychometric tests at regular intervals throughout, which range from two to four years in duration.

There is as yet no consensus on which battery of tests best detects decline. Although some reports favour immediate and delayed recall as the best predictors,[133,134] this has not been persuasively replicated.[223,224] Those studies that have found memory predictive have used tasks such as the Rey Auditory Verbal Learning Test (RAVLT),[38] the California Verbal Learning Test (CVLT),[225] the Bushke Selective Reminding Test[226] or New York University paragraph recall.[227]

As noted, EF tasks have also been informative in predicting progression of MCI. Tasks include self-monitoring, set-shifting and sequencing such as the Controlled Word Association Test, the Self-ordering Test and Trails B.[129–131,228] Albert et al followed healthy subjects ($n = 42$) and patients with very mild AD ($n = 123$) for 3 years.[134] They found that the baseline performance on memory tasks (learning score on the CVLT and immediate recall on Wechsler Memory Scale (WMS)) and EF tasks (Trails B and the self-ordering tasks), discriminated individuals who converted to AD from healthy controls (sensitivity = 89%) and from those with mild memory impairment who did not progress to AD (sensitivity = 80%).[134] EF tests can also discriminate patients with MCI, vascular cognitive impairment and AD.[229,230] Specifically, category fluency, episodic memory and CLOX test distinguished MCI from AD, and working memory, besides visuo-spatial memory and verbal learning performance best predicted progression to AD in their MCI sample.[229] Between MCI and AD, the latter perform worse on tests of EF and attention especially in subtests of initiation and perseveration of the MDRS.[231] Therefore, assessment of EF appears to be important not only to predict the progression of MCI to AD, but also to discriminate it from normal aging, even though a so-called 'isolated' memory deficit is used to define this state. Whether very early EF impairment detects a group at risk of progression, or a group with very early AD remains unclear.

The Memory Impairment Study is a multicentre, double-blind placebo-controlled study in patients with MCI that is currently being analysed by the Alzheimer Disease Cooperative Study (ADCS) to determine if donepezil or vitamin E can delay conversion to AD, as the primary outcome. A number of cognitive measures are also being followed, including immediate and delayed recall, Boston Naming Test, digit symbol, fluency, number cancellation, digits backwards, the Maze test and clock drawing.[232,233] The cognitive and functional characteristics of patients with MCI recruited in this trial have been published recently.[234] The operational criteria for the diagnosis of MCI, implemented across 69 centres in North America, included objective memory impairment documented using cut-offs in performance of delayed paragraph recall (Logical Memory II subtest of the WMS).[235] These MCI patients were compared with mild AD patients recruited to another ADCS trial on instruments common to both groups, the MMSE, ADAS-Cog,

Clinical Dementia Rating and the Hachinski scale. The MCI group had memory deficits intermediate between controls and the AD group. In addition to the memory tests, the MCI group performed worse than the controls on all non-memory tasks, indicating that, even with selection criteria emphasising memory deficits, more generalised cognitive decline may clearly be present in preclinical AD.[234] Hopefully, the sample size will permit interpretation of intervention effects on these cognitive functions, as the results from this study emerge.

Ongoing and completed trials have used the ADAS-Cog as the primary cognitive outcome measure in MCI.[232,233] As discussed however, the ADAS-Cog may not be sensitive in prodromal or early-stage AD, and concerns regarding potential ceiling effects clearly extend to the MCI population as well. It is perhaps not surprising that the study duration of some of the MCI trials has been extended to four years, because of slower than expected conversion rates as assessed by these tools.[233] While more recent trials in MCI will evaluate cognitive outcomes rather than just rate of conversion to AD, the cognitive repertoire remains limited. Aspects of memory, such as immediate and delayed recall, and of EF, such as set-shifting or sequencing, have not always been sampled. Some clinical designs have used delayed recall tasks, while others have used a modified version of the digit symbol substitution test, a test of attention.[233] The incorporation of specific EF tasks in future MCI trials may be key in detecting beneficial effects of drug intervention.

Outcome measures in advanced disease

In more advanced AD, domains such as episodic and semantic memory, EF, praxis, visuo-spatial attention and language are usually severely affected, whereas over-learned skills, implicit memory, perceptual priming and emotional modulation of cognition are relatively preserved.[236,237] Any cognitive capacity may be difficult to assess when communicative abilities are severely affected.[236] Floor effects in this stage on cognitive batteries such as the MMSE and the ADAS-Cog, may mean that relatively preserved cognitive abilities may be difficult to access. To overcome these floor effects, special scales geared to this population have been developed.

Severe Impairment Battery (SIB)

The Severe Impairment Battery (SIB) is structured in a way that does not require too much effort from the patient, is easy flowing and avoids a 'testing-environment' during its administration.[103] The SIB contains nine subscales, each of which provides a separate score: orientation, attention, language, praxis, memory, social interaction, reactivity to external stimuli, construction and visuo-spatial abilities. Though this scale is administered verbally, non-verbal responses can also be scored. Total scores range from 0 to 100 with higher scores indicating less impairment. The battery takes about 15–20 minutes to administer. Panisset et al segregated participants with AD into four groups based on their MMSE performance, ranging from less than 5 to greater than 17, and studied the performance on the SIB.[238] They found that the SIB could differentiate the low-performance MMSE severity groups at a cut-off of 11 points, and the groups that obtained fewer than five points on the MMSE were still able to obtain a mean of 46 points on the SIB.[231] Other studies also confirmed the reliability of the SIB as a cognitive assessment tool in patients with severe AD who have reached a point where conventional tests lose their sensitivity due to floor effects.[239,240] Translations of the SIB have also been validated in different countries.[241,242] A factorial analysis of the SIB identified four factors (percentage of variance): a cognitive factor (39%), praxis and visuo-spatial factor (24%), external stimuli factor (14%) and social aptitude factor (12%).[243] Longitudinal studies in severe AD using the SIB have shown a variable annual decline ranging from 17.6 points (MMSE 0–4), to 18.8 points (MMSE 0–11).[239,244] Other scales that have been used in the severe stage of AD include the Hierarchic Dementia Scale,[245] the Severe Cognitive Impairment Profile,[246] and the Test for Severe Impairment Battery.[247] The SIB, in comparison to these, has had the most reliability and validity studies and is the only tool that has been evaluated for measuring longitudinal changes in cognitive functioning in advanced stages of AD.[238,239,244]

The SIB was recently used in clinical trials with patients in more severe stages of AD, and captured changes in memory, visuo-spatial function, language and praxis.[104] In a study of donepezil in moderate to severe AD, the treatment group showed a statistically significant difference of 5.62 points on the SIB.[56] In this study, the sMMSE was also used as an additional cognitive measure, but the decline in SIB in the placebo group confirmed the greater ability of the SIB to capture cognitive change in this disease stage compared to the sMMSE.[56]

Patients in severe stages of AD comprise a large group of patients, growing in proportion, who have received less attention in clinical trials. The data regarding cognitive capacities in this stage are sparse and are based on limited data. The role of ChEIs is still to be established, though the results of the trial in moderate to severe AD patients were positive not only for the SIB but also for behavioural and functional outcomes.[56] Reliable tools such as the SIB, make it possible to track the response to interventions in advanced stage disease, though clearly, effects in function and behaviour will probably have more impact on caregiver burden. The SIB and other similar batteries do not tap into certain cognitive capacities such as perceptual priming, implicit memory, over-learned skills and emotions, which remain relatively preserved in severe AD.[237] Additionally, better measures are needed for documenting 'cognitive awareness or alertness' in the penultimate stage of dementia, between the severe stage and a terminal 'persistent vegetative state'. Scales that load on 'social interaction' and 'reactivity to external stimuli' more than the cognitive domains would be needed to appreciate 'alertness' or 'awareness' in a patient's interaction with the environment in this very advanced stage.

Cognitive outcome measures in future clinical trials: suggestions and recommendations

- Recently, computerised methods of cognitive assessment have been used in clinical trials and have been found to be successful in documenting change.[145,248] In particular, the CDR computerised assessment system,[189,190] the Computerized Neuropsychological Test Battery (CNTB) and the CANTAB have gained some popularity in this regard,[187,249] while other new batteries have been suggested for future use.[250] The CNTB has been used to study treatment effects on cognition in AD clinical trials.[251–255] The CANTAB has also been used in a wide range of age groups, including a number of studies on cognitive aging, as well as drug studies of AD.[188,256,257] The CogState, which assesses multiple cognitive domains including episodic memory, working memory and attention, is another computerized battery that reliably differentiates early MCI from normal individuals on repeated testing within a three-hour period[258] and is also more sensitive than conventional measures in detecting change in memory over a one-year period.[259] Computerised batteries capture speed of responses and may better reflect attentional processes. They may also allow better standardisation across sites in multi-centre trials, but it remains to be seen what role they will play in the design of future trials.

- Often, elderly participants are apprehensive about cognitive assessments and anxiety may accentuate deficits during testing in the clinic environment. Wide fluctuations in blood pressure readings prior to and soon after cognitive evaluation in a clinic setting provide empirical evidence for this effect. Testing done in the participant's own home may help to overcome this anxiety. This would also facilitate collection of cognitive data in those study participants who are unable to return for frequent assessments throughout the duration of the trial. For example, in an innovative follow-up of participants through a clinical trial for AD, assessments were conducted and videorecorded by a trained psychometrist in the patient's own home, and these were discretely rated by trained blinded raters for use as outcome measures.[260] Recently, telephone interviews to evaluate cognition have been reported, although this is not feasible in more severe stages.[261–265] Also milder patients may log on to specific password-protected internet sites and undergo the cognitive batteries specified for them. Web-based screening, data entry and data management in clinical trials can also lead to improved liaison and feedback between sites conducting trials and the companies sponsoring them.[266] Though data on such studies is very preliminary, testing may be problematic with this approach, because of difficulty verifying a subject's identity, and maintaining patient and drug study confidentiality. Video-conferencing can limit this drawback and provides another alternative to conventional methods of testing. The use of such telehealth methods is being evaluated extensively in countries such as Canada and Australia.[267,268] Among the available scales for dementia research, the CAMCOG has been validated for video-conferencing.[269]

- About 75% of elderly individuals have visual or hearing impairments that may preclude participation or may manifest during the course of the clinical trial. Development of cognitive tools or modifications that cater to the needs of elderly with visual and hearing impairments are needed to allow monitoring of their response to treatment, and possibly their inclusion in clinical trials.

- Most of the cognitive assessments in current practice are modulated by the participant's educational level. Development of scales that are effective in assessing cognitive impairment in the illiterate or minimally educated elderly are needed, especially in countries where illiteracy is a common reality. These are becoming available in South America and China,[67] but in developed countries, illiterate elderly are unlikely to be included.

- With the advent of functional neuroimaging and emerging concrete data on brain activation pattern in specific cognitive tasks, new avenues for assessing the effects of drugs on cognition have opened up. A pilot study by Rombouts and colleagues demonstrated an effect of rivastigmine on working memory on four patients with AD using functional magnetic resonance imaging.[166] In addition to improvement in working memory with rivastigmine, increased brain activation was reported. Such studies could not only provide objective outcome measures but could also link an intervention to changes in brain activation, implicating appropriate neural networks. Similarly, if disease-modifying drugs emerge, structural brain measures, such as whole-brain or medial temporal atrophy may be important surrogate measures of biological effects.[270]

- To assist in comparing clinical trials using different outcome measures, methods are being developed for deriving equivalent scores on one scale when scores on another are provided.[271,272] These need validation and replication in larger samples to help develop robust methods to compare outcomes using different, but often equally valid and reliable instruments in clinical trials, which would permit meta-analytic approaches to determining efficacy across many trials.

Conclusion

In addition to the traditional cognitive outcome measures that have been used in clinical trials in AD such as the MMSE, SKT and the ADAS-Cog, newer ones have recently evolved based on our expanding knowledge of cognitive deficits in dementia. These more recent outcome measures have targeted specifically executive dysfunction, a domain found to be involved in early AD as well as VaD. The extension of current therapeutic research to MCI and early AD as well as other non-AD dementias, has led to the consideration of incorporating measures such as CAMCOG, CANTAB, EXIT25 and CLOX in the design of clinical trials. Research on longitudinal sensitivity to change, applicability to ethnic differences, and normalisation, among others, is actively being pursued. Development of cognitive measures for the penultimate stages of dementia is still needed, as existing batteries such as the SIB may not capture these changes.

While developments of cognitive measures for teleconferencing and computer-based assessments is justified for conducting trials across affluent centres around the globe, we must also strive to develop tools for cognitive evaluations in underprivileged and illiterate populations, who are currently precluded from participation in therapeutic dementia research.

Acknowledgements

The authors acknowledge financial support from the Canadian Institute of Health Research, Alzheimer's Association US, Heart and Stroke Foundation Centre for Stroke Recovery and Linda C. Campbell Cognitive Neurology Research Unit.

References

1. McKhann G, Drachman D, Folstein M et al. Clinical diagnosis of Alzheimer's disease: report of the NINCDS-ADRDA Work Group under the auspices of Department of Health and Human Services Task Force on Alzheimer's Disease. Neurology 1984; 34(7):939–44.

2. Association AP. Diagnostic and Statistical Manual of Mental Disorders-DSM-IV. Washington, DC: American Psychiatric Association; 1994.

3. Almkvist O. Neuropsychological deficits in vascular dementia in relation to Alzheimer's disease: reviewing evidence for functional similarity or divergence. Dementia. 1994; 5(3–4):203–9.

4. Starkstein SE, Sabe L, Vazquez S et al. Neuropsychological, psychiatric, and cerebral blood flow findings in vascular dementia and Alzheimer's disease. Stroke 1996; 27(3):408–14.

5. Kertesz A, Clydesdale S. Neuropsychological deficits in vascular dementia vs Alzheimer's disease. Frontal lobe deficits prominent in vascular dementia. Arch Neurol 1994; 51(12):1226–31.

6. Looi JC, Sachdev PS. Differentiation of vascular dementia from AD on neuropsychological tests. Neurology 1999; 53(4):670–8.

7. Rockwood K, Howard K, MacKnight C, Darvesh S. Spectrum of disease in vascular cognitive impairment. Neuroepidemiology 1999; 18(5):248–54.

8. McPherson SE, Cummings JL. Neuropsychological aspects of vascular dementia. Brain Cogn 1996; 31(2):269–82.

9. Tierney MC, Black SE, Szalai JP et al. Recognition memory and verbal fluency differentiate probable Alzheimer disease from subcortical ischemic vascular dementia. Arch Neurol 2001; 158(10):1654–9.

10. McKeith IG, Galasko D, Kosaka K et al. Consensus guidelines for the clinical and pathologic diagnosis of dementia with Lewy bodies (DLB): report of the consortium on DLB international workshop. Neurology 1996; 47(5):1113–24.

11. Johanson A, Hagberg B. Psychometric characteristics in patients with frontal lobe degeneration of non-Alzheimer type. Arch Gerontol Geriatr 1989; 8(2):129–37.

12. Pasquier F, Grymonprez L, Lebert F, Van der Linden M. Memory impairment differs in frontotemporal dementia and Alzheimer's disease. Neurocase 2001; 7(2):161–71.

13. Perry RJ, Hodges JR. Differentiating frontal and temporal variant frontotemporal dementia from Alzheimer's disease. Neurology 2000; 54(12): 2277–84.

14. Mendez MF, Cherrier M, Perryman KM et al. Frontotemporal dementia versus Alzheimer's disease: differential cognitive features. Neurology 1996; 47(5):1189–94.

15. Pachana NA, Boone KB, Miller BL, Cummings JL, Berman N. Comparison of neuropsychological functioning in Alzheimer's disease and frontotemporal dementia. J Int Neuropsychol Soc 1996; 2(6):505–10.

16. Gregory CA, Hodges JR. Clinical features of frontal lobe dementia in comparison to Alzheimer's disease. J Neural Transm Suppl 1996; 47:103–23.

17. McKhann GM, Albert MS, Grossman M et al. Clinical and pathological diagnosis of frontotemporal dementia: report of the Work Group on Frontotemporal Dementia and Pick's Disease. Arch Neurol 2001; 58(11):1803–09.

18. Neary D, Snowden J. Fronto-temporal dementia: nosology, neuropsychology, and neuropathology. Brain Cogn 1996; 31(2):176–87.

19. Binetti G, Locascio JJ, Corkin S et al. Differences between Pick disease and Alzheimer disease in clinical appearance and rate of cognitive decline. Arch Neurol 2000; 57(2):225–32.

20. Kertesz A, Nadkarni N, Davidson W, Thomas AW. The Frontal Behavioral Inventory in the differential diagnosis of frontotemporal dementia. J Int Neuropsychol Soc 2000; 6(4):460–8.

21. Leber P. Guidelines for Clinical Evauation of Antidementia Drugs. Washington: US Food and Drug Administration, Washington; 1990.

22. Mohr E, Feldman H, Gauthier S. Canadian guidelines for the development of antidementia therapies: a conceptual summary. Can J Neurol Sci 1995; 22(1):62–71.

23. Agency EME. Note for Guidance on Medicinal Products in the Treatment of Alzheimer's Disease. Located at: London: EMEA; 1997.

24. Orgogozo JM, Abadie E. Vascular dementia: European perspectives. Alzheimer Dis Assoc Disord 1999; 13(Suppl 3):S192–200.

25. Whitehouse P. Regulatory issues in antidementia drug development. In: Gauthier S (ed). Pharmacotherapy of Alzheimer's disease. London: Martin Dunitz; 1998: 57–74.

26. Kane RL. Which outcomes matter in Alzheimer disease and who should define them? Alzheimer Dis Assoc Disord 1997; 11(Suppl 6):12–17.

27. Claman DL, Radebaugh TS. Neuropsychological assessment in clinical trials of Alzheimer disease. Alzheimer Dis Assoc Disord 1991; 5(Suppl 1): S49–56.

28. Perry RJ, Hodges JR. Relationship between functional and neuropsychological performance in early Alzheimer disease. Alzheimer Dis Assoc Disord 2000; 14(1):1–10.

29. Gauthier L, Gauthier S. Assessment of functional changes in Alzheimer's disease. Neuroepidemiology 1990; 9(4):183–8.

30. Cahn-Weiner DA, Malloy PF, Boyle PA, Marran M, Salloway S. Prediction of functional status from neuropsychological tests in community-dwelling

elderly individuals. Clin Neuropsychol 2000; 14(2):187–95.

31. Boyle PA, Malloy PF, Salloway S, Cahn-Weiner DA, Cohen R, Cummings JL. Executive dysfunction and apathy predict functional impairment in Alzheimer disease. Am J Geriatr Psychiatry 2003; 11(2): 214–21.

32. Ravetz RS. Psychiatric disorders associated with Alzheimer's disease. J Am Osteopath Assoc 1999; 99(9 Suppl):S13–16.

33. Corey-Bloom J. The ABC of Alzheimer's disease: cognitive changes and their management in Alzheimer's disease and related dementias. Int Psychogeriatr 2002; 14(Suppl 1):51–75.

34. Folstein MF, Folstein SE, McHugh PR. 'Mini-mental state'. A practical method for grading the cognitive state of patients for the clinician. J Psychiatr Res 1975; 12(3):189–98.

35. Tinklenberg J, Brooks JO, 3rd, Tanke ED et al. Factor analysis and preliminary validation of the Mini-Mental State Examination from a longitudinal perspective. Int Psychogeriatr 1990; 2(2):123–34.

36. Abraham IL, Manning CA, Snustad DG, Brashear HR, Newman MC, Wofford AB. Cognitive screening of nursing home residents: factor structures of the Mini-Mental State Examination. J Am Geriatr Soc 1994; 42(7):750–6.

37. Brooks JO 3rd, Yesavage JA, Taylor J et al. Cognitive decline in Alzheimer's disease: elaborating on the nature of the longitudinal factor structure of the Mini-Mental State Examination. Int Psychogeriatr 1993; 5(2):135–46.

38. Lezak MD. Neuropsychological Assessment. New York: Oxford University Press; 1983.

39. Cullum CM, Thompson LL, Smernoff EN. Three-word recall as a measure of memory. J Clin Exp Neuropsychol 1993; 15(2):321–9.

40. Tierney MC, Szalai JP, Snow WG, Fisher RH, Dunn E. Domain specificity of the subtests of the Mini-Mental State Examination. Arch Neurology 1997; 54(6):713–16.

41. Juby A, Tench S, Baker V. The value of clock drawing in identifying executive cognitive dysfunction in people with a normal Mini-Mental State Examination score. CMAJ 2002; 167(8):859–64.

42. Herrmann N, Kidron D, Shulman KI et al. Clock tests in depression, Alzheimer's disease, and elderly controls. Int J Psychiatry Med 1998; 28(4):437–47.

43. Aguero-Torres H, Fratiglioni L, Winblad B. Natural history of Alzheimer's disease and other dementias: review of the literature in the light of the findings from the Kungsholmen Project. Int J Geriatr Psychiatry 1998; 13(11):755–66.

44. Tombaugh TN, McIntyre NJ. The Mini-Mental State Examination: a comprehensive review. J Am Geriatr Soc 1992; 40(9):922–35.

45. Salmon DP, Thal LJ, Butters N, Heindel WC. Longitudinal evaluation of dementia of the Alzheimer type: a comparison of 3 standardized mental status examinations. Neurology 1990; 40(8):1225–30.

46. Clark CM, Sheppard L, Fillenbaum GG et al. Variability in annual Mini-Mental State Examination score in patients with probable Alzheimer disease: a clinical perspective of data from the Consortium to Establish a Registry for Alzheimer's Disease. Arch Neurol 1999; 56(7):857–62.

47. Rogers SL, Doody RS, Mohs RC, Friedhoff LT. Donepezil improves cognition and global function in Alzheimer disease: a 15-week, double-blind, placebo-controlled study. Donepezil Study Group. Arch Intern Med 1998; 158(9):1021–31.

48. Nagaraja D, Jayashree S. Randomized study of the dopamine receptor agonist piribedil in the treatment of mild cognitive impairment. Am J Psychiatry 2001; 158(9):1517–19.

49. Becker RE, Colliver JA, Markwell SJ et al. Effects of metrifonate on cognitive decline in Alzheimer disease: a double-blind, placebo-controlled, 6-month study. Alzheimer Dis Assoc Disord 1998; 12(1):54–7.

50. Morris JC, Edland S, Clark C et al. The consortium to establish a registry for Alzheimer's disease (CERAD). Part IV. Rates of cognitive change in the longitudinal assessment of probable Alzheimer's disease. Neurology 1993; 43(12):2457–65.

51. Small BJ, Viitanen M, Winblad B, Backman L. Cognitive changes in very old persons with dementia: the influence of demographic, psychometric, and biological variables. J Clin Exp Neuropsychol 1997; 19(2):245–60.

52. Ihl R, Frolich L, Dierks T, Martin EM, Maurer K. Differential validity of psychometric tests in dementia of the Alzheimer type. Psychiatry Res 1992; 44(2):93–106.

53. Galasko DR, Gould RL, Abramson IS, Salmon DP. Measuring cognitive change in a cohort of patients

with Alzheimer's disease. Stat Med 2000; 19(11–12):1421–32.

54. Bowie P, Branton T, Holmes J. Should the Mini Mental State Examination be used to monitor dementia treatments? Lancet 1999; 354(9189): 1527–8.

55. Molloy DW, Alemayehu E, Roberts R. Reliability of a Standardized Mini-Mental State Examination compared with the traditional Mini-Mental State Examination. Am J Psychiatry 1991; 148(1):102–5.

56. Feldman H, Gauthier S, Hecker J et al. A 24-week, randomized, double-blind study of donepezil in moderate to severe Alzheimer's disease. Neurology 2001; 57(4):613–20.

57. Crum RM, Anthony JC, Bassett SS, Folstein MF. Population-based norms for the Mini-Mental State Examination by age and educational level. JAMA 1993; 269(18):2386–91.

58. MRC CFA Study Group. Cognitive function and dementia in six areas of England and Wales: the distribution of MMSE and prevalence of GMS organicity level in the MRC CFA Study. The Medical Research Council Cognitive Function and Ageing Study (MRC CFAS). Psychol Med 1998; 28(2):319–35.

59. Grigoletto F, Zappala G, Anderson DW, Lebowitz BD. Norms for the Mini-Mental State Examination in a healthy population. Neurology 1999; 53(2):315–20.

60. Laks J, Batista EM, Guilherme ER et al. [Mini-mental state examination in community-dwelling elderly: preliminary data from Santo Antonio de Padua, Rio de Janeiro, Brazil]. Arq Neuropsiquiatr 2003; 61(3B):782–5.

61. Magni E, Binetti G, Cappa S, Bianchetti A, Trabucchi M. Effect of age and education on performance on the Mini-Mental State Examination in a healthy older population and during the course of Alzheimer's disease. J Am Geriatr Soc 1995; 43(8):942–3.

62. Dufouil C, Clayton D, Brayne C et al. Population norms for the MMSE in the very old: estimates based on longitudinal data. Mini-Mental State Examination. Neurology 2000; 55(11):1609–13.

63. Kabir ZN, Herlitz A. The Bangla adaptation of Mini-Mental State Examination (BAMSE): an instrument to assess cognitive function in illiterate and literate individuals. Int J Geriatr Psychiatry 2000; 15(5):441–50.

64. Commenges D, Gagnon M, Letenneur L et al. Statistical description of the Mini-Mental State Examination for French elderly community residents. Paquid Study Group. J Nerv Ment Dis 1992; 180(1):28–32.

65. Ganguli M, Chandra V, Gilby JE et al. Cognitive test performance in a community-based nondemented elderly sample in rural India: the Indo-U.S. Cross-National Dementia Epidemiology Study. Int Psychogeriatr 1996; 8(4):507–24.

66. de Silva HA, Gunatilake SB. Mini Mental State Examination in Sinhalese: a sensitive test to screen for dementia in Sri Lanka. Int J Geriatr Psychiatry 2002; 17(2):134–9.

67. Xu G, Meyer JS, Huang Y, Du F, Chowdhury M, Quach M. Adapting Mini-Mental State Examination for dementia screening among illiterate or minimally educated elderly Chinese. Int J Geriatr Psychiatry 2003; 18(7):609–16.

68. Gungen C, Ertan T, Eker E, Yasar R, Engin F. [Reliability and validity of the standardized Mini Mental State Examination in the diagnosis of mild dementia in Turkish population]. Turk Psikiyatri Derg 2002; 13(4):273–81.

69. Gimenez-Roldan S, Novillo MJ, Navarro E, Dobato JL, Gimenez-Zuccarelli M. Mini-Mental State Examination: proposal of protocol to be used. Rev Neurol 1997; 25(140):576–83.

70. Auer S, Hampel H, Moller H-J, Reisberg B. Translations of measurements and scales: opportunities and diversities. Int Psychogeriatr 2000; 12(1):391–4.

71. Mohs RC, Rosen WG, Davis KL. The Alzheimer's disease assessment scale: an instrument for assessing treatment efficacy. Psychopharmacol Bull 1983; 19(3):448–50.

72. Rosen WG, Mohs RC, Davis KL. A new rating scale for Alzheimer's disease. Am J Psychiatry 1984; 141(11):1356–64.

73. Streiner DL, Norman G. Health Measurement Scales : a Practical Guide to their Development and Use. Oxford (England), New York: Oxford University Press; 1989.

74. Streiner DL. A checklist for evaluating the usefulness of rating scales. Can J Psychiatry 1993; 38(2):140–8.

75. Stern RG, Mohs RC, Bierer LM et al. Deterioration on the Blessed test in Alzheimer's disease: longitudinal data and their implications for clinical trials and identification of subtypes. Psychiatry Res 1992; 42(2):101–10.

76. Berg G, Edwards DF, Danzinger WL, Berg L. Longitudinal change in three brief assessments of SDAT. J Am Geriatr Soc 1987; 35(3):205–12.

77. Kramer-Ginsberg E, Mohs RC, Aryan M et al. Clinical predictors of course for Alzheimer patients in a longitudinal study: a preliminary report. Psychopharmacol Bull 1988; 24(3):458–62.

78. Stern RG, Mohs RC, Davidson M et al. A longitudinal study of Alzheimer's disease: measurement, rate, and predictors of cognitive deterioration. Am J Psychiatry 1994; 151(3):390–6.

79. Torfs K, Feldman H. 12-month decline in cognitive and daily function in patients with mild-to-moderate Alzheimer's disease: two randomized, placebo-controlled studies. Neurobiol Aging 2000; 21(Suppl 1):242–3.

80. FDA. Peripheral and Central Nervous System Drugs Advisory Committee Meeting, 1989. Located at: Rockville, MD: Department of Health and Human Services, Public Health Service; 1989.

81. Baladi JF, Bailey PA, Black S et al. Rivastigmine for Alzheimer's disease: Canadian interpretation of intermediate outcome measures and cost implications. Clin Ther 2000; 22(12):1549–61.

82. Tariot PN. Maintaining cognitive function in Alzheimer disease: how effective are current treatments? Alzheimer Dis Assoc Disord 2001; 15(Suppl 1):S26–33.

83. Zec RF, Landreth ES, Vicari SK et al. Alzheimer Disease Assessment Scale: a subtest analysis. Alzheimer Dis Assoc Disord 1992; 6(3):164–81.

84. Blessed G, Tomlinson BE, Roth M. The association between quantitative measures of dementia and of senile change in the cerebral grey matter of elderly subjects. Br J Psychiatry 1968; 114(512): 797–811.

85. Mohs RC. The Alzheimer's Disease Assessment Scale. Int Psychogeriatr 1996; 8(2): 195–203.

86. Kim YS, Nibbelink DW, Overall JE. Factor structure and reliability of the Alzheimer's Disease Assessment Scale in a multicenter trial with linopirdine. J Geriatr Psychiatry Neurol 1994; 7(2):74–83.

87. Talwalker S, Overall JE, Srirama MK, Gracon SI. Cardinal features of cognitive dysfunction in Alzheimer's disease: a factor-analytic study of the Alzheimer's Disease Assessment Scale. J Geriatr Psychiatry Neurol 1996; 9(1):39–46.

88. Gauthier S, Ferris S. Outcome measures for probable vascular dementia and Alzheimer's disease with cerebrovascular disease. Int J Clin Pract Suppl 2001; 120:29–39.

89. Ferris SH. Cognitive outcome measures. Alzheimer disease and associated disorders. 1999; 13(Suppl 3):S140–2.

90. Mohs RC, Knopman D, Petersen RC et al. Development of cognitive instruments for use in clinical trials of antidementia drugs: additions to the Alzheimer's Disease Assessment Scale that broaden its scope. The Alzheimer's Disease Cooperative Study. Alzheimer Dis Assoc Disord 1997; 11(Suppl 2):S13–21.

91. Pasquier F. [Memory: therapeutic approach. Clinical evaluation]. Therapie 2000; 55(4):513–19.

92. Rockwood K, MacKnight C. Assessing the clinical importance of statistically significant improvement in anti-dementia drug trials. Neuroepidemiology 2001; 20(2):51–6.

93. Cohen J. Statistical Power Analysis for the Behavioral Scences (2nd edn). Hillsdale, NJ: Lawrence Erlbaum Associates; 1988.

94. Zakzanis KK, Leach L, Kaplan E. Dementia of Alzheimer's Type. Neuropsychological Differential Diagnosis. Lisse: Swets & Zeitlinger; 1999.

95. Corey-Bloom J, Anand R, Veach JFTEBSG. A randomized trial evaluating the efficacy and safety of ENA 713 (rivastigmine tartrate), a new acetylcholinesterase inhibitor, in patients with mild to moderately severe Alzheimer's disease. Int J Geriatr Psychopharmacol 1998; 1:55–65.

96. Rosler M, Anand R, Cicin-Sain A et al. Efficacy and safety of rivastigmine in patients with Alzheimer's disease: international randomised controlled trial. BMJ 1999; 318(7184):633–8.

97. Rockwood K. Size of the treatment effect on cognition of cholinesterase inhibition in Alzheimer's disease. J Neurol Neurosurg Psychiatry 2004; 75(5):677–85.

98. Spreen O, Strauss E. A Compendium of Neuropsychological Tests: Administration, Norms and Commentary. New York, NY: Oxford University Press, Inc.; 1991.

99. Mohs RC, Cohen L. Alzheimer's Disease Assessment Scale (ADAS). Psychopharmacol Bull 1988; 24(4):627–8.

100. Mattes JA. Can the sensitivity of the Alzheimer's Disease Assessment Scale be increased? Am J Geriatr Psychiatry 1997; 5(3):258–60.

101. Onofrj M, Thomas A, Luciano AL et al. Donepezil versus vitamin E in Alzheimer's disease: Part 2: mild versus moderate-severe Alzheimer's disease. Clin Neuropharmacol 2002; 25(4):207–15.

102. Ruether E, Alvarez XA, Rainer M, Moessler H. Sustained improvement of cognition and global function in patients with moderately severe Alzheimer's disease: a double-blind, placebo-controlled study with the neurotrophic agent Cerebrolysin. J Neural Transm Suppl 2002; 62:265–75.

103. Saxton J, Swihart AA. Neuropsychological assessment of the severely impaired elderly patient. Clin Geriatr Med 1989; 5(3):531–43.

104. Schmitt FA, Cragar D, Ashford JW et al. Measuring cognition in advanced Alzheimer's disease for clinical trials. J Neural Transm Suppl 2002; 62:135–48.

105. Standish TI, Molloy DW, Bedard M, Layne EC, Murray EA, Strang D. Improved reliability of the Standardized Alzheimer's Disease Assessment Scale (SADAS) compared with the Alzheimer's Disease Assessment Scale (ADAS). J Am Geriatr Soc 1996; 44(6):712–16.

106. Schultz RR, Siviero MO, Bertolucci PH. The cognitive subscale of the 'Alzheimer's Disease Assessment Scale' in a Brazilian sample. Braz J Med Biol Res 2001; 34(10):1295–1302.

107. Kolibas E, Korinkova V, Novotny V, Vajdickova K, Hunakova D. ADAS-cog [Alzheimer's Disease Assessment Scale-cognitive subscale] – validation of the Slovak version. Bratisl Lek Listy 2000; 101(11):598–602.

108. Tsolaki M, Fountoulakis K, Nakopoulou E, Kazis A, Mohs RC. Alzheimer's Disease Assessment Scale: the validation of the scale in Greece in elderly demented patients and normal subjects. Dement Geriatr Cogn Disord 1997; 8(5):273–80.

109. Chu LW, Chiu KC, Hui SL, Yu GK, Tsui WJ, Lee PW. The reliability and validity of the Alzheimer's Disease Assessment Scale Cognitive Subscale (ADAS-Cog) among the elderly Chinese in Hong Kong. Ann Acad Med Singapore 2000; 29(4):474–85.

110. Homma A. [Assessment and treatment of patients with dementia of the Alzheimer type]. Nippon Ronen Igakkai Zasshi 1992; 29(4):264–70.

111. Pena-Casanova J, Aguilar M, Santacruz P et al. [Adaptation and normalization of the Alzheimer's disease Assessment Scale for Spain (NORMACODEM) (II)]. Neurologia 1997; 12(2):69–77.

112. Puel M, Hugonot-Diener L. [Presentation by the GRECO group of the French adaptation of a cognitive assessment scale used in Alzheimer type dementia]. Presse Med 1996; 25(22):1028–32.

113. Youn JC, Lee DY, Kim KW et al. Development of the Korean version of Alzheimer's Disease Assessment Scale (ADAS-K). Int J Geriatr Psychiatry 2002; 17(9):797–803.

114. Hannesdottir K, Snaedal J. A study of the Alzheimer's Disease Assessment Scale-Cognitive (ADAS-Cog) in an Icelandic elderly population. Nord J Psychiatry 2002; 56(3):201–6.

115. Pena-Casanova J. Alzheimer's Disease Assessment Scale – cognitive in clinical practice. Int Psychogeriatr 1997; 9(Suppl 1):105–114.

116. Verhey FR, Houx P, Van Lang N et al. Cross-national comparison and validation of the Alzheimer's Disease Assessment Scale: results from the European Harmonization Project for Instruments in Dementia (EURO-HARPID). Int J Geriatr Psychiatry 2004; 19(1):41–50.

117. Burch EA Jr, Andrews SR. Comparison of two cognitive rating scales in medically ill patients. Int J Psychiatry Med 1987; 17(2):193–200.

118. Doraiswamy PM, Krishen A, Stallone F et al. Cognitive performance on the Alzheimer's Disease Assessment Scale: effect of education. Neurology 1995; 45(11):1980–4.

119. Doraiswamy PM, Bieber F, Kaiser L et al. Memory, language, and praxis in Alzheimer's disease: norms for outpatient clinical trial populations. Psychopharmacol Bull 1997; 33(1):123–8.

120. Stuss DT, Benson DF. The Frontal Lobes. New York: Raven Press; 1986.

121. Roman GC, Royall DR. Executive control function: a rational basis for the diagnosis of vascular dementia. Alzheimer Dis Assoc Disord 1999; 13(Suppl 3):S69–80.

122. Black S, Roman GC, Geldmacher DS et al. Efficacy and tolerability of donepezil in vascular dementia: positive results of a 24–week, multicenter, international, randomized, placebo-controlled clinical trial. Stroke 2003; 34(10):2323–30.

123. Wilkinson D, Doody R, Helme R et al. Donepezil in vascular dementia: a randomized, placebo-controlled study. Neurology 2003; 61(4):479–86.

124. Binetti G, Magni E, Padovani A et al. Executive dysfunction in early Alzheimer's disease. J Neurol Neurosurg Psychiatry 1996; 60(1):91–3.

125. Perry RJ, Hodges JR. Attention and executive deficits in Alzheimer's disease. A critical review. Brain 1999; 122(3):383–404.

126. Rizzo M, Anderson SW, Dawson J, Myers R, Ball K. Visual attention impairments in Alzheimer's disease. Neurology 2000; 54(10):1954–9.

127. Buck BH, Black SE, Behrmann M, Caldwell C, Bronskill MJ. Spatial- and object-based attentional deficits in Alzheimer's disease. Relationship to HMPAO-SPECT measures of parietal perfusion. Brain 1997; 120(7):1229–44.

128. Filoteo JV, Delis DC, Massman PJ et al. Directed and divided attention in Alzheimer's disease: impairment in shifting of attention to global and local stimuli. J Clin Exp Neuropsychol 1992; 14(6):871–83.

129. Albert MS. Cognitive and neurobiologic markers of early Alzheimer disease. Proc Natl Acad Sci U S A 1996; 93(24):13547–51.

130. Lafleche G, Albert M. Executive function deficits in mild Alzheimer's Disease. Neuropsychology 1995; 9:313–20.

131. Grady CL, Haxby JV, Horwitz B et al. Longitudinal study of the early neuropsychological and cerebral metabolic changes in dementia of the Alzheimer type. J Clin Exp Neuropsychol 1988; 10(5):576–96.

132. Artero S, Tierney MC, Touchon J, Ritchie K. Prediction of transition from cognitive impairment to senile dementia: a prospective, longitudinal study. Acta Psychiatr Scand 2003; 107(5):390–3.

133. Tierney MC, Szalai JP, Snow WG et al. Prediction of probable Alzheimer's disease in memory-impaired patients: a prospective longitudinal study. Neurology 1996; 46(3):661–5.

134. Albert MS, Moss MB, Tanzi R, Jones K. Preclinical prediction of AD using neuropsychological tests. J Int Neuropsychol Soc 2001; 7(5):631–9.

135. Locascio JJ, Growdon JH, Corkin S. Cognitive test performance in detecting, staging, and tracking Alzheimer's disease. Arch Neurol 1995; 52(11):1087–99.

136. Traykov L, Baudic S, Thibaudet MC et al. Neuropsychological deficit in early subcortical vascular dementia: comparison to Alzheimer's disease. Dement Geriatr Cogn Disord 2002; 14(1):26–32.

137. Lamar M, Podell K, Carew TG et al. Perseverative behavior in Alzheimer's disease and subcortical ischemic vascular dementia. Neuropsychology 1997; 11(4):523–34.

138. Lafosse JM, Reed BR, Mungas D et al. Fluency and memory differences between ischemic vascular dementia and Alzheimer's disease. Neuropsychology 1997; 11(4):514–22.

139. Yuspeh RL, Vanderploeg RD, Crowell TA, Mullan M. Differences in executive functioning between Alzheimer's disease and subcortical ischemic vascular dementia. J Clin Exp Neuropsychol 2002; 24(6):745–54.

140. Cannata AP, Alberoni M, Franceschi M, Mariani C. Frontal impairment in subcortical ischemic vascular dementia in comparison to Alzheimer's disease. Dement Geriatr Cogn Disord 2002; 13(2):101–11.

141. Baddeley A, Logie R, Bressi S, Della Sala S, Spinnler H. Dementia and working memory. Q J Exp Psychol A 1986; 38(4):603–18.

142. Baddeley AD, Bressi S, Della Sala S, Logie R, Spinnler H. The decline of working memory in Alzheimer's disease. A longitudinal study. Brain 1991; 114:2521–42.

143. Agid Y, Bruno D, Anand R, Gharabawi G. Efficacy and tolerability of rivastigmine in patients with dementia of the Alzheimer type. Curr Ther Res 1998; 59(12):837–45.

144. Wesnes K, Scott M, Morrison S et al. The effects if galanthamine on attention in Alzheimer's disease. J Psychopharmacol 1998; 12(Suppl A):A46.

145. Wesnes KA, McKeith IG, Ferrara R et al. Effects of rivastigmine on cognitive function in dementia with Lewy bodies: a randomised placebo-controlled international study using the

cognitive drug research computerised assessment system. Dement Geriatr Cogn Disord 2002; 13(3):183–92.

146. Rockwood K, Black SE, Robillard A, Lussier I. Potential treatment effects of donepezil not detected in Alzheimer's disease clinical trials: a physician survey. Int J Geriatr Psychiatry 2004; 19:954–60.

147. Rapport LJ, Hanks RA, Millis SR, Deshpande SA. Executive functioning and predictors of falls in the rehabilitation setting. Arch Phys Med Rehabil 1998; 79(6):629–33.

148. Grigsby J, Kaye K, Baxter J, Shetterly SM, Hamman RF. Executive cognitive abilities and functional status among community-dwelling older persons in the San Luis Valley Health and Aging Study. J Am Geriatr Soc 1998; 46(5):590–6.

149. Royall DR, Chiodo LK, Polk MJ. Correlates of disability among elderly retirees with 'subclinical' cognitive impairment. J Gerontol A Biol Sci Med Sci 2000; 55(9):M541–6.

150. Chen ST, Sultzer DL, Hinkin CH, Mahler ME, Cummings JL. Executive dysfunction in Alzheimer's disease: association with neuropsychiatric symptoms and functional impairment. J Neuropsychiatry Clin Neurosci 1998; 10(4):426–32.

151. Boyle PA, Cohen RA, Paul R, Moser D, Gordon N. Cognitive and motor impairments predict functional declines in patients with vascular dementia. Int J Geriatr Psychiatry 2002; 17(2):164–9.

152. Bell-McGinty S, Podell K, Franzen M, Baird AD, Williams MJ. Standard measures of executive function in predicting instrumental activities of daily living in older adults. Int J Geriatr Psychiatry 2002; 17(9):828–34.

153. Barberger-Gateau P, Fabrigoule C, Helmer C, Rouch I, Dartigues JF. Functional impairment in instrumental activities of daily living: an early clinical sign of dementia? J Am Geriatr Soc 1999; 47(4):456–62.

154. Schmand B, Gouwenberg B, Smit JH, Jonker C. Assessment of mental competency in community-dwelling elderly. Alzheimer Dis Assoc Disord 1999; 13(2):80–7.

155. Daigneault G, Joly P, Frigon JY. Executive functions in the evaluation of accident risk of older drivers. J Clin Exp Neuropsychol 2002; 24(2):221–38.

156. McPherson S, Fairbanks L, Tiken S, Cummings JL, Back-Madruga C. Apathy and executive function in Alzheimer's disease. J Int Neuropsychol Soc 2002; 8(3):373–81.

157. Mega MS, Cummings JL. Frontal-subcortical circuits and neuropsychiatric disorders. J Neuropsychiatry Clin Neurosci 1994; 6(4):358–70.

158. Landes AM, Sperry SD, Strauss ME, Geldmacher DS. Apathy in Alzheimer's disease. J Am Geriatr Soc 2001; 49(12):1700–7.

159. Bondi MW, Houston WS, Salmon DP et al. Neuropsychological deficits associated with Alzheimer's disease in the very-old: discrepancies in raw vs. standardized scores. J Int Neuropsychol Soc 2003; 9(5):783–95.

160. Koss E, Edland S, Fillenbaum G et al. Clinical and neuropsychological differences between patients with earlier and later onset of Alzheimer's disease: a CERAD analysis, Part XII. Neurology 1996; 46(1):136–41.

161. Binetti G, Magni E, Padovani A et al. Neuropsychological heterogeneity in mild Alzheimer's disease. Dementia 1993; 4(6):321–6.

162. Goodglass H, Kaplan E. The Assessment of Aphasia and Related Disorders. Philadelphia: Lea and Febiger; 1972.

163. Ferris SH, Lucca U, Mohs R et al. Objective psychometric tests in clinical trials of dementia drugs. Position paper from the International Working Group on Harmonization of Dementia Drug Guidelines. Alzheimer Dis Assoc Disord 1997; 11(Suppl 3):34–8.

164. Rockwood K, Mintzer J, Truyen L, Wessel T, Wilkinson D. Effects of a flexible galantamine dose in Alzheimer's disease: a randomised, controlled trial. J Neurol Neurosurg Psychiatry 2001; 71(5):589–95.

165. Wilcock GK, Lilienfeld S, Gaens E. Efficacy and safety of galantamine in patients with mild to moderate Alzheimer's disease: multicentre randomised controlled trial. Galantamine International-1 Study Group. BMJ 2000; 321 (7274):1445–9.

166. Rombouts SA, Barkhof F, Van Meel CS, Scheltens P. Alterations in brain activation during cholinergic enhancement with rivastigmine in Alzheimer's disease. J Neurol Neurosurg Psychiatry 2002; 73(6):665–71.

167. Royall DR, Mahurin RK, Gray KF. Bedside assessment of executive cognitive impairment: the executive interview. J Am Geriatr Soc 1992; 40(12):1221–6.

168. Royall DR, Mulroy AR, Chiodo LK, Polk MJ. Clock drawing is sensitive to executive control: a comparison of six methods. J Gerontol B Psychol Sci Soc Sci 1999; 54(5):328–33.

169. Royall DR, Palmer R, Chiodo LK, Polk MJ. Declining executive control in normal aging predicts change in functional status: the freedom house study. J Am Geriatr Soc 2004; 52(3):346–52.

170. Royall DR, Cordes JA, Polk M. CLOX: an executive clock drawing task. J Neurol Neurosurg Psychiatry 1998; 64(5):588–94.

171. Freedman M, Leach L, Kaplan E et al. Clock Drawing: a Neuropsychological Analysis. New York: Oxford University Press; 1994.

172. Sunderland T, Hill JL, Mellow AM et al. Clock drawing in Alzheimer's disease. A novel measure of dementia severity. J Am Geriatr Soc 1989; 37(8):725–9.

173. Mendez MF, Ala T, Underwood KL. Development of scoring criteria for the clock drawing task in Alzheimer's disease. J Am Geriatr Soc 1992; 40(11):1095–9.

174. Watson YI, Arfken CL, Birge SJ. Clock completion: an objective screening test for dementia. J Am Geriatr Soc 1993; 41(11):1235–40.

175. Shulman KI. Clock-drawing: is it the ideal cognitive screening test? Int J Geriatr Psychiatry 2000; 15(6):548–61.

176. Mattis S. Mental status examination for organic mental syndrome in the elderly patient. In: Bellack R, Karasu B (eds). Geriatric Psychiatry. New York: Grune and Stratton; 1976: 77–121.

177. Gardner R Jr, Oliver-Munoz S, Fisher L, Empting L. Mattis Dementia Rating Scale: internal reliability study using a diffusely impaired population. J Clin Neuropsychol 1981; 3(3):271–5.

178. Freidl W, Schmidt R, Stronegger WJ, Fazekas F, Reinhart B. Sociodemographic predictors and concurrent validity of the Mini Mental State Examination and the Mattis Dementia Rating Scale. Eur Arch Psychiatry Clin Neurosci 1996; 246(6):317–19.

179. Hofer SM, Piccinin AM, Hershey D. Analysis of structure and discriminative power of the Mattis Dementia Rating Scale. J Clin Psychol 1996; 52(4):395–409.

180. Woodard JL, Auchus AP, Godsall RE, Green RC. An analysis of test bias and differential item functioning due to race on the Mattis Dementia Rating Scale. J Gerontol B Psychol Sci Soc Sci 1998; 53(6):370–4.

181. Paul RH, Cohen RA, Moser D et al. Performance on the Mattis Dementia Rating Scale in patients with vascular dementia: relationships to neuroimaging findings. J Geriatr Psychiatry Neurol 2001; 14(1):33–6.

182. Schmidt R, Freidl W, Fazekas F et al. The Mattis Dementia Rating Scale: normative data from 1001 healthy volunteers. Neurology 1994; 44(5):964–6.

183. Bank AL, Yochim BP, MacNeill SE, Lichtenberg PA. Expanded normative data for the Mattis Dementia Rating Scale for use with urban, elderly medical patients. Clin Neuropsychol 2000; 14(2):149–56.

184. Gould R, Abramson I, Galasko D, Salmon D. Rate of cognitive change in Alzheimer's disease: methodological approaches using random effects models. J Int Neuropsychol Soc 2001; 7(7): 813–24.

185. Schmidt R, Hayn M, Reinhart B et al. Plasma antioxidants and cognitive performance in middle-aged and older adults: results of the Austrian Stroke Prevention Study. J Am Geriatr Soc 1998; 46(11):1407–10.

186. Dubois B, Slachevsky A, Litvan I, Pillon B. The FAB: a Frontal Assessment Battery at bedside. Neurology 2000; 55(11):1621–6.

187. Robbins TW, James M, Owen AM et al. Cambridge Neuropsychological Test Automated Battery (CANTAB): a factor analytic study of a large sample of normal elderly volunteers. Dementia 1994; 5(5):266–81.

188. Sahakian BJ, Owen AM, Morant NJ et al. Further analysis of the cognitive effects of tetrahydroaminoacridine (THA) in Alzheimer's disease: assessment of attentional and mnemonic function using CANTAB. Psychopharmacology (Berl) 1993; 110(4):395–401.

189. Simpson PM, Surmon DJ, Wesnes KA, Wilcock GK. The cognitive drug research assessment system for demented patients: A validation study. Int J Geriatr Psychiatry 1991; 6:95–102.

190. Nicholl CG, Lynch S, Kelly CA et al. The Cognitive Drug Research computerized battery system in the evaluation of early demenita – is speed of the essence? Int J Geriatr Psychiatry 1995; 10:199–206.

191. McKeith I, Del Ser T, Spano P et al. Efficacy of rivastigmine in dementia with Lewy bodies: a randomised, double-blind, placebo-controlled international study. Lancet 2000; 356(9247):2031–6.

192. Schmand B, Walstra G, Lindeboom J, Teunisse S, Jonker C. Early detection of Alzheimer's disease using the Cambridge Cognitive Examination (CAMCOG). Psychol Med 2000; 30(3):619–27.

193. Visser PJ, Scheltens P, Verhey FR et al. Medial temporal lobe atrophy and memory dysfunction as predictors for dementia in subjects with mild cognitive impairment. J Neurol 1999; 246(6):477–85.

194. Parnowski T, Gabryclewicz T, Kiedrowska A, Czyrny M. [Usefulness of CAMDEX test in the analysis of clinical picture of dementia]. Psychiatr Pol 1995; 29(5):607–18.

195. Derix MM, Hofstede AB, Teunisse S et al. [CAMDEX-N: the Dutch version of the Cambridge Examination for Mental Disorders of the Elderly with automatic data processing]. Tijdschr Gerontol Geriatr 1991; 22(4):143–50.

196. Vilalta J, Llinas J, Lopez Pousa S, Amiel J, Vidal C. [The Cambridge Mental Disorders of the Elderly Examination. Validation of the Spanish adaptation]. Neurologia 1990; 5(4):117–20.

197. Heinik J, Werner P, Mendel A, Raikher B, Bleich A. The Cambridge Cognitive Examination (CAMCOG): validation of the Hebrew version in elderly demented patients. Int J Geriatr Psychiatry 1999; 14(12):1006–13.

198. Huppert FA, Brayne C, Gill C, Paykel ES, Beardsall L. CAMCOG – a concise neuropsychological test to assist dementia diagnosis: sociodemographic determinants in an elderly population sample. Br J Clin Psychol 1995; 34(4):529–41.

199. Hobson P, Meara J. Screening for 'cognitive impairment, no dementia' in older adults. J Am Geriatr Soc 1998; 46(5):659–60.

200. Ballard C, O'Brien J, Morris CM et al. The progression of cognitive impairment in dementia with Lewy bodies, vascular dementia and Alzheimer's disease. Int J Geriatr Psychiatry 2001; 16(5):499–503.

201. Forstl H, Sattel H, Besthorn C et al. Longitudinal cognitive, electroencephalographic and morphological brain changes in ageing and Alzheimer's disease. Br J Psychiatry 1996; 168(3):280–6.

202. Cullum S, Huppert FA, McGee M et al. Decline across different domains of cognitive function in normal ageing: results of a longitudinal population-based study using CAMCOG. Int J Geriatr Psychiatry 2000; 15(9):853–62.

203. Williams JG, Huppert FA, Matthews FE et al. Performance and normative values of a concise neuropsychological test (CAMCOG) in an elderly population sample. Int J Geriatr Psychiatry 2003; 18(7):631–44.

204. de Koning I, Dippel DW, van Kooten F, Koudstaal PJ. A short screening instrument for poststroke dementia: the R-CAMCOG. Stroke 2000; 31(7):1502–8.

205. Roth M, Huppert FA, Mountjoy CQ, Tym E. The Cambridge Examination for the Mental Disorders of the Elderly (2nd edn). Cambridge: Cambridge University Press; 1999.

206. Leeds L, Meara RJ, Woods R, Hobson JP. A comparison of the new executive functioning domains of the CAMCOG-R with existing tests of executive function in elderly stroke survivors. Age Ageing 2001; 30(3):251–4.

207. de Jager CA, Milwain E, Budge M. Early detection of isolated memory deficits in the elderly: the need for more sensitive neuropsychological tests. Psychol Med 2002; 32(3):483–91.

208. Erzigkeit H. The SKT – a short cognitive performance test as an instrument for the assessment of clinical eficacy of cognition enhancers. In: Bergener M, Reisberg B (eds). Diagnosis and Treatment of Senile Dementia. Berlin Heidelberg: Springer-Verlag; 1989: 164–74.

209. Ostrosky-Solis F, Davila G, Ortiz X et al. Determination of normative criteria and validation of the SKT for use in Spanish-speaking populations. Int Psychogeriatr 1999; 11(2):171–80.

210. Lehfeld H, Rudinger G, Rietz C et al. Evidence of the cross-cultural stability of the factor structure of the SKT short test for assessing deficits of memory and attention. Int Psychogeriatr 1997; 9(2):139–53.

211. Kim YS, Nibbelink DW, Overall JE. Factor structure and scoring of the SKT test battery. J Clin Psychol 1993; 49(1):61–71.

212. Teng EL, Chui HC. The Modified Mini-Mental State (3MS) examination. J Clin Psychiatry 1987; 48(8):314–8.

213. Tuokko H, Kristjansson E, Miller J. Neuropsychological detection of dementia: an overview of the neuropsychological component of the Canadian Study of Health and Aging. J Clin Exp Neuropsychol 1995; 17(3): 352–73.

214. Bravo G, Hebert R. Age- and education-specific reference values for the Mini-Mental and modified Mini-Mental State Examinations derived from a non-demented elderly population. Int J Geriatr Psychiatry 1997; 12(10):1008–18.

215. Correa JA, Perrault A, Wolfson C. Reliable individual change scores on the 3MS in older persons with dementia: results from the Canadian Study of Health and Aging. Int Psychogeriatr 201; 13 Supp 1:71–8.

216. Bassuk SS, Murphy JM. Characteristics of the Modified Mini-Mental State Exam among elderly persons. J Clin Epidemiol 2003; 56(7): 622–8.

217. Mathuranath PS, Nestor PJ, Berrios GE, Rakowicz W, Hodges JR. A brief cognitive test battery to differentiate Alzheimer's disease and frontotemporal dementia. Neurology 2000; 55(11):1613–20.

218. Darvesh S, Leach L, et al. The Behavioural Neurology Assessment. Can J Neurol Sci 2005; 32(2):167–77.

219. Petersen RC, Smith GE, Waring SC et al. Mild cognitive impairment: clinical characterization and outcome. Arch Neurol 1999; 56(3):303–8.

220. Touchon J, Ritchie K. Prodromal cognitive disorder in Alzheimer's disease. Int J Geriatr Psychiatry 1999; 14(7):556–63.

221. Ebly EM, Hogan DB, Parhad IM. Cognitive impairment in the nondemented elderly. Results from the Canadian Study of Health and Aging. Arch Neurol 1995; 52(6):612–19.

222. Petersen RC, Smith GE, Waring SC et al. Aging, memory, and mild cognitive impairment. Int Psychogeriatr 1997; 9(Suppl 1):65–9.

223. Bowen J, Teri L, Kukull W et al. Progression to dementia in patients with isolated memory loss. Lancet 1997; 349(9054):763–5.

224. Goldman WP, Price JL, Storandt M et al. Absence of cognitive impairment or decline in preclinical Alzheimer's disease. Neurology 2001; 56(3): 361–7.

225. Delis DC, Kramer J, Kaplan E, Ober B. The California Verbal Learning Test. New York: Pschological Corp.; 1987.

226. Buschke H. Selective reminding for analysis of memory and learning. J Verb Learn Verb Behavior 1973; 12:435–550.

227. Kluger A, Ferris SH, Golomb J, Mittelman MS, Reisberg B. Neuropsychological prediction of decline to dementia in nondemented elderly. J Geriatr Psychiatr Neurol 1999; 12(4):168–79.

228. Estevez-Gonzalez A, Kulisevsky J, Boltes A, Otermin P, Garcia-Sanchez C. Rey verbal learning test is a useful tool for differential diagnosis in the preclinical phase of Alzheimer's disease: comparison with mild cognitive impairment and normal aging. Int J Geriatr Psychiatry 2003; 18(11):1021–8.

229. De Jager CA, Hogervorst E, Combrinck M, Budge MM. Sensitivity and specificity of neuropsychological tests for mild cognitive impairment, vascular cognitive impairment and Alzheimer's disease. Psychol Med 2003; 33(6):1039–50.

230. Canning SJ, Leach L, Stuss D, Ngo L, Black SE. Diagnostic utility of abbreviated fluency measures in Alzheimer's disease and vascular dementia. Neurology 2004; 62(4):556–62.

231. Ready RE, Ott BR, Grace J, Cahn-Weiner DA. Apathy and executive dysfunction in mild cognitive impairment and Alzheimer disease. Am J Geriatr Psychiatry 2003; 11(2):2228.

232. Geda YE, Petersen RC. Clinical trials in mild cognitve impairment. In: Gauthier S, Cummings JL (eds). Alzheimer's Disease and Related Disorders Annual. London: Martin Dunitz; 2001: 69–83.

233. Petersen RC. Mild cognitive impairment clinical trials. Nat Rev Drug Discov Aug 2003; 2(8): 646–53.

234. Grundman M, Petersen RC, Ferris SH et al. Mild cognitive impairment can be distinguished from Alzheimer disease and normal aging for clinical trials. Arch Neurol 2004; 61(1):59–66.

235. Wechsler D. WMS-R Wechsler Memory Scale – Revised Manual. New York: Psychological Corporation, Harcourt Brace Jovanovich Inc.; 1987.

236. Kertesz A, Mohs RC. Cognition. In: Gauthier S (ed). Clinical Diagnosis and Management of Alzheimer's Disease (2nd edn). London: Martin Dunitz Ltd.; 1999: 179–96.

237. Boller F, Verny M, Hugonot-Diener L, Saxton J. Clinical features and assessment of severe dementia. A review. Eur J Neurol 2002; 9(2):125–36.

238. Panisset M, Roudier M, Saxton J, Boller F. Severe impairment battery. A neuropsychological test for severely demented patients. Arch Neurol Jan 1994; 51(1):41–5.

239. Schmitt FA, Ashford W, Ernesto C et al. The severe impairment battery: concurrent validity and the assessment of longitudinal change in Alzheimer's disease. The Alzheimer's Disease Cooperative Study. Alzheimer Dis Assoc Disord 1997; 11(Suppl 2):S51–6.

240. Barbarotto R, Cerri M, Acerbi C, Molinari S, Capitani E. Is SIB or BNP better than MMSE in discriminating the cognitive performance of severely impaired elderly patients? Arch Clin Neuropsychol 2000; 15(1):21–9.

241. Llinas Regla J, Lozano Gallego M, Lopez OL et al. [Validation of the Spanish version of the Severe Impairment Battery]. Neurologia 1995; 10(1): 14–18.

242. Pippi M, Mecocci P, Saxton J et al. Neuropsychological assessment of the severely impaired elderly patient: validation of the Italian short version of the Severe Impairment Battery (SIB). Gruppo di Studio sull'Invecchiamento Cerebrale della Societa Italiana di Gerontologia e Geriatria. Aging (Milano) 1999; 11(4):221–6.

243. Pelissier C, Roudier M, Boller F. Factorial validation of the Severe Impairment Battery for patients with Alzheimer's disease. A pilot study. Dement Geriatr Cogn Disord 2002; 13(2):95–100.

244. Wild KV, Kaye JA. The rate of progression of Alzheimer's disease in the later stages: evidence from the Severe Impairment Battery. J Int Neuropsychol Soc 1998; 4(5):512–16.

245. Cole MG, Dastoor DP. The Hierarchic Dementia Scale: conceptualization. Int Psychogeriatr 1996; 8(2):205–12.

246. Peavy GM, Salmon DP, Rice VA et al. Neuropsychological assessment of severely demeted elderly: the severe cognitive impairment profile. Arch Neurol 1996; 53(4):367–72.

247. Albert M, Cohen C. The Test for Severe Impairment: an instrument for the assessment of patients with severe cognitive dysfunction. J Am Geriatr Soc 1992; 40(5):449–53.

248. Mohr E, Walker D, Randolph C, Sampson M, Mendis T. Utility of clinical trial batteries in the measurement of Alzheimer's and Huntington's dementia. Int Psychogeriatr 1996; 8(3):397–411.

249. Veroff AE, Cutler NR, Sramek JJ et al. A new assessment tool for neuropsychopharmacologic research: the Computerized Neuropsychological Test Battery. J Geriatr Psychiatry Neurol 1991; 4(4):211–17.

250. Dwolatzky T, Whitehead V, Doniger GM et al. Validity of a novel computerized cognitive battery for mild cognitive impairment. BMC Geriatr 2003; 3(1):4.

251. Cutler NR, Shrotriya RC, Sramek JJ et al. The use of the Computerized Neuropsychological Test Battery (CNTB) in an efficacy and safety trial of BMY 21,502 in Alzheimer's disease. Ann N Y Acad Sci 1993; 695:332–6.

252. Giacobini E, Spiegel R, Enz A, Veroff AE, Cutler NR. Inhibition of acetyl- and butyryl cholinesterase in the cerebrospinal fluid of patients with Alzheimer's disease by rivastigmine: correlation with cognitive benefit. J Neural Transm 2002; 109(7–8):1053–65.

253. Gobburu JV, Tammara V, Lesko L et al. Pharmacokinetic-pharmacodynamic modeling of rivastigmine, a cholinesterase inhibitor, in patients with Alzheimer's disease. J Clin Pharmacol 2001; 41(10):1082–90.

254. Sudilovsky A, Cutler NR, Sramek JJ et al. A pilot clinical trial of the angiotensin-converting enzyme inhibitor ceranapril in Alzheimer disease. Alzheimer Dis Assoc Disord 1993; 7(2):105–11.

255. Veroff AE, Bodick NC, Offen WW, Sramek JJ, Cutler NR. Efficacy of xanomeline in Alzheimer disease: cognitive improvement measured using the Computerized Neuropsychological Test Battery (CNTB). Alzheimer Dis Assoc Disord 1998; 12(4):304–12.

256. Eagger S, Morant N, Levy R, Sahakian B. Tacrine in Alzheimer's disease. Time course of changes in cognitive function and practice effects. Br J Psychiatry 1992; 160:36–40.

257. Sahakian BJ, Coull JT. Tetrahydroaminoacridine (THA) in Alzheimer's disease: an assessment of attentional and mnemonic function using CANTAB. Acta Neurol Scand Suppl 1993; 149:29–35.

258. Darby D, Maruff P, Collie A, McStephen M. Mild cognitive impairment can be detected by multiple assessments in a single day. Neurology 2002; 59(7):1042–6.

259. Maruff P, Collie A, Darby D et al. Subtle Memory decline over 12 months in mild cognitive impairment. Dement Geriatr Cogn Disord 2004; 18:342–8.

260. Crapper McLachlan DR, Dalton AJ, Kruck TP et al. Intramuscular desferrioxamine in patients with Alzheimer's disease. Lancet 1991; 337(8753):1304–8.

261. Wesnes KA, Ward T, Ayre G, Pincock G. Development and validation of a system for evaluating cognitive function over the telephone for use in late phase development. Eur Neuropsychopharmacol 1999; 9(Suppl 5):S368.

262. Jarvenpaa T, Rinne JO, Raiha I et al. Characteristics of two telephone screens for cognitive impairment. Dement Geriatr Cogn Disord 2002; 13(3):149–55.

263. Go RC, Duke LW, Harrell LE et al. Development and validation of a Structured Telephone Interview for Dementia Assessment (STIDA): the NIMH Genetics Initiative. J Geriatr Psychiatry Neurol 1997; 10(4):161–7.

264. de Jager CA, Budge MM, Clarke R. Utility of TICS-M for the assessment of cognitive function in older adults. Int J Geriatr Psychiatry 2003; 18(4):318–24.

265. Newkirk LA, Kim JM, Thompson JM et al. Validation of a 26–point telephone version of the mini-mental state examination. J Geriatr Psychiatry Neurol 2004; 17(2):81–7.

266. Andrews HF, Segal G, Shah C et al. Web-based screening, data entry and data management in Alzheimer's disease research. Neurobiol Aging 2000; 21(Suppl 1):160.

267. Dwyer P, Hagerman V, Ingram CA, MacFarlane R, McCourt S. Atlantic telehealth knowledge exchange. Telemed J E Health 2004; 10(1):93–101.

268. Loh PK, Ramesh P, Maher S, Saligari J, Flicker L, Goldswain P. Can patients with dementia be assessed at a distance? The use of Telehealth and standardised assessments. Intern Med J 2004; 34(5):239–42.

269. Ball C, Puffett A. The assessment of cognitive function in the elderly using videoconferencing. J Telemed Telecare 1998; 4(Suppl 1):36–8.

270. Fox NC, Rossor MN. Diagnosis of early Alzheimer's disease. Rev Neurol (Paris) 1999; 155(Suppl 4):S33–37.

271. Meiran N, Stuss DT, Guzman DA, Lafleche G, Willmer J. Diagnosis of dementia. Methods for interpretation of scores of 5 neuropsychological tests. Arch Neurol 1996; 53(10):1043–54.

272. Ihl R, Grass-Kapanke B, Janner M, Weyer G. Neuropsychometric tests in cross sectional and longitudinal studies – a regression analysis of ADAS-cog, SKT and MMSE. Pharmacopsychiatry 1999; 32(6):248–54.

9

Functional Outcomes

Serge Gauthier

Introduction

The importance of decline in activities of daily living (ADL) in dementia has been recognised in the diagnostic criteria for dementia, described as 'significant impairment in social or occupational functioning' in the *Diagnostic and Statistical Manual of Mental Disorders*.[1] Progressing over time, the loss of functional autonomy has a major impact on the quality of life of persons with dementia and their caregivers.[2] It is thus appropriate that the International Working Group on Harmonization of Dementia Drugs Guidelines (IWGHDDG) and regulatory authorities in America, Europe and Japan consider functional outcomes as part of a clinically meaningful treatment response.[3-5]

This chapter will review the natural history of functional decline in dementia, the background to current functional scales, and will highlight some of the results observed in randomised clinical trials (RCT) for the symptomatic treatment of Alzheimer's disease (AD), vascular dementia (VaD) and mixed AD/VaD, and attempts at disease modification.

Natural history of functional decline in dementia

Several longitudinal and cross-sectional studies have demonstrated the gradual loss of functional abilities in dementia over time.[2] This decline follows a hierarchical pattern best described in the Functional Assessment Staging (FAST),[6] with a range of 1 (no decrement) to 7f (ability to hold head up is lost), with involvement of instrumental activities of daily living (IADL) followed by basic or self-care activities (ADL). Examples of IADL includes leisure activities, telephoning and meal preparation, whereas ADL includes dressing, eating, toileting.

The current emphasis on very early diagnosis of AD in persons with amnestic mild cognitive impairment (MCI) has led to interest in the very early changes in IADL. Epidemiological studies suggest that impairments in four IADL items (handling medications, transportation, finances and telephone) are the most sensitive indicator of early dementia.[7] Ongoing RCT in amnestic MCI will help clarify which of the IADL are most useful for determining conversion from MCI to dementia, or alternatively, in unmasking very early dementia among persons who otherwise appear to have uncomplicated amnestic MCI.

Later in the course of dementia, functional decline can be used as a clinical milestone.[8] For instance loss of basic ADL has been delayed by alpha-tocopherol in one RCT involving patients with AD in moderate to severe stages.[9]

An unresolved issue is the hierarchy and pace of decline in non-AD dementias. It is already apparent from longitudinal studies and RCTs that the functional decline in VaD is slower than in AD (as is cognitive decline), which will impact on sample size if a functional outcome is to be used as primary outcome in a RCT.[10–12] Mixed AD/VaD populations appear to decline at a rate similar to AD populations.[13] There are no data yet published on the functional decline associated with Parkinson's disease dementia, but ongoing RCTs will be of great interest considering the prominent executive dysfunction associated with this condition.[14] Similarly, the role of the quality of performance in non-AD dementias remains to be clarified. In people with important frontal-subcortical dysfunction as, for example, in so-called 'subcortical' vascular dementia, the issue of loss of initiative in IADLs is important. Thus, for example, people can still perform certain higher-order IADL, but appear to require more cuing than before. As noted below, this also often is the case in dementia drugs trials of cholinesterase inhibitors (ChEIs) in AD.

Background to functional scales

Evidence-based reviews on the measurement of ADL in dementia have highlighted the following facts:[2,15–17] (1) dementia-specific scales are preferable over less-specific scales (such as those used in epidemiological studies of disability in elderly people) where physical disabilities such as inability to walk or climbing stairs are not necessarily related to cognitive impairment; (2) although a performance-based assessment by a trained observer would be the most objective measure of selected IADL and ADL, informant or caregiver reports are the

best available and are more reliable than self-reports; (3) some IADL items are gender- and culture-biased. In consequence, it is important that any scale has some non-arbitrary means, other than denoting the item as 'missing', for dealing with activities that while not undertaken by a patient with mild dementia, have never been in that patient's repetoire of performance.

Another issue in many scales has been the binary 'able' or 'unable' rating for individual items, whereas careful observations of decline in untreated patients as well as patterns of improvement on ChEIs suggest a more complex breakdown of abilities to initiate, plan and organise and effectively perform individual tasks. The most commonly used scales in current RCTs assess these stages of individual functions, rather than their presence or absence, giving a better picture of the capacities of the individual patient at a given point in time. These scales are the Alzheimer Disease Cooperative Study ADL scale (ADCS-ADL) and the Disability Assessment in Dementia (DAD).[18,19] These and other functional scales used in past and current RCTs are listed alphabetically in Table 9.1.

Table 9.1 Functional scales used in clinical trials for dementia

- Alzheimer Disease Cooperative Study ADL scale (ADCS-ADL)[18]
- Alzheimer's Disease Functional Assessment and Change Scale (ADFACS)[20]
- Bristol Activities of Daily Living Scale[21]
- Disability Assessment in Dementia (DAD)[19]
- Interview for Deterioration in Daily living activities in Dementia (IDDD)[22]
- Nurses' Observation Scale for Geriatric patients (NOSGER)[23]
- Progressive Deterioration Scale (PDS)[24]

Results from clinical trials aiming at symptomatic improvement

In the early days of cholinergic enhancement using tacrine, an observation had been made that initiative and interest in leisure and housework activities was often reported by families.[25] Although this fact contributed to the development of more sensitive scales to measure IADL and ADL in dementia, there was little observable improvement above baseline in the subsequent studies using therapeutic doses of ChEIs and memantine. This lack of return to previously acquired IADL abilities was termed by the IWGHDDG 'the tutoring effect',[4] meaning that caregivers were reluctant to give back the keys to the car or the cheque book, once these abilities had been lost, no matter

the regained initiative or interest by the patient on a ChEI.

Fortunately, RCTs with placebo-treated arms were able to demonstrate statistically significant differences at six months and beyond between groups, in favour of ChEIs and memantine. Examples are listed in Table 9.2. The biggest difference between drug and placebo for functional decline is in the moderate to severe AD groups, where remaining IADLs are lost and basic ADL are declining, whereas the smallest difference is in VaD, where patients are stable functionally as long as they do not have another stroke. It is unfortunate that standard errors are not systematically reported in the different publications.

Table 9.2 Examples of functional changes over time in clinical trials

Disease	Reference number	ADL scale	Drug/dose	Placebo vs baseline	Drug vs baseline	Drug vs placebo
Mild-moderate AD	26	ADCS-ADL	gal 24 mg	−3.8 (0.6)	−1.5 (0.6)	$P < 0.01$
	27	IDDD	don 10 mg	−3.0	−1.0	$P < 0.0072$
	28	PDS	riva 6–12mg	−2.18	0.05	$P < 0.1$
Moderate-severe AD	29	DAD	don 10 mg	−8.98	−0.74	$P < 0.0001$
	30	ADCS-ADL	mem 20 mg	−5.2 (6.33)	−3.1 (6.79)	$P = 0.02$
Mixed AD/VaD	11	ADFACS	don 10 mg	0.76 (0.39)	−0.23 (0.40)	NS
	12	ADFACS	don 10 mg	1.44 (0.42)	0.53 (0.38)	NS

Drugs: gal, galantamine; don, donepezil; riva, rivastigmine; mem, memantine
NS, not significant
Placebo and drug differences (standard errors) at weeks 26–28 except for reference 26 where study duration was 20 weeks
Placebo and drug differences and statistical significance from intention-to-treat populations

Results from attempts at disease modification

Although there has been no successful disease stabilisation study as yet, there have been attempts using different classes of drugs. Unfortunately, negative studies are often not published. Another issue has been the selection of scales not specific for mild to moderate stage dementia, such as the Blessed Dementia Rating Scale in RCTs comparing prednisone or oestrogen to placebo,[31–33] or the Lawton and Brody scale in a RCT comparing acetyl-L-carnitine to placebo.[34,35] It is fortunate that ongoing studies are using the newer scales based on the breakdown of initiation, planning and execution, namely the ADCS-ADL and the DAD, with supportive data from a one-year study comparing sabeluzole to placebo demonstrating linear changes over time for the DAD scale.[36] Such linearity is desirable in order to calculate slopes of decline over one year and hopefully establish divergence with the placebo group.

Another issue that longer-term trials will need to get to grips with is that caregivers' perceptions of a patient's functional ability also change, independently of changes in the patient's level of performance. For example, one study compared caregiver reports with standardised ratings (using videotapes and independent observers) of the same patients performing standardised tasks. Prior to an AD diagnosis, caregivers overestimated the extent of the ability of their affected family members. After six months, they underestimated the extent of their family members to perform tasks.[37] Particularly for longer-term studies, efforts will need to be made to assess the impact of how patients and families adapt to disease by modifying their daily roles. Similarly, studies aimed at dementia prevention will need to be careful in how functional impairment is evaluated, especially in the setting of patients with established MCI.[38]

Conclusions

Functional outcomes have been shown to be useful in proving efficacy in pivotal studies of ChEIs and of memantine. There is an expectation that measurable changes in functional autonomy will be an important component to the conversion from amnestic MCI to dementia, and to the proof that disease-modifying drugs will arrest or significantly slow down functional decline.

References

1. American Psychiatric Association. Diagnostic and Statistical Manual of Mental Disorders (4th edn). Washington: American Psychiatric Association; 2000.
2. Gélinas I, Auer S. Functional autonomy. In: Gauthier S (ed). Clinical Diagnosis and Management of Alzheimer's Disease (2nd edn). London: Martin Dunitz; 2001: 213–26.
3. Gauthier S, Bodick N, Erzigkeit E et al. Activities of daily living as an outcome measure in clinical trials of dementia drugs. Alzheimer Dis Assoc Disord 1997; 11(Suppl 3): 6–7.
4. Gauthier S, Rockwood K, Gélinas I et al. Outcome measures for the study of activities of daily living in vascular dementia. Alzheimer Dis Assoc Disord 1999; 13(Suppl 3):S143–S147.
5. Leber P. Criteria used by regulatory authorities. In: Qizilbash N, Schneider LS, Chui H et al (eds). Evidence-based Dementia Practice. Oxford: Blackwell Publishing; 2002: 376–87.
6. Reisberg B, Ferris SH, Anand R et al. Functional staging of dementia of the Alzheimer's type. Ann NY Acad Sci 1984; 435:481–3.
7. Barberger-Gateau P, Fibrigoule C, Helmer C, Rouch I, Dartigues JF. Functional impairment in Instrumental Activities of Daily Living: an early clinical sign of dementia? J Am Geriatr Soc 1999; 47:456–62.
8. Galasko D, Edland SD, Morris JC et al. The Consortium to Establish a Registry for Alzheimer's Disease (CERAD). Part XI. Clinical milestones in patients with Alzheimer's disease followed over 3 years. Neurology 1995; 45:1451–5.

9. Sano M, Ernesto C, Thomas RG et al. A controlled trial of selegiline, alpha-tocopherol, or both as treatment for Alzheimer's disease. N Engl J Med 1997; 336:1216–22.

10. Nyenhuis DL, Gorelick PB, Freels S, Garron DC. Cognitive and functional decline in African Americans with VaD, AD, and strokes without dementia. Neurology 2002; 58:56–61.

11. Wilkinson D, Doody R, Helme R et al. Donepezil in vascular dementia. A randomized, placebo-controlled study. Neurology 2003; 61:479–86.

12. Black S, Román G, Geldmacher DS et al. Efficacy and tolerability of donepezil in vascular dementia. Stroke 2003; 34:2323–32.

13. Erkinjuntti T, Kurz A, Gauthier S et al. Efficacy of galantamine in probable vascular dementia and Alzheimer's disease combined with cerebrovascular disease: a randomized trial. Lancet 2002; 359:1283–90.

14. Emre M. Dementia associated with Parkinson's disease. Lancet Neurol 2003; 2:229–45.

15. Lehfeld H, Erzigkeit H. Functional aspects of dementia. In: Gauthier S, Cummings JL (eds). Alzheimer's Disease and Related Disorders Annual. London: Martin Dunitz; 2000: 155–78.

16. Lindeboom R, Vermeulen M, Holman R, De Haan RJ. Activities of daily living instruments. Optimizing scales for neurologic assessments. Neurology 2003; 60:738–42.

17. Bullock R. Dementia rating scales. In: Qizilbash N, Schneider LS, Chui H et al (eds). Evidence-based Dementia Practice. Oxford: Blackwell Publishing; 2002: 859–69.

18. Galasko D, Bennett D, Sano M et al. An inventory to assess activities of daily living for clinical trials in Alzheimer's disease. Alzheimer Dis Assoc Disord 1997; 11:S33–S39.

19. Gélinas I, Gauthier L, McIntyre M, Gauthier S. Development of a functional measure for persons with Alzheimer's disease: the Disability Assessment of Dementia. Am J Occup Ther 1999; 53:471–81.

20. Mohs RC, Doody RS, Morris JC et al. A 1-year, placebo-controlled preservation of function survival study of donepezil in AD patients. Neurology 2001; 57:481–8.

21. Bucks RS, Ashworth DL, Wilcock GK, Siegfried K. Assessment of activities of daily living in dementia: development of the Bristol Activities of Daily Living Scale. Age Ageing 1996; 25:113–20.

22. Teunisse S, Derix MMA, van Crevel H. Assessing the severity of dementia. Patient and caregiver. Arch Neurol 1991; 48:274–7.

23. Spiegel R, Brunner C, Ermini-Fünfschilling D et al. A new behavioral assessment scale for geriatric out- and in-patients: the NOSGER. J Am Geriatr Soc 1991; 39:339–47.

24. DeJong R, Osterlund OW, Roy GW. Measurement of quality-of-life changes in patients with Alzheimer's disease. Clin Ther 1989; 11:545–54.

25. Gauthier S, Bouchard R, Lamontagne A et al. Tetrahydroaminoacridine–lecithin combination treatment in intermediate stage Alzheimer's disease: results of a Canadian double-blind cross-over multicentre study. N Eng J Med 1990; 322:1272–6.

26. Tariot PN, Solomon PR, Morris JC et al. A 5-month, randomized, placebo-controlled trial of galantamine in AD. Neurology 2000; 54:2269–76.

27. Burns A, Rossor M, Hecker J et al. The effects of donepezil in Alzheimer's disease – results from a multinational trial. Dement Geriatr Cogn Disord 1999; 10:237–44.

28. Rösler M, Anand R, Chin-Sain A et al. Efficacy and safety of rivastigmine in patients with Alzheimer's disease: international randomized controlled trial. BMJ 1999; 318:633–40.

29. Feldman H, Gauthier S, Hecker J et al. A 24-week, randomized, double-blind study of donepezil in moderate to severe Alzheimer's disease. Neurology 2001; 57:613–20.

30. Reisberg B, Doody R, Stöffler A et al. Memantine in moderate-to-severe Alzheimer's disease. N Engl J Med 2003; 348:1333–41.

31. Blessed G, Tomlinson BE, Roth M. The association between quantitative measures of dementia and of senile change in the cerebral gray matter of elderly subjects. Br J Psychiatry 1968; 114:797–811.

32. Aisen PS, Davis KL, Berg JD et al. A randomized controlled trial of prednisone in Alzheimer's disease. Neurology 2000; 54:588–93.

33. Mulnard RA, Cotman CW, Kawas C et al. Estrogen replacement therapy for treatment of mild to moderate Alzheimer Disease. A randomized controlled trial. JAMA 2000; 283:1007–15.

34. Lawton MP, Brody EM. Assessment of older people: self-maintaining and instrumental activities of daily living. Gerontologist 1969; 9:179–86.

35. Thal L, Carta A, Clarke WR et al. A 1-year multi-centre placebo-controlled study of acetyl-L-carnitine in patients with Alzheimer's disease. Neurology 1996; 47:705–11.

36. Feldman H, Sauter A, Donald A et al. The Disability Assessment for Dementia scale: a 12-month study of functional ability in mild to moderate severity Alzheimer's disease. Alzheimer Dis Assoc Disord 2001; 15:89–95.

37. Doble SE, Fisk JD, Rockwood K. Assessing the ADL functioning of persons with Alzheimer's disease: comparison of family informants' ratings and performance-based assessment findings. Int Psychogeriatr 1999; 11:399–409.

38. Fisk JD, Merry H, Rockwood K. Variations in case definition affect prevalence but not outcomes of mild cognitive impairment. Neurology 2003; 61:1179–84.

10

Behavioural Outcomes

Clive Ballard

Introduction

Behavioural and psychological symptoms (BPSD), including disturbance of affect (anxiety, depression, mania), perception (hallucinations), thought (delusions) as well as behavioural and personality changes (agitation) are commonly observed in most dementia syndromes.[1] At any one time 50% of the people with Alzheimer's disease (AD) in contact with specialist services will be experiencing at least one BPSD symptom, and more than 80% of these individuals will experience BPSD over the course of the dementia.[2] The cross-sectional frequency of BPSD is even higher amongst nursing home residents with dementia,[3] and in certain non-Alzheimer dementias such as dementia with Lewy bodies (DLB), where psychotic symptoms such as delusions and hallucinations are present in more than 80% of patients.[4]

The occurrence of BPSD leads both to distress for patients and problems for caregivers, and is associated with increased needs for care, including the precipitation of institutionalisation.[5–7] BPSD are therefore an important treatment target and, in contrast to current practice in dementia drug trials, should not be viewed as a secondary or additional feature of dementia, but rather represent a core part of the dementia syndrome.

There are relatively few longitudinal studies of BPSD, but those undertaken indicate that many symptoms tend to spontaneously recover over a period of a few months. For example, 50% of psychotic symptoms and 66% of symptoms of depression resolved without treatment over 3 months,[8,9] although more severe symptoms and those that had already been present for longer than 3 months were more likely to persist. Several studies also indicate that psychotic symptoms tend to have a good natural outcome amongst nursing home patients over a year of follow-up, although agitation symptoms such as aggression and restlessness are less likely to resolve without treatment.[10–12] Understanding the natural course of these symptoms is imperative as it allows specific treatments to be targeted optimally when required, while avoiding unnecessary interventions.

When symptoms are severe, distressing or persistent and treatment is needed, psychosocial interventions can be very effective. A detailed review of this literature is beyond the remit of the current chapter, but several key placebo-controlled trials and a large case series literature indicate significant efficacy for a variety of key symptoms.[13–15]

If good clinical management principles are followed, pharmacological intervention should only be necessary for a modest proportion of people with dementia experiencing BPSD, although in practice this is probably not the case as a number of studies indicate very high rates of psychotropic prescriptions to these individuals.[3,16] When pharmacotherapy is required, neuroleptics are probably the treatment of choice.[17] An initial meta-analysis by Schneider et al, and a more recent editorial by Ballard and O'Brien reviewing placebo-controlled trials of neuroleptics for BPSD both indicate an approximate response rate of 60% to active treatment and 40% to placebo (defined as a 30% improvement in BPSD symptoms on a standardised rating scale).[17,18] The high placebo response is probably partly explained by the benign natural course of these symptoms in many individuals, combined with a Hawthorne effect related to the study personnel and procedures. In determining the optimal management of an individual patient, this significant but modest treatment effect needs to be balanced against the potential adverse effects of the therapy. For example, in older people and those with dementia, neuroleptics are associated with a number of side-effects, some common to the class of agents and others that appear to be specific to individual drugs. These effects include an increased risk of falls and drowsiness, parkinsonism, akathisia, tar-

dive dyskinesia, stroke, risk of cardiac arrhythmias, severe neuroleptic sensitivity reactions, and the probability that neuroleptic agents may substantially accelerate cognitive decline and neuronal loss (see Figure 10.1).[19–23] As a consequence of the potentially harmful side-effects of these agents, legislation has been introduced to regulate the prescription of neuroleptics to nursing home patients in the US, and in the UK the Chief Medical Officer has recommended particular caution when prescribing neuroleptics to people with dementia.[24,25] Four of the published placebo-controlled trials are with atypical neuroleptics (three with risperidone, one with olanzapine), which appear at least in some regards to have preferable side-effect profiles with lower risks of parkinsonism and tardive dyskinesia in the context of comparable efficacy.[26–29] As all of the treatment studies are of short duration (generally 4–12 weeks); there is, however, little information regarding the long-term safety or efficacy of these agents.

On considering this evidence base it is clear that a safer and more effective pharmacological alternative is needed for the treatment of BPSD, particularly as treatment is often continued for protracted periods. Could cholinesterase

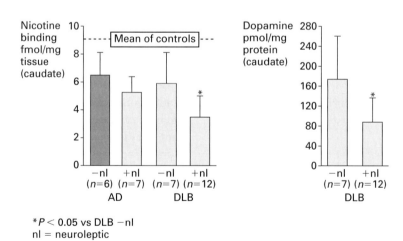

*P < 0.05 vs DLB −nl
nl = neuroleptic

Figure 10.1 Loss of nicotinic receptors in DLB patients taking neuroleptics (summarising data from Court et al 2000[23]).

inhibitors offer that alternative? The following sections will review the clinical trial evidence base and the scientific rationale for the use of cholinergic therapies to treat these symptoms.

Cholinesterase inhibitors

Acetylcholinesterase inhibitors have been introduced over the last 6–7 years as cognition-enhancing agents in the treatment of patients with mild to moderate AD. Three agents, donepezil, rivastigmine and galantamine are licensed and widely prescribed in Europe and North America. These agents exert their beneficial effect on intellectual functioning by blocking acetylcholinesterase and enhancing cholinergic function. They have been developed to improve the neuropsychological deficits of AD, as shown in most clinical trials by improved scores or superior performance compared to patients receiving placebo on standardised cognitive evaluations, and global scales.[30–32] Some studies also suggest benefit in activities of daily living (ADL).[30–32] Other than tacrine, which is associated with hepatotoxicity, the cholinesterase inhibitors are generally well tolerated, with gastrointestinal side-effects such as nausea, representing the most frequently experienced symptoms (occurring in approximately 20% of patients).[30–32] In frailer patients or those with cardiovascular disease there is some preliminary evidence to indicate that falls and syncope may ocurr.[33] Given the favourable side-effect profile compared to neuroleptic agents, there are clear potential advantages in utilising cholinesterase inhibitors for the treatment of BPSD if their efficacy is equivalent.

BPSD symptoms in double-blind placebo-controlled trials of cholinesterase inhibitors

For the purposes of the current chapter, only information from placebo-controlled trials has been considered, as the high placebo response rate seen in BPSD treatment studies renders open-label trials almost impossible to interpret. A number of placebo-controlled trials have been completed, evaluating the efficacy of cholinesterase inhibitors in people with dementia. The majority have focused upon patients with AD, although several published studies have examined the impact of cholinesterase inhibitors in other dementias. The evidence base can be difficult to evaluate, as a number of data re-analyses have been published for the majority of studies, and the number of reviews of the evidence far exceeds the original data. For the current summary of the literature I have therefore used the studies considered as part of four recent Cochrane reviews, all completed since the beginning of 2003 (donepezil in Alzheimer's disease, galantamine in Alzheimer's disease, rivastigmine in Alzheimer's disease, dementia with Lewy bodies).[30–32,34]

Although 16 placebo-controlled trials are included within the Cochrane review for donepezil in Alzheimer's disease, BPSD were only measured as a secondary outcome in two of these trials. Feldman et al reported change in Neuropsychiatric Inventory (NPI) scores as a secondary outcome measure in a cohort of 290 patients with mild to moderate Alzheimer's disease (144 assigned to active treatment with donepezil 10 mg, 146 to placebo).[35] There was no significant effect upon NPI scores at 12 weeks, but a significant advantage (-4.4 95% confidence intervals -7.9 to -0.9, $P = 0.01$) was seen for donepezil by week 24. However, in a study with a similar design, Tariot et al did not identify a significant difference between donepezil and placebo for change in BPSD symptoms.[36] Furthermore, when the two studies were combined in the Cochrane review there was no significant overall effect.

The pattern of evidence is very similar for galantamine. Two of the seven randomised placebo-controlled studies included within the Cochrane review evaluated change in the NPI as a secondary outcome measure. Tariot et al reported a 5-month trial with 978 patients randomised to galantamine or placebo (286 placebo, 140 galantamine 8 mg, 299 galantamine 16 mg, 273 galantamine 24 mg).[37] On an intention to treat analysis, the difference between galantamine and placebo was only significant at the 5-month evaluation for the 16 mg dose, with an advantage of 2.4 points for galantamine (95% confidence intervals −4.5 to −1.3, $P < 0.05$). There were no significant differences in the change of NPI score at 3 months for any of the doses compared to the placebo treatment arm. In the other study of 386 people with probable AD, there were no significant differences between galantamine and placebo, and the Cochrane meta-analysis did not identify any overall significant advantage for galantamine treatment.[38]

None of the studies included in the Cochrane review of rivastigmine in AD incorporated a standardised measure of BPSD. However, the only study considered in the Cochrane review of DLB, was a 20 week randomised placebo-controlled trial of rivastigmine in 102 patients with probable DLB in which the NPI was used as a primary outcome measure.[39] At week 20 there was a significant advantage for rivastigmine on the analysis of patients completing the study (difference in change on NPI −6.9 95% confidence intervals −11.6 to −2.3, $P = 0.003$), although there was no significant difference on the intention to treat or last observation carried forward analysis. To put this into context, the difference in the proportion of patients experiencing a 30% improvement between active treatment and placebo was comparable to that seen in trials of neuroleptics among AD patients with BPSD symptoms (see Figure 10.2). At the 3-month evaluation there was no significant difference between rivastigmine and placebo.

Taking an overview, there are hence three placebo-controlled trials where some overall effect is seen with cholinesterase inhibitor therapy on the severity of BPSD symptoms over a 5–6-month period of treatment.[35,36,39] For each of these studies, significant effects are only seen for certain analyses and the effect size is small. In addition, there are two studies showing no significant effects. None of the studies showed a significant

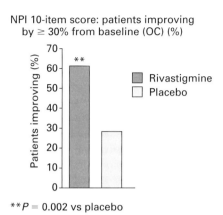

NPI 10-item score: patients improving by ≥ 30% from baseline (OC) (%)

**P = 0.002 vs placebo

Figure 10.2 Patients with a 30% improvement in NPI scores in a placebo controlled trial of rivastigmine (summarising data from McKeith et al 2000[39]).

treatment effect on BPSD with cholinesterase therapy over 3 months. While this emerging evidence is potentially interesting and should stimulate further research, it certainly cannot be taken as unequivocal evidence of efficacy. Perhaps the most clear finding is the absence of a significant treatment effect on BPSD over 3 months of cholinesterase therapy, indicating that a more prolonged period of therapy is probably needed for any significant treatment benefits to be seen.

In interpreting this evidence it is important to bear several things in mind with respect to the severity of BPSD symptoms: (1) almost all of these studies have focused upon people with mild to moderate dementia, whereas BPSD are seen predominantly in people with moderately severe to severe dementia; (2) with the exception of the rivastigmine DLB study and the report of Feldman et al,[35] the mean NPI scores are well below the threshold that would be considered as a clinically significant level of BPSD symptoms; (3) patients for these studies were not referred specifically for BPSD symptoms, and are hence unlikely to have had BPSD symptoms that were causing major problems in the day-to-day care of those individuals. For this reason, the evidence emerging from these trials cannot be considered as an equivalent of placebo-controlled trials in people with clinically significant BPSD. The preliminary indications of the possible efficacy of cholinesterase inhibitors in improving BPSD symptoms certainly highlight the importance of undertaking further trials specifically in people with clinically significant BPSD.[40] However, the value of improving subclinical non-problematic symptoms is less clear, although one possible interpretation of the data is that the emergence of BPSD may be delayed. This is quite plausible as an overall effect, as one would anticipate that the stabilisation of cognitive and global functioning seen over 6 months of treatment would also impact on BPSD, as these symptoms become more frequent with the progression of dementia severity. This hypothesis is also consistent with the lack of treatment benefit upon BPSD over a 3-month period. Although this is potentially an important secondary benefit, it does not necessarily indicate a specific mechanistic effect on BPSD related to cholinergic enhancement. The relatively low baseline NPI scores in the majority of these studies may also however have limited the statistical power to show treatment differences, and it is likely that the potential effect size may be substantially larger in people with more severe symptoms. In patients with more problematic BPSD symptoms, the significant apparent delay between the initiation of cholinesterase inhibitor therapy and the observed improvement in NPI scores (5–6 months) may be a major clinical obstacle.

It is also imperative to consider whether a global evaluation of BPSD is the best way to measure treatment response, or whether the impact of therapy on specific BPSD syndromes is a preferable outcome measure. One key issue pertains to apathy. Although a very important symptom, it is debatable whether apathy should phenomenologically be classified as a BPSD symptom. Cholinesterase inhibitor therapy appears to improve apathy, which, as one of the ten neuropsychiatric items incorporated in the NPI, has the capacity to drive modest changes in the overall score. In addition however, BPSD was developed as an umbrella term and not to describe a specific syndrome.

The following two sections will review the scientific evidence linking potential cholinergic mechanisms to specific BPSD symptoms and the phenomenological literature describing approaches to the classification of BPSD. This will then be used as a structure to review the evidence examining the impact of cholinesterase inhibitor therapy upon specific BPSD symptoms.

Scientific rationale for the use of cholinesterase inhibitors to treat BPSD

Abnormalities in cholinergic neurons (including cell loss) are prominent among the pathological changes in the brains of patients with AD and DLB. There is clear evidence that acetylcholine is an important neurotransmitter associated with cognitive (particularly attentional) deficits, but does it also play a key role in the genesis of BPSD? Several studies, focusing predominantly upon the cholinergic system, have evaluated the potential neurochemical associations of these symptoms. In DLB, two independent studies report that visual hallucinations are associated with reduced cortical choline acetyltransferase (ChAT) activity in the temporal cortex (see Table 10.1), a marker of cholinergic innervation;[41,42] whereas delusions are significantly associated with elevated M1 receptor

Table 10.1 Choline acetyl transferase and visual hallucination in dementia with Lewy bodies (from Ballard et al 2000[42])		
	M1	**ChAT**
VH (16)	122.3 ± 36.0	1.7 ± 0.6
No VH	106.3 ± 44.8	2.5 ± 0.7
	$P = 0.38$	$P = 0.02*$
*p<0.05 VH, visual hallucinations		

binding but not with reductions in ChAT.[42] Visual hallucinations and delusional misidentification (but not delusions) are also associated with lower binding to nicotinic receptors ([^{125}I]alpha bungarotoxin binding was reduced in the same area of the temporal cortex).[43] In AD, delusions also appear to be associated with upregulation of muscarinic receptors, but there is no evidence that visual hallucinations are associated with reduced ChAT activity.[44] Other risk factors such as impaired visual acuity or cataracts appear to be a more robust association of visual hallucinations in the context of AD.[45,46] The relationship of cholinergic function to restlessness and aggression in AD has also been studied, with an indication that lowered ChAT activity in the frontal and temporal cortices correlates with increasing overactivity in patients with dementia, and the ratio of cholinergic to dopaminergic function appears to be associated with aggression.[47]

While there are too few studies to establish a general consensus, the association of visual hallucinations with cholinergic defics in DLB has been replicated. The basis of BPSD symptoms may well be different in DLB and AD, and specific symptoms appear to have a different neurochemical association. There are no studies examining the relationship of cholinergic parameters to anxiety or depression in dementia patients.

Classification of BPSD

Preliminary proposals to classify BPSD have been published. For example, Jeste and Finkel suggested criteria for 'Psychosis of Alzheimer's disease and related dementias', based on the empirical evidence that psychiatric symptoms in patients with dementia tend to cluster into syndromes.[48] Lyketsos et al proposed criteria for 'Alzheimer-associated psychotic disorder' and 'Alzheimer-associated affective disorder'.[49] Both syndromes require the addition of associated symptoms to the cardinal features of the syndromes. These issues are of interest to the development of the DSM-V, which is expected to expand and revise

its taxonomy of psychiatric disorders associated with common brain diseases.

Most research of psychiatric symptoms in patients with dementia has adopted an approach with emphasis on scaled ratings of disturbances, rather than discrete diagnoses. Several rating scales have been developed for the assessment of the specific psychiatric and behavioural disturbances in dementia. However, there is emergent empirical evidence suggesting that patients with dementia exhibit multiple simultaneous symptoms, and thus the wide range of psychological and behavioural changes associated with dementia seem to cluster into psychiatric syndromes. The assumption of this approach is that underlying constructs explain the relationship between observed variables. Such an assumption is consistent with the evidence reviewed in the previous section when relating behaviour to neurochemistry.

The most widely adopted approach so far has been to utilise statistical methods to examine symptom clusters. For example, in a community-based population of 97 patients with a variety of dementias, three syndromes were identified using the Present Behavioural Examination: (1) overactivity (walking more or aimlessly, trailing the carer or checking where the carer was); (2) aggressive behaviour (physical aggression, aggressive resistance, verbal aggression); and (3) psychosis (anxiety, persecutory ideas and hallucinations).[50] Depression segregated as a separate factor, independently of the others. In another study of the prevalence and pattern of clustering of psychiatric symptoms in a representative sample of 198 persons with Alzheimer's disease, a latent class analysis indicated that the patients could be classified into three groups: (1) patients with no psychiatric symptom or with a monosymptomatic presentation (such as apathy, delusions, agitation) disturbance as measured by the NPI (40% of the sample); (2) a group of patients who exhibited a predominantly affective syndrome (28%); and (3) a psychotic syndrome (13%).[51] The results of these studies are broadly similar, but consensus is required to establish what constitute the main psychiatric syndromes, an essential process to inform meaningful treatment studies. Based upon the cluster analysis literature Ballard et al proposed three main syndromes, agitation (including aggression and restlessness), psychosis and mood disorders (including depression and anxiety),[52] although the apparent differential response of aggression and other agitation symptoms to treatment intervention perhaps indicates that they should be treated as separate syndromes.[53] In addition, the neurochemical studies indicate a different underlying pattern of cholinergic abnormalities in patients with delusions and those with visual hallucinations. Taking this on board, we are probably left with six distinct BPSD symptoms requiring separate evaluation in treatment studies: aggression, restlessness, delusions, visual hallucinations, anxiety and depression (excluding apathy which is not considered as part of the BPSD spectrum in the current review).

Evidence from clinical trials regarding the potential benefits of cholinesterase inhibitors for individual BPSD symptoms or syndromes

Of the placebo-controlled trials included in the cited Cochrane reviews,[30–32] only one of the studies specifically examines individual BPSD symptoms. Gauthier et al presented a re-analysis of the Feldman study focusing upon individual NPI items.[54] This report is particularly useful as it focuses on a group with more severe dementia (MMSE 5–17) and baseline NPI scores (mean 19), indicating that a substantial proportion of the cohort were experiencing BPSD symptoms. In this evaluation, of the BPSD symptoms, only anxiety significantly improved ($P < 0.05$), although as would be expected there was also a significant improvement in apathy. Although McKeith et al did not report the impact of rivastigmine treatment on individual NPI items in their cohort of DLB patients,[39] it is likely that improvements in visual hallucinations may have been an important

factor given the severity and persistence of this symptom in these patients. This hypothesis is supported by significant improvements in visual hallucinations in most of the open-label case series evaluating the benefit of cholinesterase inhibitors in DLB or Parkinson's disease dementia (e.g. Aarsland et al,[55] Shea et al[56]). Further evidence from placebo-controlled studies in AD is available considering earlier reports pertaining to tacrine, the unlicensed cholinesterase inhibitors metrifonate and velnacrine, and a pilot study with rivastigmine.[57–60] Raskind et al presented a re-analysis of pooled data from randomised placebo-controlled trials of tacrine, focusing upon patients who had received the maximum dose.[57] Both delusions (clinically significant improvement 67% tacrine versus 41% placebo) and pacing (clinically significant improvement 70% tacrine versus 54% placebo) significantly improved with tacrine treatment compared to placebo. In a further placebo-controlled study of another cholinesterase inhibitor (metrifonate), 273 patients with probable AD were randomised to active treatment and 135 to placebo.[58] Over the 26 weeks of treatment there was a significant treatment advantage for metrifonate with respect to visual hallucinations ($P = 0.02$), but not for the other NPI items. Antuono et al reported that patients treated with velnacrine were less likely to have emergent periods of agitation in the course of the 24-week study (1% versus 4% for patients who received placebo).[59] Our own group has just completed a pilot placebo-controlled trial comparing rivastigmine with placebo in AD patients with clinically significant agitation (31 rivastigmine, 31 placebo).[60] Over 26 weeks, the rivastigmine-treated patients had a non-significant 2-point advantage over placebo on the Cohen-Mansfield agitation inventory (CMAI), which improved to a 5-point advantage for patients with baseline CMAI scores >45.

Some additional evidence is also available from small comparator or crossover studies. Cummings and colleagues used a double-blind, active-comparator study design to investigate the relative effects of physostigmine and haloperidol in two patients with AD and psychosis.[61] Physostigmine reduced delusions and hallucinations in both patients without effects on their mood. The same group also reported a double-blind crossover trial of haloperidol and physostigmine in 13 patients with severe AD and significant behavioural disturbances.[62] Psychosis and agitation scores on the Behavioral Pathology in Alzheimer's Disease Rating Scale declined comparably in response to treatment with the two agents. In further study of long-acting, controlled-release physostigmine, Thal and colleagues found that agitation was observed in 50% fewer patients treated with the active agent than with placebo.[63]

Although somewhat inconsistent between studies, there is a building body of evidence to suggest potential benefit of cholinesterase inhibitors for the treatment of agitation (with one study indicating particular benefit for restlessness), which is consistent with scientific hypotheses generated from neurochemical studies.[64] The potential benefits with respect to other BPSD symptoms are less clear, although the impact upon anxiety in AD and visual hallucinations in DLB clearly merits further study.

Conclusion

Given the potentially harmful effects of standard pharmacological approaches for the treatment of BPSD, developing safer alternatives is a priority. The possibility that cholinesterase inhibitors may improve BPSD symptoms is therefore exciting. The current level of evidence, although sufficient to emphasise the priority of further studies does not allow any firm conclusions about the current place of cholinesterase inhibitors in the treatment of BPSD; with the possible exception of treating these symptoms in DLB patients, where cholinesterase inhibitor therapy is preferable given the considerable risk of serious detrimental consequences with neuroleptic therapy and the lack of clear evidence for any other pharmacological treatment approaches. In AD, the role of cholinesterase inhibitors in the treatment of BPSD can only be resolved by specific studies focusing upon people with specific clinically significant symptoms. The availability of preliminary data from clinical trials and initial neurochemical studies suggests that agitation is the most promising potential treatment indication, although it will be important to also undertake studies to evaluate the potential benefits for the treatment of visual hallucinations, delusions and anxiety in dementia patients. The main caveat from currently completed studies, is the time course of treatment effects, with most studies that indicate potential benefit suggesting that this only becomes evident over 6 months of treatment. This may not be a realistic time frame for the treatment of people with severe BPSD symptoms.

References

1. Finkel SI, Costa e Silva J, Cohen G, Miller S, Sartorius N. Behavioral and psychological signs and symptoms of dementia: a consensus statement on current knowledge and implications for research and treatment. Int Psychogeriatr 1996; 8(Suppl 3):497–500.

2. Howard R, Ballard C, O'Brien J, Burns A. UK and Ireland Group for Optimization of Management in dementia. Guidelines for the management of agitation in dementia. Int J Ger Psychiatry 2001; 16:714–17.

3. Margallo-Lana M, Swann A, O'Brien J et al. Prevalence and pharmacological management of behavioural and psychological symptoms amongst dementia sufferers living in care environments. Int J Geriatr Psychiatry 2001; 16:39–44.

4. Ballard C, Holmes C, McKeith I et al. Psychiatric morbidity in dementia with Lewy bodies: a prospective clinical and neuropathological comparative study with Alzheimer's disease. Am J Psychiatry 1999; 156:1039–45.

5. Ballard C, O'Brien J, James I et al. Quality of life for people with dementia living in residential and nursing home care: the impact of performance on activities of daily living, behavioral and psychological symptoms, language skills, and psychotropic drugs. Int Psychogeriatr 2001; 13:93–106.

6. Rabins PV, Mace NL, Lucas MJ. The impact of dementia on the family. JAMA 1982; 248:333–5.

7. Steele C, Rovner B, Chase GA, Folstein M. Psychiatric symptoms and nursing home placement of patients with Alzheimer's disease. Am J Psychiatry 1990; 147:1049–51.

8. Ballard C, O'Brien J, Coope B et al. A prospective study of psychotic symptoms in dementia sufferers: psychosis in dementia. Int Psychogeriatr 1997; 9:57–64.

9. Ballard CG, Patel A, Solis M, Lowe K, Wilcock G. A one year follow up study of depression in dementia sufferers. Br J Psychiatry 1996; 68:287–91.

10. Ballard C, O'Brien J, Swann A, Fossey J, Lana M. A One Year Follow-up study of behavioural and psychological symptoms in dementia (BPSD) amongst people in care environments. J Clin Psychiatry 2001; 62:631–6.

11. Haupt M, Kurz A, Janner M. A 2-year follow-up of behavioural and psychological symptoms in Alzheimer's disease. Dement Geriatr Cogn Disord 2000; 11:147–52.

12. Eustace A, Coen R, Walsh C et al. A longitudinal evaluation of behavioural and psychological symptoms of probable Alzheimer's disease. Int J Geriatr Psychiatry 2002; 17:968–73.

13. Cohen-Mansfield J, Werner P. Management of verbally disruptive behaviors in nursing home residents. J Gerontol A Biol Sci Med Sci 1997; 52:M369–77.

14. Teri L, Gibbons LE, McCurry SM et al. Exercise plus behavioral management in patients with Alzheimer disease: a randomized controlled trial. JAMA 2003; 290:2015–22.

15. Allen-Burge R, Stevens AB, Burgio LD. Effective behavioral interventions for decreasing dementia-related challenging behavior in nursing homes. Int J Geriatr Psychiatry 1999; 14:213–28.

16. McGrath AM, Jackson GA. Survey of prescribing in residents of nursing homes in Glasgow. BMJ 1996; 314:611–12.

17. Ballard CG, O'Brien J. Pharmacological treatment of behavioural and psychological signs in Alzheimer's disease: how good is the evidence for current pharmacological treatments? BMJ 1999; 319:138–9.

18. Schneider LS, Pollock VE, Lyness SA. A metaanalysis of controlled trials of neuroleptic treatment in dementia. J Am Geriatr Soc 1990; 38:553–63.

19. Reilly JG, Ayis SA, Ferrier IN, Jones SJ, Thomas SH. QTC-interval abnormalities and psychotropic drug therapy in psychiatric patients. Lancet 2000; 355:1048–52.

20. McKeith I, Fairbairn A, Perry R, Thompson P, Perry E. Neuroleptic sensitivity in patients with senile dementia of Lewy body type. BMJ 1992; 305:673–8.

21. McShane R, Keene J, Gedling K et al. Do neuroleptic drugs hasten cognitive decline in dementia? Prospective study with necropsy follow up. BMJ 1997; 314:266–70.

22. Chen CP, Alder JT, Bowen DM et al. Presynaptic serotonergic markers in community-acquired cases of Alzheimer's disease: correlations with depression and neuroleptic medication. J Neurochem 1996; 66:1592–8.

23. Court JA, Piggott MA, Lloyd S et al. Nicotine binding in human striatum: elevation in schizophrenia and reductions in dementia with Lewy bodies, Parkinson's disease and Alzheimer's disease; and in relation to neuroleptic medication. Neuroscience 2000; 98:79–87.

24. Shorr R, Fought RL, Ray WA. Changes in antipsychotic drug use in Nursing homes during implementation of the OBRA-87 regulations. J Am Med Assoc 1994; 271:358–62.

25. Chief Medical Officer. Current problems in Pharmacology. CSM update 1994; May 20–26.

26. Katz IR, Jeste DV, Mintzer JE et al. Comparison of risperidone and placebo for psychosis and behavioral disturbances associated with dementia: a randomized, double-blind trial. Risperidone Study Group. J Clin Psychiatry 1999; 60:107–15.

27. De Deyn PP, Rabheru K, Rasmussen A et al. A randomized trial of risperidone, placebo, and haloperidol for behavioral symptoms of dementia. Neurology 1999; 53:946–55.

28. Street JS, Clark WS, Gannon KS et al. Olanzapine treatment of psychotic and behavioral symptoms in patients with Alzheimer disease in nursing care facilities: a double-blind, randomized, placebo-controlled trial. The HGEU Study Group. Arch Gen Psychiatry 2000; 57:968–76.

29. Brodaty H, Ames D, Snowdon J et al. A randomized placebo-controlled trial of risperidone for the treatment of aggression, agitation, and psychosis of dementia. J Clin Psychiatry 2003; 64:134–43.

30. Birks J, Grimley Evans J, Iakovidou V, Tsolaki M. Rivastigmine for Alzheimer's disease. Cochrane Dementia and Cognitive Improvement Group. Cochrane Database Syst Rev 2005: CD 004744.

31. Olin J, Schneider L. Galantamine for Alzheimer's disease. Cochrane Dementia and Cognitive Improvement Group. Cochrane Database Syst Rev 2003; 3: CD001747.

32. Birks JS. Harvey R. Donepezil for dementia due to Alzheimer's disease. Cochrane Dementia and Cognitive Improvement Group. Cochrane Database Syst Rev 2003; 3: CD001190.

33. McLaren AT, Allen J, Murray A, Ballard CG, Kenny RA. Cardiovascular effects of donepezil in patients with dementia. Dement Geriatr Cogn Disord 2003; 15:183–8.

34. Wild R, Pettit T, Burns A. Cholinesterase inhibitors for dementia with Lewy bodies. Cochrane Dementia and Cognitive Improvement Group. Cochrane Database Syst Rev 2003; 3: CD003672.

35. Feldman H, Gauthier S, Hecker J et al. Donepezil MSAD Study Investigators Group. A 24-week, randomized, double-blind study of donepezil in moderate to severe Alzheimer's disease. Neurology 2001; 57:613–20.

36. Tariot PN, Cummings JL, Katz IR et al. A randomized, double-blind, placebo-controlled study of the efficacy and safety of donepezil in patients with Alzheimer's disease in the nursing home setting. J Am Geriatr Soc 2001; 49:1590–9.

37. Tariot PN, Solomon PR, Morris JC. A 5-month, randomized, placebo-controlled trial of galantamine in AD. The Galantamine USA-10 Study Group. Neurology 2000; 54:2269–76.

38. Rockwood K, Mintzer J, Truyen L, Wessel T, Wilkinson D. Effects of a flexible galantamine dose in Alzheimer's disease: a randomised, controlled trial. J Neurol Neurosurg Psychiatry 2001; 71:589–95.

39. McKeith I, Del Ser T, Spano P et al. Efficacy of rivastigmine in dementia with Lewy bodies: a randomised, double-blind, placebo-controlled international study. Lancet 2000; 356:2031–6.

40. Mega MS, Masterman DM, O'Connor SM, Barclay TR, Cummings JL. The spectrum of behavioral responses to cholinesterase inhibitor therapy in Alzheimer disease. Arch Neurol 1999; 56:1388–9.

41. Perry EK, Marshall E, Kerwin J et al. Evidence of a monoaminergic-cholinergic imbalance related to visual hallucinations in Lewy body dementia. J Neurochem 1990; 55:1454–6.

42. Ballard C, Piggott M, Johnson M et al. Delusions associated with elevated muscarinic binding in dementia with Lewy bodies. Ann Neurol 2000; 48:868–76.

43. Court JA, Ballard CG, Piggott MA et al. Visual hallucinations are associated with lower alpha bungarotoxin binding in dementia with Lewy bodies. Pharmacol Biochem Behav 2001; 70:571–9.

44. Lai MK, Lai OF, Keene J. Psychosis of Alzheimer's disease is associated with elevated muscarinic M2 binding in the cortex. Neurology 2001; 57:805–11.

45. Holroyd S, Rabins PV, Finkelstein D, Lavrisha M. Visual hallucinations in patients from an ophthalmology clinic and medical clinic population. J Nerv Ment Dis 1994; 182:273–6.

46. Chapman FM, Dickinson J, McKeith I, Ballard C. Association among visual hallucinations, visual acuity, and specific eye pathologies in Alzheimer's disease: treatment implications. Am J Psychiatry 1999; 156:1983–5.

47. Minger SL, Esiri MM, McDonald B et al. Cholinergic deficits contribute to behavioral disturbance in patients with dementia. Neurology 2000; 55:1460–7.

48. Jeste DV, Finkel SI. Psychosis of Alzheimer's disease and related dementias. Diagnostic criteria for a distinct syndrome. Am J Geriatr Psychiatry 2000; 8:29–34.

49 Lyketsos CG, Breitner JC, Rabins PV. An evidence-based proposal for the classification of neuropsychiatric disturbance in Alzheimer's disease. Int J Geriatr Psychiatry 2001; 16:1037–42.

50 Hope T, Keene J, Fairburn C, McShane R, Jacoby R. Behaviour changes in dementia. 2: Are there behavioural syndromes? Int J Geriatr Psychiatry 1997; 12:1074–8.

51. Lyketsos CG, Sheppard JM, Steinberg M et al. Neuropsychiatric disturbance in Alzheimer's disease clusters into three groups: the Cache County study. Int J Geriatr Psychiatry 2001; 16:1043–53.

52. Ballard CG, Bannister CL, Patel A et al. Classification of psychotic symptoms in dementia sufferers. Acta Psychiatr Scand 1995; 92:63–8.

53. Lonergan E, Luxenberg J, Colford J. Haloperdol for Agitation in Dementia. Cochrane Database Syst Rev 2002: CD002852.

54. Gauthier S, Feldman H, Hecker J et al. Donepezil MSAD Study Investigators Group. Efficacy of donepezil on behavioral symptoms in patients with moderate to severe Alzheimer's disease. Int Psychogeriatr 2002; 14:389–404.

55. Aarsland D, Hutchinson M, Larsen JP. Cognitive, psychiatric and motor response to galantamine in Parkinson's disease with dementia. Int J Ger Psychiatry 2003; 18:937–41.

56. Shea C, MacKnight C, Rockwood K. Donepezil for treatment of dementia with Lewy bodies: a case series of nine patients. Int Psychogeriatr 1998; 10:229–38.

57. Raskind MA, Sadowsky CH, Sigmund WR et al. Effect of Tacrine on language, praxis and noncognitive behavioural problems in Alzheimer's disease. Arch Neurol 1997; 54:836–40.

58. Morris JC, Cyrus PA, Orazem J et al. Metrifonate benefits cognitive, behavioral, and global function in patients with Alzheimer's disease Neurology 1998; 50:1222–30.

59. Antuono PG (Mentane Study Group): effectiveness and safety of velnacrine for the treatment of Alzheimer's disease: a double-blind, placebo-controlled study. Arch Intern Med 1995; 155: 1766–72.

60. Ballard C, Lana M, Douglas S, O'Brien JA. Placebo controlled trial of rivastigmine and quetiapine for the treatment of agitation in Alzheimer's disease. Presented at the IPA Expert Forum for Agitation, Lisbon 2003.

61. Cummings JL, Gorman DG, Shapira J. Physostigmine ameliorates the delusions of Alzheimer's disease. Biol Psychiatry 1993; 33:536–41.

62. Gorman DG, Read S, Cummings JL. Cholinergic therapy of behavioral disturbances in Alzheimer's disease. Neuropsychiatry Neuropsychol Behav Neurol 1993; 6:229–34.

63. Thal LJ, Ferguson JM, Mintzer J, Raskin A, Targum SD. A 24-week randomized trial of controlled-release physostigmine in patients with Alzheimer's disease. Neurology 1999; 52:1146–52.

64. Trinh NH, Hoblyn J, Mohanty S, Yaffe K. Efficacy of cholinesterase inhibitors in the treatment of neuropsychiatric symptoms and functional impairment in Alzheimer disease: a meta-analysis. JAMA 2003; 289:210–6.

11

Quality of Life Outcomes

Kenneth Rockwood

This chapter reviews concepts about quality of life measurement in the setting of anti-dementia drug trials for Alzheimer's disease. Often, quality of life was not measured specifically in the early trials; rather, inferences were made that if functional activity was better, then quality of life must be better too. Over time, more specific quality of life measures for Alzheimer's disease have been introduced, although the claims made for an impact on quality of life can still require a many-staged argument, with inferences from other measures.

Quality of life measurement in AD is problematic because of the nature of the construct, its reliance on some notion of the continuity of self in expression of future preferences, and the need to involve caregivers, whose own views about quality of life are susceptible to a variety of influences. Pragmatically, these objections must be overcome; not to measure quality of life will mean that it is ignored. To measure it wrongly, however, means that outcomes can be perverse. On balance, disease-specific quality of life measures that take the stage of dementia into account, as well as individualised measures that assess patient and caregiver preferences will be important. Not everyone needs to adhere to the same view of quality of life in order to appreciate it in their own lives; similarly, not everyone need adhere to the same construct of quality of life in order for it to be measured.

Introduction

That the quality of life of people with Alzheimer's disease must be considered does not mean that it is clear how to consider it. Difficulties arise both at the ontological level – getting to grips with what it is – and at the epistemological level – knowing how to measure it. In philosophy, going from how things are to how we might apprehend them, and then what to do, is to traverse well-trodden ground. Typically, the journey begins by clearing out of all the prior ontological underbrush, to discover a supposedly firm base on which all else might be constructed. That approach was undermined in the 20th century, by Wittgenstein and others, who showed how our language tricks us into believing things that are unique to us (and in that sense, not otherwise true). Still, philosophers philosophise after Wittgenstein, who if not exactly bested, can at least be compartmentalised, so that his objections notwithstanding, even in ontology it is possible to advance in knowing how it is that the world is made up of the things that comprise it.

This analogy seems to me to hold to a reasonable extent in understanding quality of life in dementia. Typically, a new contribution to the prodigious quality of life literature – there are over 1000 quality of life instruments[1] – begins by first stating what is missing in what has gone before, and proceeds by making a special expertise claim that provides the necessary solid ground. The special claim comes in many forms – from the now mostly outmoded consensus of experts to the more current – if usually shockingly ahistorical – systematic review. On this, with varying levels of methodological sophistication, the new measure is built, tested and launched.

In this chapter, we can begin somewhat ahistorically as well. First, after a very brief overview of challenges, and because of its special considerations, we can consider, as an ontologist might do a thought experiment, the case of severe dementia. Can a claim for quality of life and its measurement be made there? If so, what might we learn from it that can be applied to more mild disease? Next, we will give some brief historical perspective over the short course of modern anti-dementia trials. From this, we will move to some more substantive considerations of the problem of quality of life measurement. The chapter will conclude with the argument encapsulated above: we must measure quality of life in further anti-dementia drug trials, but we must ensure that our measures do no violence to the notion that individuals can measurably express preferences for how they would live their lives, and that by describing these preferences we can make inferences about their quality of life, even though their views will not always be the same.

The formulation of 'quality of life' as a focus for measurement in health and in dementia has been criticised on a number of grounds. Elsewhere I have objected both to its strongly reductionist bent, and to the idea that, as a focus for measurement, it detracts from the management of people whose issues do not conform to the average case.[2] Jennings has usefully drawn attention to some of the objectionable aspects of the use of the term 'quality of life'.[3] He points out that in many societies the term grates because it seems to have a strongly judgemental bent. Although he ultimately endorses being concerned about quality of life, and thus measuring it, his essay is exemplary in pointing out some of the otherwise unexamined

philosophical issues at play in attempts to measure quality of life as 'QoL'. In like spirit, I propose to simply get on with considering practical issues in quality of life measurement in drug trials.

Obviously, people with dementia can have difficulty in describing their own quality of life, and making judgements about it.[4] As will be detailed, although one has dominated, there have been several approaches to this problem. A phenomenological approach has received advocacy, but comparatively little attention.[5–7] Another approach is to ask caregivers to substitute their judgements. A third approach is to have caregivers describe how they evaluate the person's quality of life, and then how they think that the person perceives it themselves.[8] Most commonly, however, inferences are made based on ratings in multiple domains, which amongst the dominant factor is the observed behaviour of the patient.[9–11]

Quality of life and its measurement in advanced dementia

An interesting place to start with some of the dilemmas in understanding quality of life in dementia is where they are most evident, and that is in patients with severe dementia. Even though this area has only more recently been the subject of clinical trials of both anti-dementia drugs,[12–15] and non-pharmacological interventions,[16] there remains much other evidence that can be brought to bear. It is not hard to find people outside of dementia care who are sceptical about the whole notion of quality of life as a valid construct in dementia. And yet, for those who provide such care, the idea of quality of life is inherent. Volicer, for example, in considering the morass that can sometimes be end-of-life care in dementia, offers a prescriptive approach as a way out: make meaningful activity available; 'optimally' manage medical issues, and provide appropriate treatment of psychiatric symptoms, each done with due consideration to quality of life.[17] From this we clearly can conclude that quality of life still exists in severe dementia (Volicer does not view its existence as problematic, and assumes that neither will the reader), its consideration can guide the conduct of others, and intelligent people of good will can make claims about it. But will their claims agree? If they do not, will there be some non-arbitrary grounds by which these claims might be reconciled, or perhaps even adjudicated?

A useful way of conceptualising the measurement issue is between inferences about quality of life that can be made from patients' self-reports, and those that would be inferred from their behaviour, or even from the environment in which they find themselves. This is often characterised as 'objective' measurement (such as observations about the ability to perform basic functions) compared with the supposedly subjective accounts of patients.[18] Some commentators raise 'serious concerns' about the high cognitive load of some subjective rating instruments, and this would nowhere be more evident than in severe dementia.[19] If inferences about patients' quality of life in such circumstances are not just to be made from observations, what are the alternatives? One framework is that of advance directives: patients faced with the prospect of deterioration identify certain milestones ahead of time which they judge, in their present state, to be unacceptable to them.[20] Such an approach is not unproblematic: it is difficult to consider all possible outcomes (this may be a particular problem in Alzheimer's disease, where even in mild dementia, patients can have difficulty conceiving of themselves as agents for future events[21]) and preferences can change over time.

An important consideration evident from the literature on severe dementia in particular, but also on end-of-life care in dementia, is that, while quality of life needs to be considered, reconciliation of competing points of view will not be possible, as many attitudes are culturally determined (or, at least, cross-cultural differences are demonstrable).[22,23] In this context, even the approach to provide a structured process to decision making will be incompatible to those whose belief system does not allow them to engage in the process.[24] Similarly, as reviewed in detail elsewhere,[25] the experience of the SUPPORT study lends credence to the idea that structured processes for the assessment of complex issues will not result in their actual implementation, something that others too have found in dementia.[22,26] Particularly in considering these problems, the Fairhill guidelines on ethics of the care of people with Alzheimer's disease proposed that, 'in contrast with a theoretical and deductive approach to ethics' in this area, what is needed more is 'an inductive method [that] begins with attentive listening to the voices of the affected population and family members'.[27] In addition to knowing how best to conceptualise quality of life, there are conventional methodological issues, such as the validity of inferring quality of life from behaviour, and the responsiveness (sensitivity to change) of measures now in use.[4,28]

Still, the challenge must be faced. In a remarkably insightful study, that nevertheless relied on standardised instruments and structured questionnaires, Schultz et al. evaluated end-of-life care and its impact on family member caregivers of people with Alzheimer's disease, the bulk of whom showed advanced dementia.[29] They found that caregiving was highly stressful, although formal depressive symptoms showed much recovery after the death of the person with AD. Of interest, 72% of caregivers reported that the death was a relief to them, and more than 90% that it was a relief to the patient. Clearly, if anti-dementia treatment is to have some impact on this experience, it has much opportunity for demonstrable benefit.

The case can also be made that quality of life deserves measurement in advanced dementia, if only because things that are not measured tend to be ignored. For example, many nursing homes for elderly people are disgracefully inelegant in their design, a factor that appears to have little impact on their ability to be fully accredited by licensing authorities. On the other hand, innovative health designs can help alleviate behavioural problems in late-stage Alzheimer's disease.[30] Quality of life has also been a concern for treating patients with severe dementia who have behavioural and psychological symptoms.[31] It is easy to imagine making people less 'aggressive' or less 'resistive to care', but at the expense of full consciousness, such that they are unable to engage fully in their lives, whose quality is diminished in this way.[32] Such a concern appears to underlie the stance of the United States Food and Drug Administration that treatment of behavioural problems should show specific effects on behaviour, without the so-called pseudospecific effect of diminution of problems though sedation.

Even this brief overview makes clear that those who work with people with severe dementia recognise the validity of the quality of life construct. The concept informs their decision making and guides action. While quality of life in severe dementia is generally greatly compromised, grades in quality of life can be recognised. It appears to make an important target for therapeutic interventions, both pharmacological and non-pharmacological. But the question remains: how should quality of life be measured?

A growing interest in quality of life measurement in dementia

Quality of life has become an important focus for dementia trials. For example, cholinesterase inhibitors (ChEIs) are reported to have good effects on the quality of life of both patients and their caregivers,[33] and are cited as a benefit of treatment.[34] Quality of life scales are also being used as outcome measures in trials of other agents (such as neuroleptics),[31] or non-steroidal anti-inflammatory drugs,[35] and in non-pharmacological treatment, such as cognitive stimulation therapy,[36] or caregiver support.[37]

The impact of anti-dementia drug treatments on quality of life also has been disputed. For example, the 2003 Cochrane review on donepezil in dementia, while finally conceding that treatment might be useful clinically, disputes whether it is worth the price of the drugs.[38] It also concludes that the early instrumentation for quality of life is problematic. ('Although no significant changes were measured on patient-rated quality of life scales, the instrument used was crude and possibly unsuited to the task.'[38])

Economic considerations are not unrelated to quality of life measurement. An important prompt to more systematic quality of life measurement came from the desire to conduct pharmacoeconomic studies, where such measures played into so-called 'utility' analyses.[39,40] For now, it appears that the field of quality of life in dementia will be a new battleground on which those sceptical about the merit of present treatment and those more enthusiastic for it can continue to clash.[41–44]

The interest in quality of life as an outcome measure reflects both societal attitudes, and the changing quality of life metric. For example, some early studies tended, with varying degrees of explicitness, to view quality of life essentially as equivalent to the ability to perform activities of daily living.[45,46] Others included patient and caregiver quality of life as secondary measures, although, again reflecting the standards of the time, these often were crude, even single item scales.[47,48] By contrast, the most recent measures tend to be multi-factorial, and to carry out assessments of measurement properties according to widely accepted standards. Still, the tension between 'subjective' and 'objective' accounts is generally resolved in favour of the latter.[18,19,49,50]

The Quality of Life – Alzheimer's Disease (QoL-AD) represents an example of the current state of the art of quality of life scale development.[1,51] In general, the importance of the framework is emphasised, as is the source of the information (i.e. self-report, proxy report or direct observation). Interestingly, caregivers often believe that their ratings of quality of life differ from how their loved ones perceive their own quality of life.[8] Using the framework of 'the good life',[52] Logsdon et al devised a 13-item questionnaire which provides both a patient report and a caregiver report,[51] which was then assayed for content validity by an expert panel. For construct validation, they conducted interviews with 177 AD patient–caregiver dyads in their own homes during which the QoL-AD was completed, as were several other scales that measure function and cognition, mood, enjoyment of activities and caregiver burden, which were used for subsequent correlational analyses. These correlations largely proceeded as expected, as did measures of internal consistency and inter-rater reliability. Of the 177, 22 patients could not complete the interview; these were patients with more advanced dementia, as indicated by their having Mini-Mental State Examination (MMSE) scores less than 10/30. On this basis, the authors concluded that the approach was valid for patients with MMSE scores >10. Other recent examples of QoL dementia scales that conform with accepted instrument development standards include the Cornell-Brown Scale for Quality of Life in Dementia,[53] the Dementia Quality of Life scale,[54] and the Alzheimer Disease Related Quality of Life scale.[9]

Despite this growing interest and empirical study, scepticism persists

As reviewed elsewhere, much academic scepticism about ChEIs focuses on whether treatment effects are clinically relevant, and much of this has coalesced around a 'quality of life' issue.[55] This dilemma seems to have less to do with caring about patients' quality of life than it has to do with measuring quality of life in a way that it can be incorporated into the evidence of drug treatment effectiveness. Scepticism is present at several stages of this process, from knowing that a drug is available, to believing that there is a reasonable chance that it will improve first the quality of life of patients in general, and then that of the particular patient in front of you.

As noted, the construct of quality of life in dementia seems non-problematic as a starting point. Indeed, not just in severe dementia, but in many family members at higher risk for Alzheimer's disease, simply the prospect of dementia is enough to impact adversely on their lives.[56] And yet many observers are not persuaded that we have got to grips with quality of life, even

in mild dementia. Part of the dilemma has been the methodology. As noted, it is demonstrably the case that quality of life measurement, like other complex areas (such as end-of-life care) has had a strongly reductionist bent. For example, Baldwin et al concluded that 'the research literature has been dominated by surveys and studies soliciting views on predefined issues with relatively few in-depth, open-ended qualitative studies'.[50] Similarly, Bond has pointed out that different perspectives on dementia within the biomedical, psychological and social models of disability lead to radically different meanings of the concept 'quality of life' and approaches to its assessment.[54] He called for people with dementia and their informal caregivers to be involved in the development of usable outcome measures relevant to their needs and circumstances. By contrast, the traditional psychometric stance, with its emphasis on standardisation and reliability, is inimical to the idea of local standards.

Notwithstanding the issues of cultural acceptability, or the strongly reductionist bent of quality of life measurement, or the differences in perspective on whose quality of life it is anyway, or even the ability to assay quality of life in advanced dementia, one issue which must be more widely addressed if the quality of life measurement exercise is to gain ground is that of the measures' responsiveness. Responsiveness (or sensitivity to change) is often problematic in anti-dementia trials.[57] This was certainly the case in Alzheimer's disease, where the first major study of the new era showed no impact on quality of life, despite being positive in each of the two primary outcome measures.[58]

Another challenge to the use of the evidence about quality of life in dementia is its limited scope. For example, in my experience, which extends to following patients to autopsy, the treatment by cholinesterase inhibition of patients who have dementia with Lewy bodies more commonly results in dramatic clinical improvement than in any other disorder. With this comes unequivocal improvement in quality of life. Consider, for example, a case that I saw of a man who pleaded with me to kill him. He had the persistent hallucination of his daughter being gang-raped while his family stood by and did nothing. This settled within a few days of treatment to his immense relief. Such experiences tend to stay with a physician, especially when reinforced by similar, if less dramatic cases in which good results are achieved quickly. Indeed, some years ago my colleagues and I wrote this up to convey the possibility for beneficial treatment effects in an illness which had been troublingly difficult to treat.[59] (I confess to having done so, even though I am committed to scientific investigation of dementia treatment; by mitigation, it is the only clinical series of patient treatment that I have published.) For complex reasons, only one study of ChEIs in dementia with Lewy bodies has been published,[60] so that, in general, the evidence here is held to be weak.[61] But no one who has been through the usual therapeutic success with treatment is likely to be put off by weak evidence from placebo-controlled trials, especially given the alternative. Nor are people likely to be persuaded by strong evidence, if it is not detectable in their own practice.

How to move from what one reads in the studies and expert reviews to what one sees in the clinic holds a particular challenge for understanding quality of life, which, in diseases such as dementia, in turn help shape physicians' willingness to prescribe.[62,63] Future trials must therefore pay particular attention to how quality of life is to be measured, and need to proceed by some method other than validation by the assertion of favourable psychometrics: in some way, the sensibility of claims about quality of life must be addressed, and attempts to get to grips with it should be incorporated into future studies, until a satisfactory approach has been demonstrated.

The special case of quality of life in relation to caregivers

In dementia care, few precepts are more powerful than to prevent the preventable, treat the treat-

able, and care for the caregiver. While physicians are obligated to maintain the patients as the focus of their care, they must remain mindful of the impact on caregivers. For example, sometimes physicians encounter patients who, as a result of treatment, become more aware of their losses, with an anxiety that discomfits both patients and their caregivers. At the same time, it is not rare to encounter caregivers whose motives appear not always to be consistent with the patient's apparent best interest, and sometimes to encounter others who are clearly deceitful. In consequence, a clinical trails focus on caregiver outcomes must balance many considerations, as detailed elsewhere in this volume.[64]

Three considerations are particularly important to the project of assessing quality of life in dementia clinical trials: the quality of life of caregivers themselves, and how it might be improved; the role of caregivers as informants in assaying the quality of life of patients; and the role of caregivers in choosing treatment options. While recalling more hopeful messages that the caregiving experience, though stressful, is bearable, and that most caregivers show resilience,[29] there is merit in attempting to improve caregivers' quality of life. Caregiver quality of life is likely also to be an area of therapeutic intervention if there is a measurement thrust to evaluate 'hard' endpoints such as time to institutionalisation.[65]

Indeed, trials aimed at caregivers can produce effect sizes that are comparable to those of the cholinesterase inhibitors themselves.[64] Presumably, future anti-dementia trials will explore ways to combine therapeutic and social interventions.

The use of caregivers as proxies in measuring quality of life arises naturally from the scepticism that understandably can obtain when people who have been deemed to be incompetent to make choices about their care respond to questionnaires about their preferences. Still, scepticism is no grounds not to proceed with investigation. The claim has been made that, even for patients with MMSE scores as low as 13, they are able to 'respond consistently' to questions about preferences.[66] Thus some scales in this late-stage dementia use caregivers as the basis of the patient quality of life assessments.[67] Caregiver input through observable patient behaviours has also been proposed.[68] However, the hypothesis that observable behaviours give a more objective proxy assessment of quality of life has not always been borne out.[19,69] Moreover, differences between caregivers' and patients' views of quality of life are not limited to those with severe dementia.[8] Moreover, changes in patients' status are not always reflected in changes in caregivers' quality of life.[70] These differences are not unexpected: it is well accepted that caregivers are affected by the illness, and that measurement of caregiver quality of life is a subtle process that might not be susceptible to standard health-related quality of life measures.[71,72]

Quality of life in relation to how caregivers decide about treatment decisions has been a central theme of a series of studies by Karlawish and colleagues. In a very interesting study of caregiver preferences in respect of using dementia-slowing medicines, Karlawish et al found that caregivers were more likely to forgo even risk-free disease-slowing treatment when the quality of life of the person they cared for was judged to be poor.[73] In general, caregivers for people with Alzheimer's disease appear to be willing to accept risk to achieve disease slowing.[74]

Conclusion

There is no alternative to measuring quality of life measurement in future anti-dementia drug trials. For these measures to fulfil the promise of extending the scope of clinical trials measurement into clinically meaningful outcomes, two conditions must be met. In some way, the measures must include the preferences of patients. The measures also need to reflect the preferences of caregivers. This probably is best accomplished by combining disease-specific and stage-specific quality of life measures with others that assess patient and caregiver preferences. This combination would recognise that people appreciate the quality of their own lives, whether or not they adhere to the same 'average' view of quality of life so desired by most quality of life measurement instruments.

References

1. Thorgrimsen L, Selwood A, Spector A et al. Whose quality of life is it anyway?: The validity and reliability of the Quality of Life-Alzheimer's Disease (QoL-AD) scale. Alzheimer Dis Assoc Disord 2003; 17:201–8.

2. Ikuta R, Rockwood K. Quality of life in Alzheimer's dementia. In: Gauthier S (ed). Clinical Diagnosis and Management of Alzheimer's Disease (2nd edn). London: Martin Dunitz; 2001: 293–305.

3. Jennings B. A life greater than the sum of its sensations: ethics, dementia and the quality of life. In: Albert SM, Logsdon RG (eds). Assessing the Quality of Life in Alzheimer's Disease. New York: Springer; 2000: 165–78.

4. Lyketsos CG, Gonzales-Salvador T, Chin JJ et al. A follow-up study of change in quality of life among persons with dementia residing in a long-term care facility. Int J Geriatr Psychiatry 2003; 18:275–81.

5. Sabat SR. Voices of Alzheimer's disease sufferers: a call for treatment based on personhood. J Clin Ethics 1998; 9:35–48.

6. Sabat SR. The Experience of Alzheimer's Disease: Life through a Tangled Veil. Oxford: Blackwell; 2001.

7. Widdershoven GAM, Widdershoven-Heerding I. Understanding dementia: a hermeneutic perspective. In: Fulford KWM, Morris K, Sadler J, Stanghellini G. Nature and Narrative. An Introduction to the New Philosophy of Psychiatry. Oxford: Oxford University Press; 2003: 103–11.

8. Karlawish JH, Casarett D, Klocinski J, Clark CM. The relationship between caregivers' global ratings of Alzheimer's disease patients' quality of life, disease severity, and the caregiving experience. J Am Geriatr Soc 2001; 49:1066–70.

9. Rabins PV, Kasper JD, Kleinmann L et al. Concepts and methods in the development of the ADRQL: an instrument for assessing health-related quality of life in persons with Alzheimer's disease. J Mental Health Aging 1999; 5:33–48.

10. Albert SM, Logsdon RG (eds). Assessing Quality of Life in Alzheimer's Disease. New York: Springer; 2000.

11. Burgener S, Twigg P. Relationships among caregiver factors and quality of life in care recipients with irreversible dementia. Alzheimer Dis Assoc Disord 2002; 16:88–102.

12. Feldman H, Gauthier S, Hecker J et al. A 24-week, randomized, double-blind study of donepezil in moderate to severe Alzheimer's disease. Neurology 2001; 57:613–20.

13. Tariot PN, Cummings JL, Katz IR et al. A randomized, double-blind, placebo-controlled study of the efficacy and safety of donepezil in patients with Alzheimer's disease in the nursing home setting. J Am Geriatr Soc 2001; 49:1590–9.

14. Reisberg B, Doody R, Stoffler A et al. Memantine Study Group. Memantine in moderate-to-severe Alzheimer's disease. N Engl J Med 2003; 348:1333–41.

15. Tariot PN, Farlow MR, Grossberg GT et al, Memantine Study Group. Memantine treatment in patients with moderate to severe Alzheimer disease already receiving donepezil: a randomized controlled trial. JAMA 2004; 291:317–24.

16. Politis AM, Vozzella S, Mayer LS et al. A randomized, controlled, clinical trial of activity therapy for apathy in patients with dementia residing in long-term care. Int J Geriatr Psychiatry 2004; 19:1087–94.

17. Volicer L. Management of severe Alzheimer's disease and end-of-life issues. Clin Geriatr Med 2001; 17:377–91.

18. Weyerer S, Schäufele M. The assessment of quality of life in dementia. Int Psychogeriatr 2003; 15:213–18.

19. Coucill W, Bryan S, Bentham P, Buckley A, Laight A. EQ-5D in patients with dementia: an investigation of inter-rater agreement. Med Care 2001; 39:760–71.

20. Singer PA, Thiel EC, Naylor CD et al. Life-sustaining treatment preferences of hemodialysis patients: implications for advance directives. J Am Soc Nephrol 1995; 6:1410–7.

21. Rockwood K, Wallack M, Tallis R. The treatment of Alzheimer's disease: success short of cure. Lancet Neurol 2003; 2:630–3.

22. Mitchell SL, Berkowitz RE, Lawson FM, Lipsitz LA. A cross-national survey of tube-feeding decisions in cognitively impaired older persons. J Am Geriatr Soc 2000; 48:391–7.

23. Chee YK, Levkoff SE. Culture and dementia: accounts by family caregivers and health professionals for dementia-affected elders in South Korea. J Cross Cult Gerontol 2001; 16:111–25.

24. Karlawish JH, Quill T, Meier DE. A consensus-based approach to providing palliative care to patients who lack decision-making capacity. ACP-ASIM End-of-Life Care Consensus Panel. American College of Physicians-American Society of Internal Medicine. Ann Intern Med 1999; 130:835–40.

25. Rockwood K, Powell C. End-of-life decision-making. In: Marrie TJ (ed). Community-Acquired Pneumonia. New York: Raven Press; 2001, Chapter 13: 205–38.

26. Schonwetter RS, Walker RM Solomon M, Indurkhya A, Robinson BE. Life values, resuscitation preferences, and the applicability of living wills in an older population. J Am Geriatr Soc 1996; 44:954–8.

27. Post SG, Whitehouse PJ. Fairhill guidelines on ethics of the care of people with Alzheimer's disease: a clinical summary. Center for Biomedical Ethics, Case Western Reserve University and the Alzheimer's Association. J Am Geriatr Soc 1995; 43:1423–9.

28. Albert SM, Jacobs DM, Sano M et al. Longitudinal study of quality of life in people with advanced Alzheimer's disease. Am J Geriatr Psychiatry 2001; 9:160–8.

29. Schulz R, Mendelsohn AB, Haley WE et al. End-of-life care and the effects of bereavement on family caregivers of persons with dementia. N Engl J Med 2003; 349:1936–42.

30. Zeisel J, Silverstein NM, Hyde J et al. Environmental correlates to behavioral health outcomes in Alzheimer's special care units. Gerontologist 2003; 43:697–711.

31. Fontaine CS, Hynan LS, Koch K et al. A double-blind comparison of olanzapine versus risperidone in the acute treatment of dementia-related behavioral disturbances in extended care facilities. J Clin Psychiatry 2003; 64:726–30.

32. Volicer L, Hurley AC. Management of behavioural symptoms in progressive degenerative dementias. J Gerontol A Biol Sci Med Sc 2003; 58:M837–45.

33. Wynn ZJ, Cummings JL. Cholinesterase inhibitor therapies and neuropsychiatric manifestations of Alzheimer's disease. Dement Geriatr Cogn Disord 2004; 17:100–8.

34. DeKosky S. Early intervention is key to successful management of Alzheimer disease. Alzheimer Dis Assoc Disord 2003; 17(Suppl 4):S99–104.

35. Aisen PS, Schafer KA, Grundman M et al. Effects of rofecoxib or naproxen vs placebo on Alzheimer disease progression: a randomized controlled trial. JAMA 2003; 289:2819–26.

36. Spector A, Thorgrimsen L, Woods B et al. Efficacy of an evidence-based cognitive stimulation therapy programme for people with dementia: randomised controlled trial. Br J Psychiatry 2003; 183:248–54.

37. Kurz X, Scuvee-Moreau J, Rive B, Dresse A. A new approach to the qualitative evaluation of functional disability in dementia. Int J Geriatr Psychiatry 2003; 18:1050–5.

38. Birks JS, Harvey R. Donepezil for dementia due to Alzheimer's disease. Cochrane Database Syst Rev 2003; 3:CD001190.

39. Whitehouse PJ, Winblad B, Shostak D et al. First International Pharmacoeconomic Conference on Alzheimer's Disease: report and summary. Alzheimer Dis Assoc Disord 1998; 12:263–5.

40. Kurz X, Scuvee-Moreau J, Vernooij-Dassen M, Dresse A. Cognitive impairment, dementia and quality of life in patients and caregivers. Acta Neurol Belg 2003; 103:24–34.

41. Winblad B, Brodaty H, Gauthier S et al. Pharmacotherapy of Alzheimer's disease: is there a need to redefine treatment success? Int J Geriatr Psychiatry 2001; 16:653–66.

42. Clegg A, Bryant J, Nicholson T et al. Clinical and cost-effectiveness of donepezil, rivastigmine, and galantamine for Alzheimer's disease. A systematic review. Int J Technol Assess Health Care 2002; 18:497–507.

43. Thal LJ. How to define treatment success using cholinesterase inhibitors. Int J Geriatr Pyschaitry 2001; 17: 388–9.

44. Trinh NH, Hoblyn J, Mohanty S, Yaffe K. Efficacy of cholinesterase inhibitors in the treatment of neuropsychiatric symptoms and functional impairment in Alzheimer disease: a meta-analysis. JAMA 2003; 289:210–16.

45. Wilcock GK, Surmon DJ, Scott M et al. An evaluation of the efficacy and safety of tetrahydroaminoacridine (THA) without lecithin in the treatment of Alzheimer's disease. Age Ageing 1993; 22:316–24.

46. Rockwood K, Beattie BL, Eastwood MR et al. Randomized, controlled trial of linopirdine in the treatment of Alzheimer's disease. Can J Neurol Sci 1997; 24:140–5.

47. Knapp MJ, Knopman DS, Solomon PR et al. A 30-week randomized controlled trial of high-dose tacrine in patients with Alzheimer's disease. The Tacrine Study Group. JAMA 1994; 271:985–91.

48. Rogers SL, Friedhoff LT. The efficacy and safety of donepezil in patients with Alzheimer's disease: results of a US multicentre, randomized, double-blind, placebo-controlled trial. The Donepezil Study Group. Dementia 1996; 7:293–303.

49. Bond J. Quality of life for people with dementia: approaches to the challenge of measurement Aging Society 1999; 19:561–79.

50. Baldwin C, Hughes J, Hope T, Jacoby R, Ziebland S. Ethics and dementia: mapping the literature by bibliometric analysis. Int J Geriatr Psychiatry 2003; 18:41–54.

51. Logsdon RG, Gibbons LE, McCurry SM, Teri L. Assessing quality of life in older adults with cognitive impairment. Psychosom Med 2002; 64:510–19.

52. Lawton MP. A multidimensional view of quality of life in frail elders. In: Birren JE, Lubben JE, Rowe JC, Deutchman DE (eds). The Concept and Measurement of Quality of Life in the Frail Elderly. New York: Academic Press; 1991.

53. Ready RE, Ott BR, Grace J, Fernandez I. The Cornell-Brown Scale for Quality of Life in Dementia. Alzheimer Dis Assoc Disord 2002; 16:109–15.

54. Brod M, Stewart AL, Sands L, Walton P. Conceptualisation and measurement of quality of life in dementia: the dementia quality of life instrument (DQoL). Gerontologist 1999; 39:25–35.

55. Rockwood K, MacKnight C. Assessing the clinical importance of statistically significant changes in anti-dementia drug trials. Neuroepidemiol 2001; 20:51–6.

56. Axelman K, Lannfelt L, Almkvist O, Carlsson M. Life situation, coping and quality of life in people with high and low risk of developing Alzheimer's disease. Dement Geriatr Cogn Disord 2003; 16:220–8.

57. Rockwood K, Stolee P. Responsiveness of outcome measures used in an antidementia drug trial. Alzheimer Dis Assoc Disord 2000; 14:182–5.

58. Rogers SL, Farlow MR, Doody RS, Mohs R, Friedhoff LT. A 24-week, double-blind, placebo-controlled trial of donepezil in patients with Alzheimer's disease. Donepezil Study Group. Neurology 1998; 50:136–45.

59. Shea C, MacKnight C, Rockwood K. Donepezil for treatment of dementia with Lewy bodies: a case series of nine patients. Int Psychogeriatr 1998; 10:229–38.

60. McKeith I, Del Ser T, Spano P et al. Efficacy of rivastigmine in dementia with Lewy bodies: a randomised, double-blind, placebo-controlled international study. Lancet 2000; 16; 356:2031–6.

61. Wild R, Pettit T, Burns A. Cholinesterase inhibitors for dementia with Lewy bodies. Cochrane Database Syst Rev 2003; 3:CD003672.

62. The AM, Pasman R, Onwuteaka-Philipsen B, Ribbe M, van der Wal G. Withholding the artificial administration of fluids and food from elderly patients with dementia: ethnographic study. BMJ 2002; 325:1326.

63. Shah SN, Sesti AM, Copley-Merriman K, Plante M. Quality of life terminology included in package inserts for US approved medications. Qual Life Res 2003; 12:1107–17.

64. Brodaty H, Green A, Koschera A. Meta-analysis of psychosocial interventions for caregivers of people with dementia. J Am Geriatr Soc 2003; 51:657–64.

65. Mittelman MS, Ferris SH, Shulman E, Steinberg G, Levin B. A family intervention to delay nursing home placement of patients with Alzheimer disease. A randomized controlled trial. JAMA 1996; 276:1725–31.

66. Feinberg LF, Whitlatch CJ. Are persons with cognitive impairment able to state consistent choices? Gerontologist 2001; 41:374–82.

67. Weiner MF, Martin-Cook K, Svetlik DA et al. The quality of life in late-stage dementia (QUALID) scale. J Am Med Dir Assoc 2000; 1:114–16.

68. Albert SM, Del Castillo-Castaneda C, Sano M et al. Quality of life in patients with Alzheimer's disease as reported by patient proxies. J Am Geriatr Soc 1996; 44:1342–7.

69. Ankri J, Beaufils B, Novella JL et al. Use of the EQ-5D among patients suffering from dementia. J Clin Epidemiol 2003; 56(11):1055–63.

70. Martikainen J, Valtonen H, Pirttila T. Potential cost-effectiveness of a family-based program in mild Alzheimer's disease patients. Eur J Health Econ 2004; 5:136–42.

71. Bell CM, Araki SS, Neumann PJ. The association between caregiver burden and caregiver health-related quality of life in Alzheimer disease. Alzheimer Dis Assoc Disord 2001; 15:129–36.

72. Canadian Study of Health and Aging Working Group. Patterns and health effects of caring for people with dementia: the impact of changing cognitive and residential status. Gerontologist 2002; 42:643–52.

73. Karlawish JH, Casarett DJ, James BD et al. Why would caregivers not want to treat their relative's Alzheimer's disease? J Am Geriatr Soc 2003; 51:1391–7.

74. Karlawish JH, Klocinski JL, Merz J, Clark CM, Asch DA. Caregivers' preferences for the treatment of patients with Alzheimer's disease. Neurology 2000; 55:1008–14.

12

Caregiver Burden Outcomes

Henry Brodaty and Claire Thompson

Introduction

Most people with dementia live at home with their family, friends or another person who assists them in managing day to day life. In the USA approximately 49% of these caregivers are spouses of the patient, 45% are children, 77% are women and 73% are over 50 years of age.[1]

The responsibility or 'burden' of caring for a person with dementia is immense, and it has become common to refer to the caregiver as the second patient. The time demands of caregiving for a person with dementia have been referred to as the '36 hour day'.[2] Estimates range from 50 to 286 hours per month, with time demands increasing as the illness progresses.[3,4]

Caregiver burden may be divided into objective and subjective burden. Objective burden results from the direct effects of the illness: the patient's loss of cognitive skills, impaired functioning in activities of daily living (ADL) and instrumental activities of daily living (IADL), increasing behavioural and psychiatric problems in dementia and consequent loss of companionship for the caregiver. Subjective burden or strain is the caregiver's response to the objective burden, including feelings of stress, depression, overload, resentment and inability to cope. There is a large body of evidence demonstrating the psychological, physical, social and financial impact of the burden of caring for a patient with dementia.[5–9] Although the negative aspects of caregiving are most apparent and most often researched, there are also positive aspects of caregiving, such as feelings of satisfaction and a sense of meaning.[10]

Psychological morbidity, such as depression, is common in caregivers.[1,5] In a meta-analysis of research into caregiver's psychological wellbeing, Pinquart and Sorensen found that caregivers had higher levels of stress and depression as well as lower levels of subjective wellbeing and self-efficacy than non-caregiver controls.[11] The psychological impact can persist even after the person with dementia is placed in residential care, though it usually is ameliorated.[5,12]

Physically, caregivers tend to have more health problems than non-caregivers. Caregivers have more diagnosed chronic illnesses, poorer self-rated health, more visits to doctors, more hospitalisations, and more use of prescription medications.[6,13,14] In the meta-analysis of Pinquart and Sorensen, caregivers had lower levels of physical health than non-caregiver controls, although they note that this difference was smaller than the difference in prevalence of depression.[11] Another meta-analysis focusing on physical health found that caregivers reported more health-related problems than did non-caregivers.[15] Pre-existing conditions such as hypertension can be exacerbated by the caregiving role.[13] Further, caregiving has been shown to be associated with increased mortality.[13]

Social isolation can add to the burden of caregiving. Caregivers tend to give up many of their usual interests and employment as the demands of caring for the patient increase.[16,17] In addition, friends and family may feel uncomfortable with the person with the dementia, with the patient's poor communication or disinhibited behaviour, or even with the topic of dementia.[16] Isolation can be so profound that many caregivers see a person from outside the home once a week or less.[5] This can lead to a cycle of increasing burden as caregiver burden increases when the caregiver has poorer health or fewer social supports.[18]

The financial burden of caregiving can be considerable. There are direct costs such as medical consultations, investigations, medications, provision of personal and nursing care and respite or residential care. In addition, there are indirect costs in the loss of income that could have been earned by the patient and the caregiver. Reported annual costs per person of dementia in the USA vary from US$ 4900 to US$46 700.[19-21] In Sweden, costs have been estimated to be equivalent to US$14 500 to US$22 200, and in Ireland equivalent to US$12 000.[22,23] Costs increase with disease severity, with the yearly cost of providing care in the USA estimated to be US$3630 for mild dementia, and US$7420 and US$17 700 for moderate and severe dementia respectively.[24]

Patient factors, such as the severity of the disease, behaviour problems and functional impairment, all contribute to burden. Patients commonly are placed in nursing homes once they have declined in function to a level that their caregivers find unmanageable, but more so when their caregivers have reached a 'breaking point'. This may be because of the amount of time required to care for the patient, the patient's misidentifications of significant people, especially of the caregiver, the patient's clinical fluctuations and nocturnal deterioration and poor health in the caregiver.[25]

When it comes to the decision to place a person with dementia in nursing home care, characteristics of and factors relating to the caregivers themselves, especially the burden they bear, tend to be more predictive than patient variables. Institutionalisation is more likely among caregivers experiencing greater psychological distress, and by caregivers who are adult children compared to spouses.[26-29] A recent study on predictors of institutionalisation of people with dementia found a 20-fold protective effect of having a co-resident caregiver.[30] Higher ratings of behavioural problems in the person with dementia were also statistically associated with transition into residential care, as was the psychological domain of quality of life of the caregiver.[30]

A model of caregiver burden and strain has been described (see Figure 12.1).[31] This includes factors that protect or exacerbate the effects of caregiving. Protective factors include knowledge, mature coping skills, supportive relationships and good physical health in the caregiver. Exacerbating factors include lack of these attributes and behavioural problems in the patient.

Is it important to measure caregiver burden in drug trials?

The enrolment of a patient in a drug trial for dementia is often at the instigation of the caregiver. In the early stages of the disease, patients are likely to seek the advice of their caregivers, and as the severity of the disease progresses, the weight of this decision falls more and more on the caregiver. The consent of caregivers is important as regards their own participation in addition to that of the patient. Caregivers tend to be the driving force behind the patient's participation in the trial, in terms of providing or organising transportation, accompanying the patient to appointments, supervising medication, as well as the crucial role of providing valuable information about the cognitive and behavioural functioning of the patient and any adverse events.

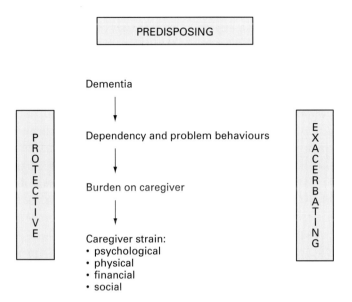

Figure 12.1 Model of caregiver burden and strain.

While the impact of caregiver burden has been well established and some clinical trials are using measures of at least some aspects of burden, the question remains as to whether caregiver burden is relevant in the context of clinical drug trials. Trials are conducted in order to establish the efficacy of a drug in alleviating symptoms of illness or altering disease progression. If a trial shows a drug to be effective, this evidence is used to obtain approval to manufacture and market the drug. Regulatory authorities approve drugs on the basis of evidence of improvement in the patient.

How useful is evidence that the prescription of a drug to a patient is associated with improvement in the caregiver? Benefits to second parties alone cannot be grounds for prescribing medication. As caregiver outcomes can only be an adjunct to measures of patient outcomes, mainly regarding cognition and function, the value of evaluating effects on caregivers needs to be justified against the extra time and expense to the researcher and the caregiver.

Measurement of caregiver outcomes is relevant to drug trials for several reasons. Medications can have benefits in reducing cognitive decline, functional impairment and behavioural disturbances, which in turn have been shown to be linked to caregiver burden. Reduced time spent caregiving or supervising, and decreased caregiver burden have been reported in trials of cholinesterase inhibitors (see below). Caregiver burden is linked to caregiver psychological stress,[11] which in turn is correlated with depression scores in patients.[31] It is probable that this relationship is bidirectional – patient mood influencing caregiver distress and vice versa. Finally, cost-effectiveness is a major consideration for regulatory authorities in approving subsidisation of medication. The single greatest driver of costs is institutional care, for which caregiver burden is the strongest determinant. Reduction in caregiver burden is thus highly relevant to patient outcome.

A variety of measures of burden and time are used

In recognition of the clinical and heuristic importance of the burden of caregiving, a number of useful measures have been developed to quantify levels of burden. Using the model shown in Figure 12.1, measures can be categorised as objective stressors, the subjective strain of caregiving, time spent in caregiving activities and resource utilisation. Choice of measures hinges on what outcomes are likely to be achieved, whether the measures are valid and reliable in the type of population under investigation, and whether the measures are sensitive to change.

Objective stressors and strain

The *Burden Interview* was designed to identify stressors for caregivers and quantify burden.[32] The Burden Interview is based on the caregiver's self-report during a structured interview and was one of the first instruments used in research to highlight the importance of providing support for caregivers of people with dementia. The 29 items assess the caregiver's feelings about aspects of the caregiving situation. Reliability has been reported to be adequate (from 0.88 to 0.91), and validity of the total burden score was established through correlation with another measure of burden ($r = 0.71$). Factor analysis confirmed two subscales: personal strain and role strain.[33]

The *Screen for Caregiver Burden* is a measure designed to identify distressing caregiving experiences in spouse caregivers of patients with Alzheimer's disease (AD), with 25 items measuring objective experiences and subjective distress.[34] This measure has sound psychometric properties, with coefficient reliabilities of 0.84 and 0.88 for objective burden and subjective burden respectively, and test–retest reliabilities of 0.70 and 0.64. Construct validity has been established by correlations with measures of care recipient functioning for objective burden, and of caregiver variables for subjective burden. The measure has internal consistency reliabilities of 0.89 and 0.81 for objective and subjective burden respectively, and has been found to be sensitive to changes in caregiver burden over time.[34]

The *Revised Memory and Behaviour Problems Checklist* (RMBPCL) is a 24-item caregiver report of behavioural problems that provides a total score and scores for three patient-related subscales (memory, depression, disruptive behaviour) and corresponding scores for the caregiver's reaction to each behaviour.[35] Overall reliability is 0.84 for patients' behaviours and 0.90 for caregivers' reactions. Validity has been assessed through correlations with other established measures of patient mood and cognitive functioning.

In the *Neuropsychiatric Inventory* (NPI), a clinician conducts a brief standardised interview with an informant and rates the frequency and severity of 12 specific behavioural and psychiatric symptoms of dementia.[36] There is also a rating of caregiver distress for each problem identified. The NPI has been shown to be sensitive to stage-specific behavioural and psychological symptoms in AD and also sensitive to effects of cholinergic drugs.[37,38] A briefer version, the NPI-Q is a self-administered informant questionnaire screen for the 12 symptoms of the NPI, and rates severity and caregiver distress. Test–retest reliability was 0.80 for total symptoms and 0.94 for caregiver distress. Validity was established through interscale correlations with the NPI of 0.84 for total scores and measures of distress.[38]

Caregiving time

The *Caregiver's Activity Time Survey* (CATS) records information on time caregivers spend in patient care activities such as feeding, toileting, bathing, dressing, giving drugs, and providing supervision.[39] The CATS was developed in 1990 and was an early measure of caregiving to be used in clinical drug trials. Validity, reliability and sensitivity to change were not reported.[40]

The *Caregiver Activity Survey* is a very brief (six-item) measure of the time caregivers spend in caregiving activities during a typical 24-hour period, including ADLs, IADLs as well as communication and supervision time.[41] Reliability was established through test–retest intraclass correlation coefficient of 0.88 over a 3-week period and convergent validity with the Alzheimer's Disease Assessment Scale–Cognitive Subscale (ADAS-Cog)

$(r = 0.61)$, Mini-Mental State Examination (MMSE) $(r = 0.57)$ and Physical Self Maintenance Scale (PSMS)[42] $(r = 0.43)$.

Resource use

Resource Utilization in Dementia (RUD) focuses specifically on use of formal and informal resources by the patient and the caregiver.[43,44] Validity, reliability and sensitivity to change have not been reported, however the RUD was based on direct observation of caregiving activities and has shown a clear connection between dementia severity and caregiver time.[45] In recognition of the complexity of clinical trials with multiple outcome measures, a shortened version, the 'RUD Lite' has been developed.[44]

What is not being measured?

Where drug trials consider caregiver outcomes they usually focus on objective burden, i.e. the time and activities involved in caregiving,[46] and external resource use.[47] Caregivers' subjective burden is assessed less often,[48] and more distal measures of psychological effects on caregivers are simply not rated, despite the extensive evidence of the prevalence of depression in caregivers,[11] and the many reliable and valid measures of depression available, such as the Centre for Epidemiologic Studies Depression Scale,[49] the Beck Depression Inventory[50] and the Hamilton Rating Scale for Depression.[51] Similarly, measures of anxiety and mediating variables such as coping styles and premorbid relationship quality are not used, despite the evidence of their contribution to caregiver burden and the availability of reliable valid measurement tools. Finally, the ultimate measure perhaps for both caregiver and patient is quality of life, a difficult construct to define, especially in people with dementia. No clinical drug trial study that we could locate measured quality of life in caregivers, despite the implications for health economics.[52]

Clinical meaningfulness

Clinical drug trial research has not explored how clinically meaningful are the changes in caregiver burden (symptomatology of disease and caregiver psychological, physical, social and financial impact) or the impact on caregivers of residential care placement. The findings noted above are statistically significant, however research findings are increasingly being judged on their 'clinical significance' or the extent to which the effects of a treatment are meaningful to the patient, caregiver and society. Schulz et al suggested asking firstly whether the assessed outcome is important to the individual or society, and secondly, how large an effect is required in order to be considered clinically meaningful.[53] Schulz et al conducted a meta-analysis of studies that reported caregiver outcomes from psychological and social interventions as well as clinical drug trials.[53] Of the 33 studies which included measures of caregiver burden, 16 reported positive impacts (and a further five found benefits for subgroups of subjects). Schulz et al also focused on the clinical significance of the interventions, which had percentile changes ranging from 1.5% to 14% on standard measures used.[53] Brodaty et al, in their meta-analysis of caregiver interventions, excluding drug trials and respite care, concluded that psychosocial interventions had an average effect size on caregiver outcomes of 0.32 but no influence on patient cognition or function.[54]

Empirical data of caregiver outcomes from drug trials

Caregivers of patients treated with metrifonate, a cholinesterase inhibitor, reported significantly reduced burden. The magnitude of treatment effects was 5–6% and 1.8% on two measures of caregiver distress, 2–4% for caregiver stress and a saving of approximately half an hour (a 2% improvement in a 24 hour day, or 12.5% improvement from the 4 hours at baseline) in the caregiver's subjective impression of unpaid caregiving time.[55] Similarly, a *post hoc* survey of caregivers of donepezil-treated patients reported less difficulty with caregiving compared to caregivers of patients who did not receive donepezil, with magnitude of treatment effects of 4.6% on the Caregiver Burden Scale.[56] Caregivers of patients treated with galantamine experienced significantly reduced distress (compared with controls) associated with behavioural symptoms of AD.[57] The difference was primarily due to an increase in distress in the caregivers of the placebo group. The percentage difference was not reported. A comparison of galantamine and donepezil found caregivers of patients on either medication reported reduced caregiving burden on the Screen for Caregiver Burden, however the authors provide the percentage of caregivers who report improvement greater than zero, but not the magnitude of the change. There was no significant change in NPI scores from baseline between groups at endpoint.[42]

With regard to time spent in caregiving activities, a placebo-controlled trial of velnacrine found that unpaid professional caregiver time measured with the CATS correlated with AD severity, and that time spent in caregiving was reduced by about 45% (or 3.2 hours per week) from baseline among patients taking the medication, which represents approximately half an hour a day or 2% more free time in a 24-hour day.[46] A trial of venacrine that significantly increased cognitive function in patients relative to placebo, was associated with a trend towards a reduction in caregiver time, with a decrease in 3.3 hours per day caregiving, or 14% less out of a 24-hour day, for caregivers of patients on a high dose.[40] A multi-centre trial found that after a year, compared to caregivers of patients on placebo, caregivers of patients treated with donepezil spent an average of an hour less providing care, a 4.2% saving in a 24-hour day.[58] Similarly, treatment with galantamine decreased the time caregivers spent assisting patients by 21 minutes from baseline to 6-month follow-up, whereas the time spent assisting those who received placebo increased by 34 minutes.[59] This difference

of 55 minutes represents a magnitude of treatment of 3.8%.

Conversely, the need to supervise and monitor the patient's medication or the failure of medication to help may increase caregiver burden. In a study of 31 family caregivers (where the person cared for was not necessarily affected by dementia), 7.7% of total caregiving time was spent on medication-related tasks. Further to this, 10 (32%) caregivers reported problems directly related to medications, 6 (19%) had current problems in managing drug regimens, and 16 (52%) had problems within the past year.[60] If no benefit to the person with dementia is observed, this can lead to feelings of disappointment, anger and disillusionment. In placebo-controlled clinical trials, it is as likely as not that the patient has not in fact been taking the active medication and therefore receiving no benefit of medication whilst the caregiver still has the added burden of medication-related tasks.

Methodological issues in measuring caregiver outcomes

Brodaty et al[54] found four studies that reported delay in institutionalisation.[54] Caregivers participating in drug trials often receive more support than they would have done otherwise. Theoretically, this could influence rates of placement although placebo randomisation should even out any such effect. However in open-label studies there is no correction for such an effect. Thus the delay in nursing home placement associated with cholinesterase inhibitor use, reported by Lopez et al, may have other claimants for the credit.[61]

Residential placement is a particularly attractive outcome measure to pharmaceutical companies seeking to demonstrate cost-effectiveness of medications in order to qualify for subsidisation. The costs of nursing home care are so large that any delay in placement can offset the payment for the medication under consideration. However, residential placement is complex and depends on many factors which vary considerably by country, for example, financial burden. Also, while delayed institutionalisation is cited as a socially significant outcome, debate exists about whether this is necessarily beneficial for the caregiver.[53] For the family caregiver, burden usually decreases after placement, although a minority of caregivers have difficulty adjusting.[62–64] The Canadian Study of Health and Aging found that over five years, caregivers of people with dementia who remained in the community had higher Zarit burden scores than did caregivers of people with dementia who were institutionalised.[63]

Once in residential care, the patient is attended to by a number of professional caregivers who are usually nurses or nursing assistants, and who are caring for many residents. Measurement of their levels of stress is not very meaningful in determining effects of a drug unless the stress can be linked to particular patients. Also as most of the staff in nursing homes are untrained in the process of clinical trials, care must be taken to ensure the reliability and validity of their reports.

Patients not caregivers are selected for trials. While patients can be guaranteed by definition to have measurable impairment, burden, depression and other outcome measures will be at low levels in over half of their caregivers. Low base rates on these variables limit possible improvement.

Caregivers are a heterogeneous group; spouses and adult children, men and women, relatives and friends differ in how they care for someone and how much they are affected by the caregiving role. Drug trials cannot practically control for this variability, which needs to be dealt with statistically.

Severity of dementia will affect the choice of instruments appropriate to measure caregiver outcome. Trials usually recruit patients with mild to moderately severe Alzheimer's disease, typically with MMSE scores ranging between 14 and 25. Yet many of the scales used are more relevant to later stage disease. Time to supervise ADL and use of respite care will be minimal for people with early disease and is unlikely to change much over 24 weeks. Caregivers' reactions to behavioural problems are only relevant in the presence of these symptoms which characteristically occur in the middle and later stages of dementia.

As trials focus on milder forms of disease, such as mild cognitive impairment (MCI), more subtle measures are required. These might be strain scales, e.g. the Caregiver Stress Scale,[65] or more distal direct measures of caregiver psychological state, be it of morbidity such as depression or distress in general, or of quality of life.

Measurement error bedevils all scales; there are extra challenges in asking caregivers to recall the amount of time they spend performing various tasks. Anecdotally, we note that many caregivers have difficulty in estimating the time involved in caregiving activities, as often (particularly in the earlier stages of illness) these may be hard to distinguish from the regular pattern of household tasks. Many spouse caregivers do not consider their caregiving activities to be any different from those they performed prior to the onset of the illness. Caregivers will often tell us they can not estimate time spent in supervision as they are engaged in other tasks at the same time, or that it is difficult to add up the 'five minutes here and there' they devote to supervising the patient. Many caregivers take over responsibilities such as driving, but find this saves them time and stress compared to being driven by the patient. Many caregivers tell us they cannot describe a typical day or estimate hours spent in caregiving activities in a typical day, as some days involve excursions, appointments, or other activities whereas other days are uneventful. In later stages of the disease, a patient may be receiving nursing care or day care services thereby freeing some of the caregiver's time. Caregivers may also engage other resources such as cleaning or meal preparation services, thereby lessening the total time spent in caring for the patient. Other factors that may influence time spent in caregiving include assistance from other family members or the presence of physical disability in the caregiver or the patient, and financial resources available to employ help. Trials rely on the randomisation process equalising these variables, but this needs to be checked. If time caring is to be an outcome variable it is important to try and capture all the time spent caring for the person with dementia, not just that provided by the principal caregiver. Table 12.1 summarises the issues involved in measuring caregiver outcomes.

Mediating variables can be crucial in determining caregiver outcomes. These can be variables associated with the patient or with the caregiver. There is a strong association between behavioural disturbance and caregiver distress. Behavioural disturbance accounts for about 25% of the variance in caregiver psychological distress.[66–69] Certain types of behaviour are particularly likely to be associated with distress – incontinence, immobility, nocturnal wandering, proneness to fall, inability to engage in meaningful activities, difficulties with communication, sleep disturbance, loss of companionship, disruptiveness, constant demands and aggression.[64,70–72] Caregiver variables include presence of psychiatric disorder, an exclusion criterion for patients but not asked about with caregivers, physical illness, other caregiving responsibilities and level of support from family and friends.

Qualitative data are not captured by trial designs. Caregivers will often use the assessment sessions to unburden themselves, to talk about their feelings and to seek help. Descriptive, qualitative information about the patient's condition is imparted, but in a way that is not quantifiable or standardised, and therefore not useful to the trial. Interviewers vary in how long they allow unbur-

dening to continue, which may influence caregiver outcome too. The notion that the assessment visits merely allow for recording of data is artificial. Just participating in studies can be beneficial and will dilute any differences between drug and placebo groups.[73] On the other hand, the process of being in a trial may add to the burden due to extra appointments, though this has been rare in our experience. The division between routine medical care and trial appointments may not be apparent to the caregiver who expects more from study visits.

Caregivers as informants – methodological issues

Caregivers are the main informants for rating patient ADL and IADL functioning and behaviour, and also for rating the CIBIC-Plus, a standard instrument in most drug trials. But not all caregivers are reliable, some maximising their family member's cognitive abilities and others minimising these.[74] Doble et al compared caregivers' reports of patients' ADL to performance-based assessments conducted in patients' homes.[75] While there was 77% agreement of ratings, in all cases of disagreement the patient's functioning was overrated by the caregiver. Informants were more likely to overestimate the patient's functioning (against the criterion of occupational therapist assessment) when the patient had MCI rather than dementia.

What directions are needed for future trials?

Caregivers are essential to drug studies, with most trials stipulating a minimum period of daily contact in order to be eligible for inclusion. Caregivers instigate participation, motivate patients, organise schedules, administer medication, ensure compliance and provide data on symptoms, function and on adverse reactions. A maxim in old age care is that when a person is diagnosed with dementia there is almost always a second

---

Table 12.1 Methodological issues in measuring caregiver outcomes

1. Low base rates
2. Heterogeneity of caregivers
3. Severity of dementia
4. Measurement error
5. Mediating variables not accounted for
6. Qualitative data lost
7. Non-specific effects of trial participation
8. Residential placement

patient, the caregiver. Prescription of drugs for people with dementia can also benefit their caregivers.

Two main directions are anticipated for the future. Firstly, trials should include, as secondary endpoints, measures to ascertain the effects on caregivers. The choice of measure depends on the stage of dementia and the anticipated effect. Measures that could be taken need to go beyond the proximal measures of patient functional impairment and caregiver time, to caregiver burden scales and those which measure reactions to patient's behaviour (Revised Memory and Behaviour Problems Checklist[35]), to those which rate caregiver depression (Beck Depression Inventory, Hamilton Rating Scale for Depression, Geriatric Depression Scale), anxiety (Beck Anxiety Inventory) and psychological morbidity (General Health Questionnaire). Quality of life measures are increasingly employed in studies, e.g. Dementia Quality of Life Instrument (DQoL[76]) and the Quality of Life in Alzheimer's Disease: Patient and Caregiver Report (QoL-AD[77] and Assessment of Quality of Life (AQoL[78]) however these are measures of the patient's quality of life, not that of the caregiver. Another consideration is to ensure the questionnaire is not so large that measurement itself becomes burdensome to caregivers. As regulatory authorities generally require demonstration of cost benefit before authorising subsidisation, resource use data are essential. Finally it is wise to record known major contributors to caregiver outcomes such as caregiver demographics, relationship to patient, previous psychological and physical health and other caregiving responsibilities, as well as other persons in the house and levels of support.

A second new direction is the investigation of the effects of combining drug treatments for people with dementia and psychosocial interventions aimed at caregivers. The '3-Country Study' conducted in Manchester, New York and Sydney, has tested the hypothesis that the concurrent patient and caregiver interventions may have additive or even synergistic effects. The 3-country study

enrolled 156 patients with mild to moderately severe Alzheimer's disease and their spouses. All patients received donepezil, and caregivers were randomly allocated to receive standard information pack and standard care or a counselling package, based on two caregiver intervention programmes.[79,80] Caregivers who received the counselling package had less depression after six months than caregivers who received standard care, and over two year follow-up, this difference from the control group became greater.[81]

Now that the efficacy of cholinesterase inhibitors has been established, double-blind, randomised, placebo-controlled trials are being approved by few if any research ethics committees. Future research is likely to be aimed at establishing equal (or better) effectiveness of new drugs compared to the existing drugs, or an incremental effect of adding a new drug to an existing drug. Caregiver outcomes could profitably be included in such studies.

References

1. Brodaty H, Green A. Defining the role of the caregiver in Alzheimer's disease treatment. Drugs & Aging 2002; 19(12):891–8.

2. Mace NK, Rabins PV. The 36-hour day. Baltimore: The Johns Hopkins University Press; 1991.

3. Leon J, Neumann PJ, Hermann RC et al. Health-related quality-of-life and service utilization in Alzheimer's disease a cross sectional study. Am J Alzheimer's Dis 2000; 152:94–108.

4. Max W, Webber PA, Fox PJ. Alzheimer's disease: the unpaid burden of caring. J Aging Health 1995; 7:179–99.

5. Brodaty H, Hadzi-Pavlovic D. Psychosocial effects on caregivers of living with persons with dementia. Aust N Z J Psychiatry 1990; 24:351–61.

6. Baumgarten M, Battista RN, Infante-Rivard C et al. The psychological and physical health of family members caring for an elderly person with dementia. J Clin Epidemiol 1992; 45(1):61–70.

7. Grafström M, Winblad B. Family burden in the care of the demented and nondemented elderly – a longitudinal study. Alzheimer Dis Assoc Disord 1995; 92:78–86.

8. Gonzalez-Salvador MT, Arango C, Lyketsos CG et al. The stress and psychological morbidity of the Alzheimer patient caregiver. Int J Geriatr Psychiatry 1999; 149:701–10.

9. Wright LK, Hickey JV, Buckwalter KC, Hendrix SA, Kelechi T. Emotional and physical health of spouse caregivers of persons with Alzheimer's disease and stroke. J Adv Nurs 1999; 30:552–63.

10. Cohen CA, Colantonio A, Vernich L. Positive aspects of caregiving: rounding out the caregiver experience. Int J Geriatr Psychiatry 2002; 17:184–8.

11. Pinquart M, Sorensen S. Differences between caregivers and noncaregivers in psychological health and physical health: A meta-analysis. Psychol Aging 2003; 18(2):250–67.

12. Stephens MA, Ogrocki PK, Kinney JM. Sources of stress for family caregivers of institutionalized patients. J Appl Gerontol 1991; 103:328–42.

13. Schulz R, Williamson GM. The measurement of caregiver outcomes in Alzheimer disease research. Alzheimer Dis Assoc Disord 1997; 11(6):117–24.

14. Haley WE, Levine EG, Brown SL et al. Stress appraisal coping and social support as predictors of adaptational outcome among dementia caregivers. Psychol Aging 1987; 2:323–30.

15. Vitaliano PP, Zhang J, Scanlan J. Is caregiving hazardous to one's physical health? A meta-analysis. Psychol Bull 2003; 129(6):946–72.

16. Brodaty H. The family and drug treatments for Alzheimer's disease. In: Gauthier S. (ed) Pharmacotherapy of Alzheimer's Disease. London: Martin Dunitz; 1998.

17. Braekhus A, Oksengard AR, Engedal K, Laake K. Social and depressive stress suffered by spouses of patients with mild dementia. Scand J Prim Health Care 1998; 16:242–6.

18. Vitaliano PP, Russo J, Young HM, Teri L, Maiuro RD. Predictors of burden in spouse caregivers of individuals with Alzheimer's disease. Psychol Aging 1991; 6(3):392–402.

19. Wimo A, Ljunggren G, Winblad B. Costs of dementia and dementia care: a review. Int J Geriatr Psychiatry 1997; 12:841–56.

20. Huang LF, Cartwright WS, Hu TW. The economic cost of senile dementia in the United States, 1985. Pub Health Rep 1988; 103:3–7.

21. Ernst RL, Hay JW. The US economic and social costs of Alzheimer's disease revisited. Am J Public Health 1994; 84:1–4.

22. Wimo A, Karlsson G, Sandman PO, Corder L, Winblad B. Cost of illness due to dementia in Sweden. Int J Geriatr Psychiatry 1997; 12:857–61.

23. O'Shea E, O'Reilly S. The economic and social cost of dementia in Ireland. Int J Geriatr Psychiatry 2000; 15:208–18.

24. Langa KM, Chernew ME, Kabeto MU et al. National estimates of the quantity and cost of informal caregiving for the elderly with dementia. J Gen Intern Med 2001; 16:770–8.

25. Annerstedt L, Elmstahl S, Ingvad B, et al. Family caregiving in dementia – an analysis of the caregiver's burden and the 'breaking-point' when home care becomes inadequate. Scand J Public Health 2000; 28:23–31.

26. Colerick EJ, George LK. Predictors of institutionalization among caregivers of patients with Alzheimer's disease. J Am Geriatr Soc 1986; 34:493–8.

27. Lieberman MA, Kramer JH. Factors affecting decisions to institutionalise demented elderly. Gerontologist 1991; 313:371–4.

28. Brodaty H, McGilchrist C, Harris L, Peters KE. Time until institutionalization and death in patients with dementia. Role of caregiver training and risk factors. Arch Neurol 1993; 50(6):643–50.

29. Yaffe K, Fox P, Newcomer R et al. Patient and caregiver characteristics and nursing home placement in patients with dementia. JAMA 2002; 187:2090–7.

30. Banerjee S, Murray J, Foley B, Atkins L, Schneider J, Mann A. Predictors of institutionalisation in people with dementia. J Neurol Neurosurg Psychiatry 2003; 74(9):1315–16.

31. Brodaty H, Luscombe G. Psychological morbidity in caregivers is associated with depression in patients with dementia. Alzheimer Dis Assoc Disord 1998; 12:62–70.

32. Zarit S, Reever K, Bach-Peterson J. Relatives of the impaired elderly: correlates of feelings of burden. Gerontologist 1980; 20:649–55.

33. Kuhlenschmidt S. Review of the Memory and Behaviour Problems Checklist and the Burden interview. In: Plake BS, Impara JC (eds). The Fourteenth Mental Measurements Yearbook. Lincoln NE: Buros Institute of Mental Measurements; 2001. Retrieved 23 March 2004 from Mental Measurements Yearbook database.

34. Vitaliano PP, Russo J, Young HM, Becker J, Maiuro RD. The screen for caregiver burden. Gerontologist 1991; 31:76–83.

35. Teri L, Truax P, Logsdon RG, Uomoto J, Zarit S, Vitaliano PP. Assessment of behavioral problems in dementia: the Revised Memory and Behavior Problems Checklist. Psychol Aging 1992; 7:622–31.

36. Cummings JL, Mega M, Gray K et al. The Neuropsychiatric Inventory comprehensive assessment of psychopathology in dementia. Neurology 1994; 44:2308–14.

37. Mega MS, Cummings JL, Fiorello T et al. The spectrum of behavioral changes in Alzheimer's disease. Neurology 1996; 46;130–5.

38. Kaufer DI, Cummings JL, Ketchel P et al. Validation of the NPI-Q, a brief clinical form of the Neuropsychiatric Inventory. J Neuropsychiatry Clin Neurosci 2000; 12(2):233–9.

39. Moore MJ, Clipp EC. Alzheimer's disease and caregiver time. Lancet 1994; 343:239–40.

40. Clipp E, Moore M. Caregiver time use: an outcome measure in clinical trial research on Alzheimer's disease. Clin Pharmacol Ther 1995; 58(2):228–36.

41. Davis K, Marin D, Kane R et al. The caregiver activity survey: CAS Development and validation of a new measure for caregivers of persons with Alzheimer's disease. Int J Geriatr Psychiatry 1997; 12:978–88.

42. Lawton MP, Brody EM. Assessment of older people: self-maintaining and instrumental activities of daily living. Gerontologist 1969; 9(3):179–86.

43. Wimo A, Wetterholm A, Mastey V, Winblad B. Evaluation of the resource utilization and caregiver time in anti-dementia drug trials – a quantitative battery. In: Wimo A, Jönsson B, Karlsson G et al (eds). The Health Economics of Dementia. London: John Wiley & Sons; 1998: 465–99.

44. Wimo A, Winblad B. Resource utilization in dementia: 'RUD Lite'. Brain Aging 2003; 3(1):48–59.

45. Wimo A, Nordberg G, Jansson W, Grafstrom M. Assessment of informal services to demented people with the RUD instrument. Int J Geriatr Psychiatry 2000; 15:969–71.

46. Antuono P. Effectiveness and safety of venlacrine for the treatment of Alzheimer's disease: a double-blind placebo-controlled study. Arch Intern Med 1995; 155:1766–72.

47. Winblad B, Wimo A, Mastey V et al. Caregiver health benefits and associated reductions in health-care costs as a consequence of treating patients with donepezil. Value In Health 2003; 6(3):276.

48. Wilcock G, Howe I, Coles H et al. A long-term comparison of galantamine and donepezil in the treatment of Alzheimer's disease. Drugs Aging 2003; 20(10):777–89.

49. Radloff LS. The CES-D Scale: a self-report depression scale for research in the general population. Applied Psychological Measurement 1977; 1:385–401.

50. Beck AT, Rush J, Shaw B, Emery G. Cognitive Therapy of Depression. New York: Guilford; 1961.

51. Hamilton M. Development of a rating scale for primary depressive illness. Br J Soc Clin Psychol 1967; 6:278–96.

52. Whitehouse PJ, Patterson MB, Sami SA. Quality of life in dementia: Ten years later. Alzheimer Dis Assoc Disord 2003; 17(4):199–214.

53. Schulz R, O'Brien A, Czaja S et al. Dementia caregiver intervention research: In search of clinical significance. Gerontologist 2002; 42(5):589–602.

54. Brodaty H, Green A, Koschera A. Meta-analysis of psychosocial interventions for caregivers of people with dementia. J Am Geriatr Soc 2003; 51:657–64.

55. Shikiar R, Shakespeare A, Sagnier PP et al. The impact of metrifonate therapy on caregivers of patients with Alzheimer's disease: results from the MALT clinical trial. J Am Geriatr Soc 2000; 48:268–74.

56. Fillit HM, Gutterman EM, Brooks RL. Impact of donepezil on caregiving burden for patients with Alzheimer's disease. Int Psychogeriatr 2000; 123:389–401.

57. Tariot PN, Truyen L. Reminyl (galantamine) reduces caregiver distress. In: 10th Congress of the International Psychogeriatric Association; 9–14 September 2001, Nice, France: IPA; 2001.

58. Mastey V, Wimo A, Winblad B et al. Donepezil reduces the time caregivers spend providing care results of a one-year double-blind randomized trial in patients with mild to moderate Alzheimer's disease. In: 14th Annual Meeting of the American Association for Geriatric Psychiatry; 23–26 February 2001, San Francisco, CA USA; 2001.

59. Lilienfeld S, Gaens E. Galantamine alleviates caregiver burden in Alzheimer's disease: a 12 month study. In: 5th Congress of the European Federation of Neurological Societies; October 14–18 2000; Copenhagen Denmark; 2000.

60. Ranelli P, Aversa S. Medication-related stressors among family caregivers. Am J Hosp Pharm 1994; 51(1):75–9.

61. Lopez OL, Becker JT, Wisniewski S et al. Cholinesterase inhibitor treatment alters the natural history of Alzheimer's disease. J Neurol Neurosurg Psychiatry 2002; 72:310–14.

62. Gold D, Reis M, Markiewicz D, Andres D. When home caregiving ends a longitudinal study of outcomes for caregivers of relatives with dementia. J Am Geriatr Soc 1995; 431:10–16.

63. The Canadian Study of Health and Aging Working Group. Patterns and health effects of caring for people with dementia: the impact of changing cognitive and residential status. Gerontologist 2002; 42:643–52.

64. Brodaty H, Hadzi-Pavlovic D. Psychosocial effects on carers of living with persons with dementia. Aust N Z J Psychiatry 1990; 24(3):351–61.

65. Pearlin LI, Mullan JT, Semple SJ et al. Caregiving and the stress process: an overview of concepts and their measures. Gerontologist 1990; 30:583–94.

66. Brodaty H. Caregivers and behavioral disturbances: effects and interventions. Int Psychogeriatr 1996; 8(Suppl 3):455–8.

67. Mangone CA, Sanguinetti RM, Baumann PD et al. Influence of feelings of burden on the caregiver's perception of the patient's functional status. Dementia 1993; 4:287–93.

68. Cohen R, Swanwick G, O'Boyle C, Coakley D. Behaviour disturbance and other predictors of carer burden in Alzheimer's disease. Int J Geriatr Psychiatry 1997; 12:331–6.

69. Bond J, Buck D. Long-term psychological distress among informal caregivers of frail older people in England: a longitudinal study. In: Paper presented at The 6th International Conference on Alzheimer's Disease and Related Disorders, 18–23 July, Amsterdam, 1998.

70. Gilleard CJ. Problems posed for supporting relatives of geriatric and psychogeriatric day patients. Acta Psychiatr Scand 1984; 70:198–208.

71. Greene JG, Smith R, Gardiner M, et al. Measuring behavioural disturbance of elderly demented patients in the community and its effects on relatives: a factor analytic study. Age Ageing 1982; 11:121–6.

72. Morris RG, Morris LW, Britton PG. Factors affecting the emotional wellbeing of the caregivers of dementia sufferers. Br J Psychiatry 1988; 153:147–56.

73. Schneider L. Designing phase III trials of anti-dementia drugs with a view towards pharmacoeconomic considerations. In: Wimo A, Jonsson B, Karlsson G, Winblad B (eds). The Health Economics of Dementia. London: John Wiley and Sons; 1998: 451–464.

74. Kemp NM, Brodaty H, Pond D, Luscombe G. Diagnosing dementia in primary care: the accuracy of informant reports. Alzheimer Dis Assoc Disord 2002; 16(3):171–6.

75. Doble S, Fisk J, Rockwood K. Assessing the ADL functioning of persons with Alzheimer's disease: Comparison of family informants' ratings and performance-based assessment findings. International Psychogeriatrics 1999; 11(4):399–409.

76. Brod M, Stewart AL, Sands L, Walton P. Conceptualisation and measurement of quality of life in dementia: the dementia quality of life instrument (DQoL). Gerontologist 1999; 39:25–35.

77. Logsdon RG, Gibbons LE, McCurry SM, Teri L. Assessing quality of life in older adults with cognitive impairment. Psychosomatic Med 2002; 64:510–19.

78. Wlodarczyk J, Brodaty H, Hawthorne G. The relationship between quality of life Mini-Mental State Examination and the Instrumental Activities of Daily Living in patients with Alzheimer's disease. Arch Gerontol Geriatr; 2004; 39:25–33.

79. Mittelman MS, Ferris SH, Shulman E et al. A family intervention to delay nursing home placements of patients with Alzheimer's disease: a randomized controlled trial. JAMA 1996; 276(21):1725–31.

80. Brodaty H, Gresham M, Luscombe G. The Prince Henry Hospital Dementia Caregivers' Training Programme. Int J Geriatr Psychiatry 1997; 12:183–92.

81. Mittelman MS, Brodaty H, Burns A. A multinational clinical trial of the effectiveness of a combined psychosocial and pharmacological intervention for AD patients and family caregivers. In: International Psychogeriatric Association Congress; August; Chicago; 2003.

13

Structural Neuroimaging Outcomes

Philip Scheltens and Frederik Barkhof

Introduction

While the role of structural neuroimaging (computed tomography/magnetic resonance imaging (CT/MRI)) in the diagnostic work up in dementia is recognised by international and national consensus committees, its role in clinical trials in Alzheimer's disease (AD) or its predecessor mild cognitive impairment (MCI) is less clear, although trials have been and are being conducted that incorporate MRI.

The focus of this chapter is to outline the different roles of MRI in a clinical trial with an anti-dementia compound in dementia, and to provide practical advice with regard to image acquisition, quality control and analysis. We will cover the ability of MRI to distinguish between dementia subgroups, monitor disease progression and measure treatment effects.

The ultimate aim is to improve the effectiveness of future trials and to facilitate comparisons between different trials.

The main potential roles of MRI in trials are to:

- *define the study population* – provide exclusion and inclusion criteria for selection of patients and determine markers for sample stratification
- *measure outcome* – provide surrogate markers of disease progression using serial MRI;

may also be done in clinical or MRI subgroups.

Each of these activities requires consideration of optimal acquisition and analysis for the particular subject group studied.

MRI is a feasible technology to use in that it is widely disseminated, and relatively inexpensive in the context of clinical trial costs. MRI is non-invasive, and even with repeated imaging no adverse effects are known, as long as care is taken to exclude subjects with pacemakers or certain metallic implants, and to wear appropriate ear protection. MRI research studies often require patients to have a Mini-Mental State Examination (MMSE) >10 in order to comply with instructions, and to remain still during the scan. More severely affected subjects may however still be scanned as long as short scan times (<10 minutes) are used. Although sedation (usually benzodiazepines) is sometimes used in clinical practice, for example in claustrophobic patients, this seems inappropriate in the context of a trial. The demented patient group raises numerous ethical questions particularly related to issues of informed consent in research in dementia; the non-invasive nature of MRI is therefore an important consideration in this regard.

Define the study population

MCI and AD

Due to the lack of specificity of clinical criteria and the heterogeneity of AD and MCI,[1–11] MRI has a role to play in defining and homogenising study populations, by adding MRI exclusion and inclusion criteria to existing clinical criteria. Excluding patients with significant small vessel cerebrovascular disease as well as large vessel strokes will narrow but also homogenise the study population. Including patients with a minimal degree of medial temporal atrophy (MTA) will increase the proportion of subjects that will progress from MCI to fulfil a diagnosis of AD.[12–19] The challenge in the MCI group is to determine which patients will progress to fulfil criteria for AD, as some of these patients may have fixed deficits, and will proceed to other diagnoses (e.g. vascular dementia (VaD) or fronto-temporal dementia (FTD)), or will turn out to be 'worried well'. In the correct clinical setting, the presence of MTA may enhance the likelihood of conversion to AD (and other features may rule out VaD and FTD, at the risk of reducing the generalisability of the study to the entire MCI population).[20–27] This strategy is attractive since it enhances the likelihood of reaching clinical endpoints and thus increases the statistical power to detect a treatment effect.

Vascular dementia

Trials in VaD are now planned and carried out widely. The most often used clinical criteria are those of the NINDS-AIREN work group, published in 1993.[28] The NINDS-AIREN criteria for VaD have three main features: a patient can fulfil the criteria if he or she is demented, has evidence on clinical examination and imaging of cerebrovascular disease and if a temporal relationship between the dementia and the cerebrovascular disease can be established. In order to assess the cerebrovascular disease on the brain scan, a list of radiological features has been formulated. In short, the radiological part of the criteria requires that vascular lesions should fulfil criteria for topography

as well as severity. In the case of large vessel stroke, the locations that meet criteria are: bilateral anterior cerebral artery, paramedian thalamic, inferior medial temporal lobe, parietotemporal and temporo-occipital association areas and angular gyrus, superior frontal and parietal watershed areas, as long as they involve the dominant hemisphere. In the case of small vessel disease, lesions that fulfil criteria are: white matter hyperintensities (WMH) more than 25% of the total white matter, multiple basal ganglia and frontal white matter lacunes and bilateral thalamic lesions. Interobserver agreement of these criteria has been reported to be low, even after the use of operational criteria.[29,30] Current use of these criteria in a clinical trial setting requires the use of a central reader to minimise variance in the included population and secure fulfilment of the radiological criteria of the NINDS-AIREN.

Recommendations for MRI inclusion and exclusion criteria

We recommend the use of MRI for all dementia trials.

Imaging can be used to:

- exclude non-degenerative, non-vascular, pathology: e.g. tumour, subdural haematoma, hydrocephalus, etc.
- exclude cerebrovascular disease as the major pathological substrate for the cognitive problems, for instance in MCI or AD
- rule out or include specific degenerative diseases, such as FTD (presenting with focal, lobar atrophy), corticobasal degeneration and progressive supranuclear palsy
- establish a minimum amount of MTA as a safeguard of correct diagnosis for AD
- secure fulfilment of radiological criteria in case of VaD.

Even if a given therapy is considered non-specific for one sort of pathology and licensing sought for a non-specific indication such as 'dementia', then the use of MRI could still be considered since there is potential value in determining which subgroups show response to the therapy (assessments of vascular load or MTA may be useful covariates in outcome analyses).

Measuring progression: surrogate endpoints

Whole brain atrophy

The advent of potential disease-modifying agents in AD has made the development of imaging markers of progression increasingly important. Such markers will not only increase our knowledge of the disease process and help provide prognostic information to patients, but importantly may also provide cost-effective ways of identifying those therapies that slow AD as opposed to providing temporary symptomatic benefit. Ideally, a surrogate marker of disease progression should relate directly to the extent of the underlying molecular pathology, e.g. synaptic loss, amyloid load or abnormal tau deposition. Such measures have been sought, but to date are not available *in vivo*. A downstream event, which is nevertheless central to pathological progression, is cerebral atrophy secondary to neuronal destruction. MRI can measure rates of atrophy that can serve as *in vivo* markers of disease progression. Progressive atrophy can be assessed by repeat MRI scanning non-invasively, blind to treatment allocation and to time point within a trial. Care should be taken to exclude (identify) other factors that may produce alterations in brain volume unrelated to atrophy (e.g. steroid treament or dehydration),[31] and these potential confounding factors need to be considered in relation to specific interventions. There have been several large multi-centre clinical trials in MCI and AD that used MRI measures of atrophy as outcome measures, and many are under way. Some of the published results of these trials will be reviewed below. The outcome measures that have been chosen in these studies are measures of regional (medial temporal lobe) atrophy and/or whole brain atrophy rates.

Measures of rates of atrophy based on manual outlining of regions of interest (e.g. hippocampus, entorhinal cortex), and semi-automated whole brain atrophy rates from serial MRI are the first choices as outcome measures. Rates of atrophy of other structures should also be investigated as possible surrogate markers. Alternative manual and automated image-analysis techniques are in development and merit comparison with these outcome measures in future multi-centre studies to determine the most powerful markers of disease progression. A central site for standardised analysis should be used; if multiple independent measures are being chosen these may be performed at different central sites.

Whole brain registration-based methods

Registration allows semi-automated measurement of atrophy rates, for example by determining the deformation of a brain contour, which has shown annualised mean (SD) rates of atrophy in AD of 1.4 (±1.1) to 2.4%/year (±1.1) versus controls 0.5% (±0.4).[32] These methods can incorporate correction for scanner geometry variability. There are some published data on sample sizes for the so-called boundary shift integral (BSI),[33] and this method is currently being used in several multi-centre trials in AD and MCI (see Figure 13.1). A related measure, called SIENA, has been used succesfully in MS trials,[34,35] and performs equally well as the BSI in subjects with AD.[36]

Ventricular CSF measurements

Ventricular measures are simple, albeit indirect markers of global atrophy but have rarely been used in longitudinal studies to date.[37]

High-dimensional non-linear registration methods

These novel techniques have the potential to warp a template or a baseline image onto follow-up images – allowing 'compression maps' to provide rates of atrophy in different regions.[38,39] These methods have considerable potential in reducing user input especially with multiple scans per subject; however more research is needed since there is presently only limited validation in AD or clinical trials.

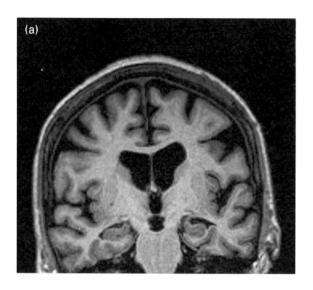
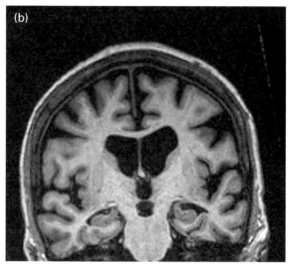

Figure 13.1 (a) Baseline T1 weighted coronal view of early-onset AD patient; (b) same patient scanned 1 year later; note progressive medial temporal and diffuse cortical atrophy and verntricular widening.

Medial temporal lobe atrophy
Hippocampus (HC)

The HC is the most extensively studied structure in AD; large numbers of cross-sectional and a small number of longitudinal studies have shown increased rate of atrophy (4–6% per annum) in patients with AD relative to controls (1–2% per annum).[18,40] Manual tracing of the HC on digital image data is the most validated and recommended method. Alternatives that require further longitudinal evaluation include: visual assessment using rating scales;[41] stereological measures;[42] and automated deformation-based methods either from standard template or from a baseline segmentation.[43,44]

Entorhinal cortex (EC)

The EC should on pathological grounds be even more sensitive than the HC as a measure of progression in AD. However, measurement reliability has been lower for the EC than for the hippocampus. It is presently unclear if there is any practical advantage in using EC measures over those from the HC; both measures appear to provide similar power for clinical trials in AD.[45] Advances in image

acquisition and analysis techniques may in the future mean that EC atrophy rates will prove to be superior to HC atrophy rates in MCI of the AD type and possibly also in AD.[46]

Amygdala

The lower reproducibility of amygdala measurements means that currently amygdala measures provide a slightly lower power than the HC to detect possible disease modification effects.[47]

White matter changes

In contrast to the above where white matter/ vascular changes were excluded, the goal of a trial in AD/MCI may also be to focus on vascular changes, of which small vessel changes are the most prominent and prevalent. One could think of a trial on treatment of arterial hypertension in AD to influence both cognitive and MRI endpoints. Earlier trials did not include MRI and missed the opportunity to document the underlying mechanism of the beneficial effect, for example in the setting of mixed dementia.[48] In a trial of VaD, such a goal is self-evident and in this case even changes in the development of large vessel strokes or small vessel disease may be the surrogate endpoint. In terms of causing cognitive impairment, however, cerebral small vessel changes are the most important.[49]

Several studies have reported longitudinal data on the progression of white matter lesions. The main findings are summarised in Table 13.1. At first glance all these data suggest a highly variable and often 'benign' course of white matter lesions. However, stratification of data by the baseline grade of white matter abnormalities in the 6-year data from the Austrian Stroke Prevention Trial study demonstrate that in the study participants with a baseline finding of early confluent or confluent changes, there was a remarkably rapid increase in lesion volume.[50] A baseline finding of early confluent and confluent abnormalities resulted in a median (interquartile range) volume increase of 2.7 (0.5; 5.9) cm^3 and 9.3 (7.1; 21.0) cm^3 after 6 years. Almost two-third of study participants with early confluent, and all subjects with confluent lesions demonstrated progression beyond measurement error (1.81 cm^3) after 6 years. In contrast, this was seen in none of the subjects with a normal baseline MRI scan and in only 14.6% of those with punctate foci. These findings indicate selection of patients with significant white matter abnormalities will increase the possibility to detect change over time (see Figure 13.2).

Power calculations

In general, power calculations rely on several simple input variables; the rate of atrophy in the placebo group, and its variance (which is often assumed to be equal in both treatment arms); the magnitude of the treatment effect; the number of subjects and the observation period. The most uncertain aspect in the setting of an MRI endpoint is the estimated treatment effect size (which may be completely different from the clinically expected effect size). Currently, our knowledge about the statistical power of MRI-derived endpoints is derived from a very limited number of studies in different and small patient groups; the comparison of the figures presented below should therefore be made with great caution.

Whole brain atrophy

As noted above, the annual rate of whole brain atrophy may be on the order of 2.4% per annum (SD 1.1%) and that in controls is 0.5% (0.4%). Based

Table 13.1 Listing of studies with MRI documenting progression of white matter changes

Authors	n	Interval (years)	Method	Progression
Wahlund et al 1996[51]	13	5	Visual rating	92% (12/13)
Veldink et al 1998[52]	14	2	Visual rating	55% (8/14)
Schmidt et al 1999[53]	273	3	Visual rating	17.9%/8.1% marked
Whitman et al 2001[54]	70	4	Stereology	Mean 1.1 cm^3/4 years
De Carli et al 2002[55]	168	4	Volumetry	0.38cm^3/year
Schmidt et al 2003[50]	296	6	Volumetry	17.2 % > 1.81 cm^3
Prins et al 2004[56]	20	3	Volumetry	0.42 cm^3/year (PVH) 0.15 cm^3/year (WMH)

PVH denotes periventricular hyperintensities; WMH denotes white matter hyperintensities

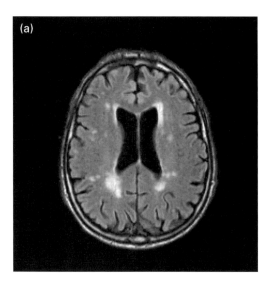
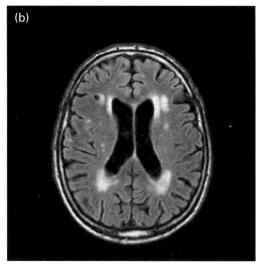

Figure 13.2 (a) and (b) Two axial FLAIR sequences showing progression of white matter lesions and occurrence of lacunar infarction in a 68-year-old probable AD patient.

on these assumptions, in a placebo-controlled trial of 12 months' duration, the power to detect a 20% decrease in the rate of atrophy was estimated to be 90% for 2 × 207 patients;[33] it should be noted that these estimates are based on very limited numbers of highly selected familial cases of dementia. For example, in the study by Jack et al, the yearly rate of atrophy was between 0.8 and 1.4%/year in AD;[32] nevertheless, in a more typical sample of MCI patients with and without conversion, sample size estimates (90% power, 25% treatment effect) were as low as 2 × 130 patients.

Hippocampal atrophy

The annual rate of hippocampal atrophy typically is around 2–4% per year for (MCI patients converting to) AD, with an SD of 2%, compared to a yearly rate of 1.4% (SD 1.2) per year for non-declining controls. Based on these assumptions, the sample size to detect a 25% treatment effect with a power of 90% is 2 × 102 patients, which is only slightly smaller than what is obtained using whole brain volume analysis in the same sample.[32]

Other volumetric atrophy measures

Less is known about the natural history of other imaging markers, especially EC. It has been suggested that ventricular volume is quite sensitive to change, and would provide even better power than whole brain or hippocampal volume, although the measurement definitions may be less well developed and the interpretation less obvious from a clinical point-of-view.[32,57]

White matter changes

In a recent 'Medical hypothesis' paper in *Neurology*, Schmidt and colleagues argued for including MRI white matter lesion volumes as surrogate marker in trials on patients with small vessel disease, with or without dementia.[58] As can be seen from the sample size calculations presented in Table 13.2, actually such a trial would need low numbers of patients to demonstrate treatment effects in these patients. A total of 227 patients with early confluent and confluent lesions per treatment arm would be needed to show a 30% therapeutic effect in a 3-year study. If one

Table 13.2 Sample sizes per treatment arm according to lesion grade at baseline and effect size in a 3-year interventional trial with the outcome measure being change in lesion volume

Treatment effect (%)	Early confluent (%)	Confluent (%)	Early confluent and confluent combined (%)
100	27	9	22
90	33	11	26
80	46	13	33
70	53	17	43
60	73	23	58
50	104	32	83
40	159	50	129
30	289	87	227

Samples sizes were calculated for a 100% reduction in the rate of increase in MRI white matter lesion volume (i.e. complete stabilisation of lesion load in the treatment arm) and for less marked effect sizes down to 30% reductions in volume increase over 3 years. All calculations assumed a power of 0.80 and a two-sided significance level of 5%

focuses on subjects with confluent abnormalities alone, which is the most likely scenario in a subcortical vascular dementia trial, a 30% therapeutic effect can already be detected with 87 patients per treatment arm, and a 40% effect with 50 in each arm.

General requirements for including MRI in clinical trials

Acquisition

Scanner field strength should be at least 0.5 T, and preferably 1.5 T, and ideally a single field strength should be used for all patients in a particular trial to allow comparison between sites and scanners. Centres should be selected for not planning an upgrade in the foreseeable future and each patient should be examined on the same scanner, using the same receiver-coil, and exactly the same sequence-parameters throughout the study.

T2-weighted or FLAIR (fluid attenuated inversion recovery) techniques should be used, particularly to assess white matter disease, and acquisition time should be tailored to patient tolerability.

Whole brain coronal 3D T1-weighted imaging is ideally suited to assess both whole brain and local medial temporal lobe atrophy. Volumetric T1-weighted imaging (e.g. SPGR, MPRAGE) ideally isotropic with voxel dimensions (1 mm or less) should be used, since these lend themselves quite well for image registration and reslicing. Interestingly, it has been suggested that small technical perturbations may not have great impact on group differences, when whole brain volume techniques are being employed.[36]

Analysis

Central assessment of all scans is key, based on visual inspection to exclude tumour and subdural haematoma, and to include patients with a significant amount of MTA and vascular damage, checking fulfilment of NINDS-AIREN criteria, etc.

Baseline analysis should include

- exclusion of surgically treatable disorders of other exclusion criteria
- assessment of hippocampal atrophy using established visual scales or using a region-of-interest-based volumetric analysis[40–42]

- assessment of white matter load using established visual scales[29,56] or automated or semi-automated method for quantifying vascular load.

For analysis of change, automated whole brain techniques are quite appealing, given their low measurement error and limited sensitivity to technical perturbations.

Manufacturer

While the use of only one MRI manufacturer in a multi-centre study may simplify acquisition and Quality Control (QC) and thereby improve consistency, this is unrealistic for large, multi-centre phase II(I) studies. The opportunity exists for MRI manufacturers to establish themselves as having a particular interest in consistent MRI acquisition techniques, specifically for multi-centre quantitative studies; and there is a need to improve the standards for QC for quantitative as opposed to clinical work. Investigators should strive for standardisation of acquisition techniques across trials to enhance the comparability of results. We recommend that setting standards for consistent acquisition and continuous QA should be the responsibility of, and handled by, an independent image-analysis coordinating centre.

Conclusions

It has become evident and already routine practice that MRI is included in clinical trials with anti-dementia drugs. As outlined above, MRI may be used in several ways, ranging from baseline assessment for diagnostic purposes only to multiple scans to measure progression and treatment effects. Clearly, the field has moved already into this application, while comparative studies using the various available techniques are lacking, rendering adequately powering studies using MRI as an outcome measure impossible.

While including MRI as a diagnostic screen seems logical and definitely improves the quality of included patient data (the large number of excluded patients in VaD trials using central reading may be taken as an example here), using it as a surrogate marker has still to be proven useful by showing positive results in a trial.

Ackowledgement

The authors are indebted to Professor Nick Fox for help and advice.

References

1. Knopman DS, DeKosky ST, Cummings JL et al. Practice parameter: diagnosis of dementia (an evidence-based review). Report of the Quality Standards Subcommittee of the American Academy of Neurology. Neurology 2001; 56(9):1143–53.

2. McKhann G, Drachman D, Folstein M et al. Clinical diagnosis of Alzheimer's disease: report of the NINCDS-ADRDA Work Group under the auspices of Department of Health and Human Services Task Force on Alzheimer's disease. Neurology 1984; 34:939–44.

3. American Psychiatric Association. Diagnostic and statistical manual of mental disorders, Fourth Edition, Washington, DC: American Psychiatric Association; 1997.

4. Chui H, Young Lee, A. Clinical criteria for various dementia subtypes: Alzheimer disease (AD), vascular dementia (VaD) Dementia with Lewy bodies (DLB), and Frontotemporal dementia (FTD). In: Qizilbash N, Schneider L, Chui H et al (eds). Evidence Based Dementia. Oxford: Blackwell Science, Ltd.; 2002: 106–19.

5. Petrovitch H, White LR, Ross GW et al. Accuracy of clinical criteria for AD in the Honolulu-Asia aging study, a population based study. Neurology 2001; 57:226–34.

6. Varma AR, Snowden JS, Lloyd JJ et al. Evaluation of the NINCDS-ADRDA criteria in the differentiation of Alzheimer's disease and frontotemporal dementia. J Neurol Neurosurg Psychiatry 1999; 66:184–8.

7. Braak H, Braak E. Frequency of stages of Alzheimer-related lesions in different age categories. Neurobiol Aging 1997; 18(4):351–7.

8. Linn RT, Wolf PA, Bachman DL et al. The 'pre-clinical phase' of probable Alzheimer's disease. A

13-year prospective study of the Framingham cohort. Arch Neurol 1995; 52:485–90.

9. Bozoki A, Giordani B, Heidebrink JL et al. Mild cognitive impairments predict dementia in non-demented elderly patients with memory loss. Arch Neurol 2001; 58(3):411–16.

10. Bookheimer SY, Strojwas MH, Cohen MS, Saunders AM, Pericak-Vance MA, Mazziotta JC et al. Patterns of brain activation in people at risk for Alzheimer's disease. N Engl J Med 2000; 343(7):450–6.

11. Morris JC, Storandt M, Miller JP et al. Mild cognitive impairment represents early-stage Alzheimer disease. Arch Neurol 2001; 58(3):397–405.

12. Visser PJ, Scheltens P, Verhey FR et al. Medial temporal lobe atrophy and memory dysfunction as predictors for dementia in subjects with mild cognitive impairment. J Neurol 1999; 246(6): 477–85.

13. Visser PJ, Verhey FR, Hofman PA, Scheltens P, Jolles J. Medial temporal lobe atrophy predicts Alzheimer's disease in patients with minor cognitive impairment. J Neurol Neurosurg Psychiatry 2002; 72(4):491–7.

14. Convit A, De Leon MJ, Tarshish C et al. Specific hippocampal volume reductions in individuals at risk for Alzheimer's disease. Neurobiol Aging 1997; 18(2):131–8.

15. Korf ESC WL-O, Visser PJ, Scheltens P. Medial temporal lobe atrophy on MRI predicts dementia in subjects with mild cognitive impairment. Neurology 2004; 63:94–101.

16. Convit A, de Asis J, de Leon MJ et al. Atrophy of the medial occipitotemporal, inferior, and middle temporal gyri in non-demented elderly predict decline to Alzheimer's disease. Neurobiol Aging 2000; 21(1):19–26.

17. Jack CR Jr, Petersen RC, Xu YC et al. Prediction of AD with MRI-based hippocampal volume in mild cognitive impairment. Neurology 1999; 52(7): 1397–1403.

18. Jack CR Jr, Petersen RC, Xu Y et al. Rates of hippocampal atrophy correlate with change in clinical status in aging and AD. Neurology 2000; 55(4):484–9.

19. Dickerson BC, Goncharova I, Sullivan MP et al. MRI-derived entorhinal and hippocampal atrophy in incipient and very mild Alzheimer's disease. Neurobiol Aging 2001; 22(5):747–54.

20. Laakso MP, Soininen H, Partanen K et al. MRI of the hippocampus in Alzheimer's disease: sensitivity,

specificity, and analysis of the incorrectly classified subjects. Neurobiol Aging 1998; 19(1):23–31.

21. Laakso MP, Frisoni GB, Kononen M et al. Hippocampus and entorhinal cortex in frontotemporal dementia and Alzheimer's disease: a morphometric MRI study. Biol Psychiatry 2000; 47(12):1056–63.

22. Laakso MP, Lehtovirta M, Partanen K, Riekkinen PJ, Soininen H. Hippocampus in Alzheimer's disease: a 3-year follow-up MRI study. Biol Psychiatry 2000; 47(6):557–61.

23. Barber R, Gholkar A, Scheltens P et al. Medial temporal lobe atrophy on MRI in dementia with Lewy bodies. Neurology 1999; 52(6):1153–8.

24. Barber R, Gholkar A, Scheltens P et al. MRI volumetric correlates of white matter lesions in dementia with Lewy bodies and Alzheimer's disease. Int J Geriatr Psychiatry 2000; 15(10):911–16.

25. Barber R, Ballard C, McKeith IG, Gholkar A, O'Brien JT. MRI volumetric study of dementia with Lewy bodies: a comparison with AD and vascular dementia. Neurology 2000; 54(6):1304–9.

26. Barber R, Panikkar A, McKeith IG. Dementia with Lewy bodies: diagnosis and management. Int J Geriatr Psychiatry 2001; 16(Suppl 1):S12–18.

27. Barber R, McKeith IG, Ballard C, Gholkar A, O'Brien JT. A comparison of medial and lateral temporal lobe atrophy in dementia with Lewy bodies and Alzheimer's disease: magnetic resonance imaging volumetric study. Dement Geriatr Cogn Disord 2001; 12(3):198–205.

28. Roman GC, Tatemichi TK, Erkinjuntti T et al. Vascular dementia: diagnostic criteria for research studies. Report of the NINDS-AIREN International Workshop. Neurology 1993; 43(2):250–60.

29. Scheltens P, Erkinjuntti T, Leys D et al. White matter changes on CT and MRI: an overview of visual rating scales. European Task Force on Age-Related White Matter Changes. Eur Neurol 1998; 39(2):80–9.

30. Van Straaten ECW, Scheltens P, Knol D et al. Operational definitions for the NINDS-AIREN criteria for vascular dementia. An interobserver study. Stroke 2003; 34:1907–12.

31. Walters RJ, Fox NC, Crum WR, Taube D, Thomas DJ. Haemodialysis and cerebral oedema. Nephron 2001; 87(2):143–7.

32. Jack CR Jr, Shiung MM, Gunter JL et al. Comparison of different MRI brain atrophy rate measures

with clinical disease progression in AD. Neurology 2004; 62(4):591–600.

33. Fox NC, Cousens S, Scahill R, Harvey RJ, Rossor MN. Using serial registered brain magnetic resonance imaging to measure disease progression in Alzheimer disease: power calculations and estimates of sample size to detect treatment effects. Arch Neurol 2000; 57(3):339–44.

34. Smith SM, De Stefano N, Jenkinson M, Matthews P. Normalized accurate measurement of longitudinal brain change. J Comput Assist Tomogr 2001; 25:466–75.

35. Filippi M, Rovaris M, Inglese M et al. Interferon beta-1a for brain tissue loss in patients at presentation with syndromes suggestive of multiple sclerosis: a randomised, double-blind, placebo-controlled trial. Lancet 2004; 364:1489–96.

36. Gunter JL, Shiung MM, Manduca A, Jack CR Jr. Methodological considerations for measuring rates of brain atrophy. J Magn Reson Imaging 2003; 18:16–24.

37. Shear PK, Sullivan EV, Mathalon DH et al. Longitudinal volumetric computed tomographic analysis of regional brain changes in normal aging and Alzheimer's disease. Arch Neurol 1995; 52:392–402.

38. Thompson PM, Mega MS, Wood RP et al. Cortical change in Alzheimer's disease detected with a disease-specific population-based brain atlas. Cerebral Cortex 2001; 11:1–16.

39. Csernansky JG, Wang L, Joshi S et al. Early DAT is distinguished from aging by high-dimensional mapping of the hippocampus. Neurology 2000; 55:1636–43.

40. Jack CR Jr, Petersen RC, Xu Y et al. Rate of medial temporal lobe atrophy in typical aging and Alzheimer's disease. Neurology 1998; 51(4):993–9.

41. Scheltens P, Fox N, Barkhof F, De Carli C. Structural magnetic resonance imaging in the practical assessment of dementia: beyond exclusion. Lancet Neurology 2002; 1:13–21.

42. Wahlund LO, Julin P, Lindqvist J, Scheltens P. Visual assessment of medical temporal lobe atrophy in demented and healthy control subjects: correlation with volumetry. Psychiatry Res 1999; 90(3):193–9.

43. Haller JW, Banerjee A, Christensen GE et al. Three-dimensional hippocampal MR morphometry with high-dimensional transformation of a neuroanatomic atlas. Radiology 1997; 202(2):504–10.

44. Crum WR, Scahill RI, Fox NC. Automated hippocampal segmentation by regional fluid registration of serial MRI: validation and application in Alzheimer's disease. Neuroimage 2001; 13(5): 847–55.

45. Du AT, Schuff N, Zhu XP et al. Atrophy rates of entorhinal cortex in AD and normal aging. Neurology 2003; 60(3):481–6.

46. Du AT, Schuff N, Amend D et al. Magnetic resonance imaging of the entorhinal cortex and hippocampus in mild cognitive impairment and Alzheimer's disease. J Neurol Neurosurg Psychiatry 2001; 71(4):441–7.

47. Chan D, Fox NC, Scahill RI et al. Patterns of temporal lobe atrophy in semantic dementia and Alzheimer's disease. Ann Neurol 2001; 49(4):433–42.

48. Birkenhager WH, Forette F, Staessen JA. Dementia and antihypertensive treatment. Curr Opin Nephrol Hypertens 2004; 13(2):225–30.

49. Scheltens PH, Kittner B. Preliminary results from an MRI/CT-based database for vascular dementia and Alzheimer's disease. Ann NY Acad Sci 2000; 903:542–7.

50. Schmidt R, Enzinger C, Ropele S, Schmidt H, Fazekas F. Progression of cerebral white matter lesions: 6-year results of the Austrian Stroke Prevention Study. Lancet 2003; 361:2046–8.

51. Wahlund L, Almkvist O, Basin H, Julin P. MRI in successful ageing, a 5-year follow-up study from eighth to ninth decade of life. Magn Res Imag 1996; 14:601–8.

52. Veldink J, Scheltens P, Jonker C, Launer LJ. Progression of cerebral white matter hyperintensities on MRI is related to diastolic blood pressure. Neurology 1998; 51:319–20.

53. Schmidt R, Fazekas F, Kapeller P, Schmidt H, Hartung H-P. MRI white matter hyperintensities: three-year follow-up of the Austrian Stroke Prevention Study. Neurology 1999; 53:132–9.

54. Whitman G, Tang T, Lin MA, Baloh RW. A prospective study of cerebral white matter abnormalities in older people with gait dysfunction. Neurology 2001; 57:990–4.

55. DeCarli C, Swan GE, Park M et al. Longitudinal changes in brain and white matter hyperintensity

volumes among elderly male twins from the NHLBI twin study. Neurology 2002; 58(Suppl 3):A399.

56. Prins ND, van Straaten EC, van Dijk EJ et al. Measuring progression of cerebral white matter lesions on MRI: visual rating and volumetrics. Neurology 2004; 62:1533–9.

57. Bradley KM, Bydder GM, Budge MM et al. Serial brain MRI at 3–6 month intervals as a surrogate marker for Alzheimer's disease. Br J Radiol 2002; 75(894):506–13.

58. Schmidt R, Scheltens P, Erkinjuntti T et al. for the European Task Force on Age-Related White Matter Changes. White matter lesion progression: a surrogate endpoint for trials in cerebral small-vessel disease. Neurology 2004; 63:139–44.

14

Guidelines for Randomised Clinical Studies in Parkinson's Disease with Dementia and Dementia with Lewy Bodies

Richard Camicioli and Serge Gauthier

Introduction

Parkinson's disease with dementia (PDD) and dementia with Lewy bodies (DLB) are common causes of dementia. Overlapping pathological features include subcortical and cortical Lewy bodies, and co-existent Alzheimer disease (AD) pathology in some cases.[1]

Although the prominent pathological finding in PDD and DLB is alpha-synuclein-positive Lewy bodies, amyloid, tau and synuclein pathology may be synergistic.[2] Neurochemical systems possibly affected in Lewy body-related dementias include dopamine, acetylcholine, noradrenaline (norepinephrine), serotonin and glutamate.[3] Potential outcomes for clinical trials include progression to dementia (e.g. conversion rate), cognitive dysfunction (especially executive dysfunction), behavioural and psychiatric problems (e.g. depression and psychosis), functional status (e.g. activities of daily living (ADL) and instrumental ADL (IADL)), global impairment and quality of life, among others (see Table 14.1). Motor symptoms of parkinsonism represent an added dimension that requires specific investigation and monitoring in clinical trials,[4] as do sleep disorders.

Definitions

Parkinsonism precedes dementia by at least one year for PDD, but co-occurs within 12 months in DLB. Consensus criteria exist for DLB (dementia with parkinsonism, hallucinations and cognitive fluctuations),[5] and are in development for PDD. Some DLB cases may not have parkinsonism. Currently, generic criteria for dementia such as DSM-IV (which specifies memory and another cognitive domain to be affected along with functional

Table 14.1 Targets of treatments
• Cognitive impairment
• Behavioural impairment
• Motor function
• Functional impairment
• Global outcomes
• Onset of dementia
• Caregiver burden
• Quality of life
• Sleep disorder
• Falls
• Hospitalisation
• Nursing home placement
• End-of-life care
• Mortality
• Economic impact

impairment) or ICD-10 criteria are applied to diagnose PDD.[6] Clinical features to be kept in mind for the definition of PDD include the stipulation that PD onset should precede dementia (by at least 1–2 years). Definitions of PDD should recognise that memory deficits may improve with cueing and that non-amnestic features can be prominently affected (e.g. attention/fluctuations, executive function/working memory/verbal fluency, and visuo-spatial function) with relative preservation of other domains such as language and praxis. Behavioural symptoms such as apathy, depression, psychosis and personality changes may also be prominently affected.

Outcome measures (see Table 14.2)

Neuropsychological measures

Both PDD and DLB patients can exhibit greater executive dysfunction and cognitive slowing with less severe episodic memory impairment than seen in AD.[7–10] Visuo-spatial deficits are prominent in DLB,[11] whereas semantic memory may be equally affected in DLB and AD.[12] Whether patients with mild cognitive impairment (e.g. with memory, executive, visuo-spatial, or mild multidomain impairment) can be defined by analogy to the elderly who progress to AD remains to be established.[13] Studies that target specific subgroups or target symptoms may be relevant, since such treatment response in various cognitive domains may differ. Studies that directly compare PDD and DLB with respect to cognitive profile mostly show overlap.[8,14–16] Patients with REM sleep behaviour disorder also show this cognitive profile, and these patients can evolve into DLB.[17,18]

Global cognitive measures

The Mini-Mental State Examination (MMSE) is widely used as a cognitive assessment tool in dementia. A recent study examining patients with PD, including those who developed dementia, found that patients with PDD decline on the MMSE at a rate similar to patients with AD (PDD: 2.3, 95% confidence interval (CI) 2.1–2.5 points

Table 14.2 Examples of assessment instruments – see text for citation details
Cognitive scales and batteries
• Global
– Mini-Mental State Examination (MMSE)
– Alzheimer Disease Assessment Scale-Cognitive (ADAS-Cog)
• Comprehensive
– Repeatable Battery for Assessment of Neuropsychological Status (RBANS)
– Wechsler Adult Intelligence Scale-Revised Neuropsychological Instrument (WAIS-R NI)
– Mattis Dementia Rating Scale (DRS)
• Executive Function
– Delis–Kaplan Executive Function Scale (D-KEFS)
– Executive Interview (EXIT25)
– Scales for Outcomes of Parkinson's disease-cognition (SCOPA-Cog)
Non-cognitive domains
• Global
– Clinicians Interview-Based Impression of Change (CIBIC-Plus)
– Clinical Dementia Rating Scale (CDR)
• Function
– Disability Assessment in Dementia Scale (DADS)
• Behaviour
– Neuropsychiatric Inventory (NPI)
– Cornell Depression Rating Scale for Dementia

per year).[19] A recent clinical trial of rivastigmine in PDD reported a 2.9-point difference in performance between the treated group compared to the placebo group using the 70-point Alzheimer Disease Assessment Scale – cognitive subscale (ADAS-Cog), an instrument used in Alzheimer's disease trials.[20] Interestingly, similar modest effect sizes were observed for the MMSE, a computer-based test of attention, frontal measures (DKEFS, see next section) and clock drawing. (This trial is also remarkable for employing a retrieved drop out strategy, so that patients who withdrew early could still be evaluated close to the time of the end point.)

Executive function

Executive function was the cognitive domain that was most susceptible to decline in a recent longitudinal Parkinson's disease study, consistent with prior results.[21] Some global neuropsychological batteries and scales (Repeatable Battery for Assessment of Neuropsychological Status (RBANS),[22] Dementia Rating Scale (DRS),[23] Wechsler Adult Intelligence Scale – Revised Neuropsychological Instrument (WAIS-R NI),[24] Scales for Outcomes of Parkinson's disease–cognition (SCOPA-Cog)[25]) incorporate executive function measures, and others exclusively target executive functions (Delis–Kaplan Executive Function System (DKEFS),[26] Executive Interview (EXIT25)).[27] These would be appropriate cognitive outcome measures in studies of PDD and DLB.[28]

Attention and fluctuation

Measurement of attention is important, given that fluctuating attention is one of the core criteria for DLB, evident in PDD. A computerised cognitive battery has been used in clinical trials of DLB and PDD, which may be useful for monitoring fluctuating attention.[29] Others have utilised similar tests of reaction time in discriminating demented from non-demented PD patients.[30] Several questionnaire-based approaches to measuring fluctuations have been validated and should be considered in trials of PDD and DLB.[31,32]

Non-cognitive domains
Global assessment

The use of a Clinician's Global Assessments of Change (CIBIC) with (CIBIC-Plus) or without caregiver input is not without controversy in Alzheimer's disease trials.[33] Nevertheless such global impressions are commonly used in clinical trials in dementia. The Clinical Dementia Rating Scale is a semi-structured approach that allows assessment of global function in that it incorporates both cognitive and functional measures as assessed by interview and testing.[34] It has been used to rate dementia severity in PD.[35]

Functional status

Functional status is an important outcome, reflecting caregiver burden. It is determined by cognitive, motor and behavioural features. Scales that reflect the executive dysfunction evident in this patient population would ideally be used in clinical trials. The Disability Assessment in Dementia Scale (DADS) is one such instrument.[36]

Behavioural symptoms

Behavioural symptoms are overlapping between DLB and PDD.[37] Psychiatric symptoms including psychosis (hallucinations and delusions), depression and anxiety can be distressing to patients and family members.[38] Depression and apathy are distinct phenomena that can confound and contribute to cognitive dysfunction.[39–42] Behavioural problems such as agitation are difficult to manage and can trigger nursing home placement. The Neuropychiatric Inventory (NPI) has been used to measure behaviour problems in DLB,[43,44] and was sensitive to change in clinical trials.[20,45] Specific measures with a wider range of responses should be included in trials in which specific target symptoms are relevant; however, most scales have not been validated in parkinsonian patients with dementia, which necessitates using scales that have been used in other dementias, such as the Cornell Depression Rating Scale,[46] or the development of new instruments for specific target symptoms.[47]

Biomarkers

There are no established biomarkers in PDD and DLB. Volumetric magnetic resonance brain imaging studies have shown atrophy for both PDD and DLB, generally to a lesser degree than in AD, in medial temporal, frontal, occipital, parietal lobes as well as in subcortical structures.[48–53] Progressive brain atrophy has been reported.[54] Magnetic resonance spectroscopy has shown decreased NAA/Cr

(N-acetylaspartate/creatine) and normal mI/Cr (myoinositol/Cr) in Parkinson's disease, in contrast to Alzheimer's disease.[55–57] Blood flow studies (PET (positron emission tomography) and SPECT (single photon emission computed tomography) have shown cerebral hypoperfusion in Lewy body dementias.[58,59] Findings in PET and SPECT studies examining dopaminergic terminals show decreased binding, distinct from AD.[60–63] Cholinergic changes have been shown in PDD and DLB using PET.[64] Cerebrospinal fluid changes in DLB may be distinct from those in AD.[65] These biomarkers may hold promise for defining patient groups in future clinical trials, and potentially for tracking the course of disease, a concept that requires longitudinal validation.

Clinical trials in PDD and DLB

Cholinesterase inhibitors and cholinergic agents

Defining DLB has enabled the development of targeted clinical trials. Initial case series were encouraging with the cholinesterase inhibitors donepezil,[66–70] rivastigmine[71–73] and galantamine.[74,75] A Cochrane review identified one placebo-controlled trial showing modest efficacy.[45,76] In some series, subgroups of patients appear more likely to benefit from therapy.[77,78]

Until recently most published trials in PDD have been open label.[79,80] Two small placebo-controlled trials of donepezil,[81,82] and a recent large placebo-controlled trial of rivistigmine have been published.[20] In one study, withdrawal of medication led to deterioration in a series of patients with PDD and DLB treated with donepezil.[83] Both rivastigmine and donepezil have been associated with marked deterioration in case studies.[84,85] One study suggested that hallucinations were prone to improvement with rivastigmine,[86] consistent with results of a study of donepezil.[87] An open-label study of galantamine in PDD has been reported.[88] While cholinesterase inhibitors remain promising, confirmatory studies, and trials of different medications, which may differ in their efficacy in DLB

and PDD, will be needed. Cholinergic agonists may also hold promise, but few studies are currently available.

Anti-psychotic agents

A clinical feature highlighting the distinction between DLB and other dementias is marked sensitivity to typical anti-psychotic medications.[89] This has led to important practice changes, including the avoidance of these medications. Psychotic symptoms are also associated with poor outcomes in PD. Clozapine, an atypical anti-psychotic drug, has been shown to be effective in PD in placebo-controlled trials that included patients with mild PDD.[90–93] Olanzapine has been shown to worsen motor function in placebo-controlled studies, including one comparing olanzapine to clozapine.[94,95] Recent studies comparing anti-psychotic agents provide a plausible model for studies in this frail population.[96,97] Agents such as quetiapine, aripiprazole and ziprasidone have been studied in open-label series or case reports,[98–101] but are awaiting placebo-controlled trials. Anti-psychotic agents have not been subjected to appropriate trials in DLB, though open-label studies have been reported.[102]

Dopaminergic medications

Since motor impairment is significant in patients with PDD, and can be significant in patients with DLB, studies targeting its treatment are relevant. One study has shown that patients with PDD and DLB respond to dopaminergic therapy to a lesser degree than patients without dementia.[103] Studies examining dopaminergic therapy in advanced PD should monitor cognitive outcomes.

Other potential approaches

Noradrenergic and serotonergic systems may be involved in DLB and PDD and offer potential therapeutic approaches. Anti-glutaminergic medications merit consideration, especially in light of studies that suggest that such drugs may confer cognitive benefit in AD.[104,105] Unlike the situation for PD and AD, there is currently no animal model for DLB and PDD that can be used to understand its pathophysiology and develop new therapies.

General clinical trial design considerations

Non-randomised and open-label studies only provide a basis for randomised trials, and cannot form the basis for treatment choices. The gold standard for clinical trials remains the parallel-group randomised double-blind placebo-controlled design. If there are no accepted treatments, it remains ethical to consider a placebo arm. Given the potential to worsen patients, stopping rules and careful safety monitoring should be built into clinical trials. When treatments are available, equivalency, inferiority or superiority trials can be considered, where interventions are compared. Generally these studies require larger sample sizes than placebo-controlled studies. If the duration of effect of a treatment is in question, or if symptoms might resolve spontaneously, randomised start/stop designs are considerations, but these may be impractical in this frail population. Fluctuating or transitory symptoms, which are common in PD, DLB and PDD, make crossover studies (where patients act as their own controls) difficult to interpret.

A major challenge in all dementia studies is to obtain a representative sample of subjects. Data suggest that patients with Alzheimer's disease in clinical trials are not representative of the overall patient population.[106] Studies of Parkinson's disease have excluded elderly subjects.[107] Nevertheless, subgroups of patients might be targeted for intervention. For example, executive dysfunction, fluctuating attention, depression, apathy and psychosis are specific, disturbing symptoms that might be targeted in clinical trials. For the present, PDD and DLB should be considered distinct entities. In the future, clinical trials aimed at common problems might include patients with both PDD and DLB. If that is done, preplanned stratification would have to be done from the start to allow assessment of potential differential treatment response. Ethical concerns are paramount in all clinical research. Patients with PDD may have difficulty with comprehending treatment options, highlighting the importance of establishing surrogate decision makers and taking extreme care in obtaining informed consent.[108] This concern applies equally to DLB.

Conclusions

Clinical trials for PDD and DLB are in their early stages compared to studies in AD and those for the motor aspects of PD. In part this is related to barriers that can be overcome with additional research (see Table 14.3). Defining these entities and their natural history remain important ongoing issues. Natural history studies would ideally be population based and with autopsy confirmation. Separating patients with co-existent AD may be important for future trials, and may be possible using biomarkers. Cognitive outcome measures designed for AD may not be ideal for PDD and DLB in that they emphasise memory and not frontal/executive measures. Measures that are valid in the setting of PDD/DLB need to be developed. Non-cognitive behaviours are also important targets. Biomarkers in PDD and DLB are critically needed, both to monitor change in clinical trials and to exclude co-existent pathologies. Clearly, motor function and its progression must be taken into account in these disorders. Conversely, studies in advanced Parkinson's disease that target motor symptoms should also assess cognitive function.

Acknowledgements

We thank Sheri Foster for assisting with manuscript preparation and Dr Wendy Johnston for reviewing the manuscript.

Table 14.3 Critical barriers to clinical trials in Parkinson's disease with dementia (PDD) and dementia with Lewy bodies (DLB)

- Lack of consensus definition for PDD
- Inaccuracies in diagnosis of DLB
- Overlapping pathology (cortical and subcortical Lewy bodies, amyloid plaques, neurofibrillary tangles, cerebral infarctions, white matter disease)
- Subgroups of patients within disease groups who may have treatment response:
  - cognitive: amnestic, executive dysfunction, multiple cognitive domains
  - fluctuating
  - psychotic, hallucinating
  - apathetic, depressed
- Natural history of cognitive impairment in PD is not fully defined
- Assessment instruments available for clinical trials have been designed for AD, specific challenges include:
  - longitudinal measurement of executive function
  - measurement of attention and fluctuations
  - ADL instruments with executive components
- Imaging/other biomarker correlates of dementia in PDD and DLB are needed
- Most current studies in advanced PD target motor symptoms only
- Motor impairment confounds cognitive and functional assessments

References

1. Camicioli R, Fisher N. Progress in clinical neurosciences: Parkinson's disease with dementia and dementia with Lewy bodies. Can J Neurol Sci 2004; 31:7–21.
2. Lee VM, Giasson BI, Trojanowski JQ. More than just two peas in a pod: common amyloidogenic properties of tau and alpha-synuclein in neurodegenerative diseases. Trends Neurosci 2004; 27:129–34.
3. Perry EK, McKeith I, Thompson P et al. Topography, extent, and clinical relevance of neurochemical deficits in dementia of Lewy body type, Parkinson's disease, and Alzheimer's disease. Ann N Y Acad Sci 1991; 640:197–202.
4. Burn DJ, Rowan EN, Minett T et al. Extrapyramidal features in Parkinson's disease with and without dementia and dementia with Lewy bodies: a cross-sectional comparative study. Mov Disord 2003; 18:884–9.
5. McKeith IG, Galasko D, Kosaka K et al. Consensus guidelines for the clinical and pathologic diagnosis of dementia with Lewy bodies (DLB): report of the consortium on DLB international workshop. Neurology 1996; 47:1113–24.
6. American Psychiatric Association. Diagnostic and statistical manual of mental disorders: DSM-IV (4th edn). Washington, DC: American Psychiatric Association; 1994.
7. Stern Y, Tang MX, Jacobs DM et al. Prospective comparative study of the evolution of probable Alzheimer's disease and Parkinson's disease dementia. J Int Neuropsychol Soc 1998; 4:279–84.
8. Noe E, Marder K, Bell KL et al. Comparison of dementia with Lewy bodies to Alzheimer's disease and Parkinson's disease with dementia. Mov Disord 2004; 19:60–7.
9. Hamilton JM, Salmon DP, Galasko D et al. A comparison of episodic memory deficits in neuropathologically-confirmed Dementia with Lewy bodies and Alzheimer's disease. J Int Neuropsychol Soc 2004; 10:689–97.
10. Simard M, van Reekum R, Myran D et al. Differential memory impairment in dementia with Lewy bodies and Alzheimer's disease. Brain Cogn 2002; 49:244–9.
11. Collerton D, Burn D, McKeith I, O'Brien J. Systematic review and meta-analysis show that dementia with Lewy bodies is a visual-perceptual and attentional-executive dementia. Dement Geriatr Cogn Disord 2003; 16:229–37.
12. Lambon RMA, Powell J, Howard D et al. Semantic memory is impaired in both dementia with Lewy bodies and dementia of Alzheimer's type: a comparative neuropsychological study and literature review. J Neurol Neurosurg Psychiatry 2001; 70:149–56.

13. Davis HS, Rockwood K. Conceptualization of mild cognitive impairment: a review. Int J Geriatr Psychiatry 2004; 19:313–19.

14. Cormack F, Aarsland D, Ballard C, Tovee MJ. Pentagon drawing and neuropsychological performance in Dementia with Lewy Bodies, Alzheimer's disease, Parkinson's disease and Parkinson's disease with dementia. Int J Geriatr Psychiatry 2004; 19:371–7.

15. Ballard CG, Aarsland D, McKeith I et al. Fluctuations in attention: PD dementia vs DLB with parkinsonism. Neurology 2002; 59:1714–20.

16. Downes JJ, Priestley NM, Doran M et al. Intellectual, mnemonic, and frontal functions in dementia with Lewy bodies: A comparison with early and advanced Parkinson's disease. Behav Neurol 1998; 11:173–83.

17. Boeve BF, Silber MH, Ferman TJ et al. Association of REM sleep behavior disorder and neurodegenerative disease may reflect an underlying synucleinopathy. Mov Disord 2001; 16:622–30.

18. Boeve BF, Silber MH, Ferman TJ. REM sleep behavior disorder in Parkinson's disease and dementia with Lewy bodies. J Geriatr Psychiatry Neurol 2004; 17:146–57.

19. Aarsland D, Andersen K, Larsen JP et al. The rate of cognitive decline in Parkinson disease. Arch Neurol 2004; 61:1906–11.

20. Emre M, Aarsland D, Albanese A et al. Rivastigmine for dementia associated with Parkinson's disease. N Engl J Med 2004; 351:2509–18.

21. Azuma T, Cruz RF, Bayles KA et al. A longitudinal study of neuropsychological change in individuals with Parkinson's disease. Int J Geriatr Psychiatry 2003; 18:1115–20.

22. Beatty WW, Ryder KA, Gontkovsky ST et al. Analyzing the subcortical dementia syndrome of Parkinson's disease using the RBANS. Arch Clin Neuropsychol 2003; 18:509–20.

23. Brown GG, Rahill AA, Gorell JM et al. Validity of the Dementia Rating Scale in assessing cognitive function in Parkinson's disease. J Geriatr Psychiatry Neurol 1999; 12:180–8.

24. Peavy GM, Salmon D, Bear PI et al. Detection of mild cognitive deficits in Parkinson's disease patients with the WAIS-R NI. J Int Neuropsychol Soc 2001; 7:535–43.

25. Marinus J, Visser M, Verwey NA et al. Assessment of cognition in Parkinson's disease. Neurology 2003; 61:1222–8.

26. Delis DC, Kramer JH, Kaplan E, Holdnack J. Reliability and validity of the Delis-Kaplan Executive Function System: an update. J Int Neuropsychol Soc 2004; 10:301–3.

27. Royall DR, Mahurin RK, Gray KF. Bedside assessment of executive cognitive impairment: the executive interview. J Am Geriatr Soc 1992; 40:1221–6.

28. Royall DR, Lauterbach EC, Cummings JL et al. Executive control function: a review of its promise and challenges for clinical research. A report from the Committee on Research of the American Neuropsychiatric Association. J Neuropsychiatry Clin Neurosci 2002; 14:377–405.

29. Wesnes KA, McKeith IG, Ferrara R et al. Effects of rivastigmine on cognitive function in dementia with lewy bodies: a randomised placebo-controlled international study using the cognitive drug research computerised assessment system. Dement Geriatr Cogn Disord 2002; 13:183–92.

30. Pate DS, Margolin DI. Cognitive slowing in Parkinson's and Alzheimer's patients: distinguishing bradyphrenia from dementia. Neurology 1994; 44:669–74.

31. Bradshaw J, Saling M, Hopwood M et al. Fluctuating cognition in dementia with Lewy bodies and Alzheimer's disease is qualitatively distinct. J Neurol Neurosurg Psychiatry 2004; 75:382–7.

32. Ferman TJ, Smith GE, Boeve BF et al. DLB fluctuations: specific features that reliably differentiate DLB from AD and normal aging. Neurology 2004; 62:181–7.

33. Quinn J, Moore M, Benson DF et al. A videotaped CIBIC for dementia patients: validity and reliability in a simulated clinical trial. Neurology 2002; 58:433–7.

34. Morris JC. The Clinical Dementia Rating (CDR): current version and scoring rules. Neurology 1993; 43:2412–14.

35. Goldman WP, Baty JD, Buckles VD et al. Cognitive and motor functioning in Parkinson disease: subjects with and without questionable dementia. Arch Neurol 1998; 55:674–80.

36. Gelinas I, Gauthier L, McIntyre M, Gauthier S. Development of a functional measure for persons with Alzheimer's disease: the disability assessment for dementia. Am J Occup Ther 1999; 53:471–81.

37. Aarsland D, Ballard C, Larsen JP, McKeith I. A comparative study of psychiatric symptoms in dementia with Lewy bodies and Parkinson's disease with and without dementia. Int J Geriatr Psychiatry 2001; 16:528–36.

38. Weintraub D, Moberg PJ, Duda JE et al. Effect of psychiatric and other nonmotor symptoms on disability in Parkinson's disease. J Am Geriatr Soc 2004; 52:784–8.

39. Norman S, Troster AI, Fields JA, Brooks R. Effects of depression and Parkinson's disease on cognitive functioning. J Neuropsychiatry Clin Neurosci 2002; 14:31–6.

40. Cubo E, Bernard B, Leurgans S, Raman R. Cognitive and motor function in patients with Parkinson's disease with and without depression. Clin Neuropharmacol 2000; 23:331–4.

41. Pluck GC, Brown RG. Apathy in Parkinson's disease. J Neurol Neurosurg Psychiatry 2002; 73:636–42.

42. Isella V, Melzi P, Grimaldi M et al. Clinical, neuropsychological, and morphometric correlates of apathy in Parkinson's disease. Mov Disord 2002; 17:366–71.

43. Cummings JL, Mega M, Gray K et al. The Neuropsychiatric Inventory: comprehensive assessment of psychopathology in dementia. Neurology 1994; 44:2308–14.

44. Del Ser T, McKeith I, Anand R et al. Dementia with lewy bodies: findings from an international multicentre study. Int J Geriatr Psychiatry 2000; 15:1034–45.

45. McKeith I, Del Ser T, Spano P et al. Efficacy of rivastigmine in dementia with Lewy bodies: a randomised, double-blind, placebo-controlled international study. Lancet 2000; 356:2031–6.

46. Alexopoulos GS, Abrams RC, Young RC, Shamoian CA. Cornell Scale for Depression in Dementia. Biol Psychiatry 1988; 23:271–84.

47. Robert PH, Clairet S, Benoit M et al. The apathy inventory: assessment of apathy and awareness in Alzheimer's disease, Parkinson's disease and mild cognitive impairment. Int J Geriatr Psychiatry 2002; 17:1099–105.

48. Burton EJ, McKeith IG, Burn DJ et al. Cerebral atrophy in Parkinson's disease with and without dementia: a comparison with Alzheimer's disease, dementia with Lewy bodies and controls. Brain 2004; 127:791–800.

49. Camicioli R, Moore MM, Kinney A et al. Parkinson's disease is associated with hippocampal atrophy. Mov Disord 2003; 18:784–90.

50. Cousins DA, Burton EJ, Burn D et al. Atrophy of the putamen in dementia with Lewy bodies but not Alzheimer's disease: an MRI study. Neurology 2003; 61:1191–5.

51. Almeida OP, Burton EJ, McKeith I et al. MRI study of caudate nucleus volume in Parkinson's disease with and without dementia with Lewy bodies and Alzheimer's disease. Dement Geriatr Cogn Disord 2003; 16:57–63.

52. Middelkoop HA, van der Flier WM, Burton EJ et al. Dementia with Lewy bodies and AD are not associated with occipital lobe atrophy on MRI. Neurology 2001; 57:2117–20

53. Barber R, McKeith I, Ballard C, O'Brien J. Volumetric MRI study of the caudate nucleus in patients with dementia with Lewy bodies, Alzheimer's disease, and vascular dementia. J Neurol Neurosurg Psychiatry 2002; 72:406–7.

54. O'Brien JT, Paling S, Barber R et al. Progressive brain atrophy on serial MRI in dementia with Lewy bodies, AD, and vascular dementia. Neurology 2001; 56:1386–8.

55. Molina JA, Garcia-Segura JM, Benito-Leon J et al. Proton magnetic resonance spectroscopy in dementia with Lewy bodies. Eur Neurol 2002; 48:158–63.

56. Summerfield C, Gomez-Anson B, Tolosa E et al. Dementia in Parkinson disease: a proton magnetic resonance spectroscopy study. Arch Neurol 2002; 59:1415–20.

57. Camicioli RM, Korzan JR, Foster SL et al. Posterior cingulate metabolic changes occur in Parkinson's disease patients without dementia. Neurosci Lett 2004; 354:177–80.

58. Firbank MJ, Colloby SJ, Burn DJ et al. Regional cerebral blood flow in Parkinson's disease with and without dementia. Neuroimage 2003; 20:1309–19.

59. Colloby S, O'Brien J. Functional imaging in Parkinson's disease and dementia with Lewy bodies. J Geriatr Psychiatry Neurol 2004; 17:158–63.

60. Gilman S, Koeppe RA, Little R et al. Striatal monoamine terminals in Lewy body dementia and Alzheimer's disease. Ann Neurol 2004; 55:774–80.

61. Plotkin M, Amthauer H, Klaffke S et al. Combined (123)I-FP-CIT and (123)I-IBZM SPECT for the diagnosis of parkinsonian syndromes: study on 72 patients. J Neural Transm 2004. DOI: 10.1007/s00702–004–0208–x.

62. O'Brien JT, Colloby S, Fenwick J et al. Dopamine transporter loss visualized with FP-CIT SPECT in the differential diagnosis of dementia with Lewy bodies. Arch Neurol 2004; 61:919–25.

63. Walker Z, Costa DC, Walker RW et al. Striatal dopamine transporter in dementia with Lewy bodies and Parkinson disease: a comparison. Neurology 2004; 62:1568–72.

64. Bohnen NI, Kaufer DI, Ivanco LS et al. Cortical cholinergic function is more severely affected in parkinsonian dementia than in Alzheimer disease: an in vivo positron emission tomographic study. Arch Neurol 2003; 60:1745–8.

65. Gomez-Tortosa E, Gonzalo I, Fanjul S et al. Cerebrospinal fluid markers in dementia with lewy bodies compared with Alzheimer disease. Arch Neurol 2003; 60:1218–22.

66. Kaufer DI, Catt KE, Lopez OL, DeKosky ST. Dementia with Lewy bodies: response of delirium-like features to donepezil. Neurology 1998; 51:1512.

67. Shea C, MacKnight C, Rockwood K. Donepezil for treatment of dementia with Lewy bodies: a case series of nine patients. Int Psychogeriatr 1998; 10:229–38.

68. Aarsland D, Bronnick K, Karlsen K. Donepezil for dementia with Lewy bodies: a case study. Int J Geriatr Psychiatry 1999; 14:69–72.

69. Samuel W, Caligiuri M, Galasko D et al. Better cognitive and psychopathologic response to donepezil in patients prospectively diagnosed as dementia with Lewy bodies: a preliminary study. Int J Geriatr Psychiatry 2000; 15:794–802.

70. Querfurth HW, Allam GJ, Geffroy MA et al. Acetylcholinesterase inhibition in dementia with Lewy bodies: results of a prospective pilot trial. Dement Geriatr Cogn Disord 2000; 11:314–21.

71. McKeith IG, Grace JB, Walker Z et al. Rivastigmine in the treatment of dementia with Lewy bodies: preliminary findings from an open trial. Int J Geriatr Psychiatry 2000; 15:387–92.

72. Grace J, Daniel S, Stevens T et al. Long-Term use of rivastigmine in patients with dementia with Lewy bodies: an open-label trial. Int Psychogeriatr 2001; 13:199–205.

73. Maclean LE, Collins CC, Byrne EJ. Dementia with Lewy bodies treated with rivastigmine: effects on cognition, neuropsychiatric symptoms, and sleep. Int Psychogeriatr 2001; 13:277–88.

74. Edwards KR, Hershey L, Wray L et al. Efficacy and safety of galantamine in patients with dementia with Lewy bodies: a 12-week interim analysis. Dement Geriatr Cogn Disord 2004; 17(Suppl 1):40–8.

75. Holm AC. Alleviation of multiple abnormalities by galantamine treatment in two patients with dementia with Lewy bodies. Am J Alzheimers Dis Other Dement 2004; 19:215–18.

76. Wild R, Pettit T, Burns A. Cholinesterase inhibitors for dementia with Lewy bodies. Cochrane Database Syst Rev 2003; CD003672.

77. McKeith IG, Wesnes KA, Perry E, Ferrara R. Hallucinations predict attentional improvements with rivastigmine in dementia with lewy bodies. Dement Geriatr Cogn Disord 2004; 18:94–100.

78. Pakrasi S, Mukaetova-Ladinska EB, McKeith IG, O'Brien JT. Clinical predictors of response to acetyl cholinesterase inhibitors: experience from routine clinical use in Newcastle. Int J Geriatr Psychiatry 2003; 18:879–86.

79. Werber EA, Rabey JM. The beneficial effect of cholinesterase inhibitors on patients suffering from Parkinson's disease and dementia. J Neural Transm 2001; 108:1319–25.

80. Hutchinson M, Fazzini E. Cholinesterase inhibition in Parkinson's disease. J Neurol Neurosurg Psychiatry 1996; 61:324–5.

81. Aarsland D, Litvan I, Salmon D et al. Performance on the dementia rating scale in Parkinson's disease with dementia and dementia with Lewy bodies: comparison with progressive supranuclear palsy and Alzheimer's disease. J Neurol Neurosurg Psychiatry 2003; 74:1215–20.

82. Leroi I, Brandt J, Reich SG et al. Randomized placebo-controlled trial of donepezil in cognitive impairment in Parkinson's disease. Int J Geriatr Psychiatry 2004; 19:1–8.

83. Minett TS, Thomas A, Wilkinson LM et al. What happens when donepezil is suddenly withdrawn? An open label trial in dementia with Lewy bodies and Parkinson's disease with dementia. Int J Geriatr Psychiatry 2003; 18:988–93.

84. Richard IH, Justus AW, Greig NH et al. Worsening of motor function and mood in a patient with Parkinson's disease after pharmacologic challenge with oral rivastigmine. Clin Neuropharmacol 2002; 25:296–9.

85. Onofrj M, Thomas A. Severe worsening of parkinsonism in Lewy body dementia due to donepezil. Neurology 2003; 61:1452.

86. Bullock R, Cameron A. Rivastigmine for the treatment of dementia and visual hallucinations associated with Parkinson's disease: a case series. Curr Med Res Opin 2002; 18:258–64.

87. Bergman J, Lerner V. Successful use of donepezil for the treatment of psychotic symptoms in patients with Parkinson's disease. Clin Neuropharmacol 2002; 25:107–10.

88. Aarsland D, Hutchinson M, Larsen JP. Cognitive, psychiatric and motor response to galantamine in Parkinson's disease with dementia. Int J Geriatr Psychiatry 2003; 18:937–41.

89. Ballard C, Grace J, McKeith I, Holmes C. Neuroleptic sensitivity in dementia with Lewy bodies and Alzheimer's disease. Lancet 1998; 351:1032–3.

90. The French Clozapine Parkinson Study Group. Clozapine in drug-induced psychosis in Parkinson's disease. Lancet 1999; 353:2041–2.

91. Pollak P, Tison F, Rascol O et al. Clozapine in drug induced psychosis in Parkinson's disease: a randomised, placebo controlled study with open follow up. J Neurol Neurosurg Psychiatry 2004; 75:689–95.

92. The Parkinson Study Group. Parkinson. Low-dose clozapine for the treatment of drug-induced psychosis in Parkinson's disease. N Engl J Med 1999; 340:757–63.

93. Factor SA, Friedman JH, Lannon MC et al. Clozapine for the treatment of drug-induced psychosis in Parkinson's disease: results of the 12 week open label extension in the PSYCLOPS trial. Mov Disord 2001; 16:135–9.

94. Ondo WG, Levy JK, Vuong KD et al. Olanzapine treatment for dopaminergic-induced hallucinations. Mov Disord 2002; 17:1031–5.

95. Breier A, Sutton VK, Feldman PD et al. Olanzapine in the treatment of dopamimetic-induced psychosis in patients with Parkinson's disease. Biol Psychiatry 2002; 52:438–45.

96. Goetz CG, Blasucci LM, Leurgans S, Pappert EJ. Olanzapine and clozapine: comparative effects on motor function in hallucinating PD patients. Neurology 2000; 55:789–94.

97. Morgante L, Epifanio A, Spina E et al. Quetiapine and clozapine in parkinsonian patients with dopaminergic psychosis. Clin Neuropharmacol 2004; 27:153–6.

98. Juncos JL, Roberts VJ, Evatt ML et al. Quetiapine improves psychotic symptoms and cognition in Parkinson's disease. Mov Disord 2004; 19:29–35.

99. Targum SD, Abbott JL. Efficacy of quetiapine in Parkinson's patients with psychosis. J Clin Psychopharmacol 2000; 20:54–60.

100. Fernandez HH, Trieschmann ME, Burke MA et al. Long-term outcome of quetiapine use for psychosis among Parkinsonian patients. Mov Disord 2003; 18:510–14.

101. Fernandez HH, Trieschmann ME, Friedman JH. Aripiprazole for drug-induced psychosis in Parkinson disease: preliminary experience. Clin Neuropharmacol 2004; 27:4–5.

102. Fernandez HH, Trieschmann ME, Burke MA, Friedman JH. Quetiapine for psychosis in Parkinson's disease versus dementia with Lewy bodies. J Clin Psychiatry 2002; 63:513–15.

103. Bonelli SB, Ransmayr G, Steffelbauer M et al. L-dopa responsiveness in dementia with Lewy bodies, Parkinson disease with and without dementia. Neurology 2004; 63:376–8.

104. Reisberg B, Doody R, Stoffler A et al. Memantine in moderate-to-severe Alzheimer's disease. N Engl J Med 2003; 348:1333–41.

105. Tariot PN, Farlow MR, Grossberg GT et al. Memantine treatment in patients with moderate to severe Alzheimer disease already receiving donepezil: a randomized controlled trial. JAMA 2004; 291:317–24.

106. Schneider LS, Olin JT, Lyness SA, Chui HC. Eligibility of Alzheimer's disease clinic patients for clinical trials. J Am Geriatr Soc 1997; 45:923–8.

107. Mitchell SL, Sullivan EA, Lipsitz LA. Exclusion of elderly subjects from clinical trials for Parkinson disease. Arch Neurol 1997; 54:1393–8.

108. Dymek MP, Atchison P, Harrell L, Marson DC. Competency to consent to medical treatment in cognitively impaired patients with Parkinson's disease. Neurology 2001; 56:17–24.

15

Clinical Trials for Vascular Cognitive Impairment

Kenneth Rockwood and Gordon Gubitz

Introduction

No therapeutic effort in medicine is free from its own history, but it is easy to believe that the treatment of vascular dementia is encumbered more than most others. Understanding treatment efficacy is made challenging by the shifting status of vascular dementia as an entity (reflected in varying terminology and criteria), its long status as an 'also ran' not just in dementia, but so too in stroke, and by the lack of a compound with even the equivocal success accorded the cholinesterase inhibitors in Alzheimer's disease. Add to this that the entity is a grouping of potentially many illnesses – giving a more variable natural history – that its conceptualisation has been influenced by the comparatively more specific phenomenon of Alzheimer's disease and that its major deficits often go unmeasured by standard tests, and the scope of the problem is easy to imagine.

Some historical considerations

To understand where we are now, it is perhaps best to review how we came to be here. Many physicians will know of an era in which it was held that 'multi-infarct dementia' was the second most common cause of dementia after Alzheimer's disease. Over time, as reviewed elsewhere, the deficiencies of this convenient story became clear.[1-3] First, many people with vascular dementia did not have multiple strokes, but rather had a picture of chronic ischaemia, often manifest more by cognitive disorders than by the usual stroke syndromes. Thus the 'multi-infarct' part of the description was seen to be insensitive. Second, unlike the memory-dominant conceptualisation of the dementia of Alzheimer's disease, a dysexecutive syndrome was appreciated to be the chief source of disability. In consequence, the 'dementia' needed to be distinguished from that of Alzheimer's disease. Following a pioneering re-conceptualisation by Hachinski and Erkinjuntti in 1993, the term 'vascular cognitive impairment' (VCI) has come to be the umbrella term for cognitive disorders that present as a result of cerebrovascular disease.[4]

This evolution has spanned many years, during which the drive for better therapy has not stood still. Therapeutic effort in the modern era is highly constrained by public regulatory authorities; however, the need for standardisation cannot always accommodate the more rapidly changing scientific understanding of disease. In consequence, although the field in general is moving to the VCI terminology, even recent reports of clinical trials use the term 'vascular dementia' (VaD). The term VaD has not come without its own history, however. The conceptualisation of VaD employed in recent trials was developed from a consensus-based process that gave rise to the highly influential VaD criteria of the National Institutes of Neurological Disorders and Stroke/ Association Internationale pour la Recherche et l'Enseignement en Neurosciences (NINDS-AIREN). The NINDS-AIREN model of VaD, as it turns out, has been none other than that of 'multi-infarct dementia'.

Challenges that arise from a multi-infarct dementia conceptualisation of vascular dementia

The journey from multi-infarct dementia (MID) to VaD and back has posed its own challenges, which can be grouped as those that arise as a consequence of the entity itself, and those that arise as a consequence of how we understand that entity. The former, ontological challenges are those that faced MID. The entity is only a small part of the VCI spectrum (indeed, in some clinic settings, MID is not even the second most common cause of VCI, let alone dementia[5]). This is also likely to be the case in population settings, given that dementia occurs most commonly in those who are very old, and that group usually has mixed dementia.[6] It does not give rise to a specific syndrome in the manner of the progressive memory impairment seen with Alzheimer's disease. The natural history shows periods of plateaus and periods of acute exacerbations, sometimes with improvement or it may be indistinguishable from Alzheimer's disease, even with careful patient selection.[7–10] Narrowing the focus to post-stroke patients, in the manner of criteria that use the multi-infarct model of VaD, does not diminish this variability.[11]

In consequence of these considerations, how we understand this entity has challenges for how we carry out and then interpret clinical trials in VaD. In the first instance, MID-modelled VaD criteria, although highly specific,[12,13] have been markedly insensitive (in the order of 0.10) to the entity of VCI, especially that seen in memory clinic settings.[5] Indeed the NINDS-AIREN criteria are insensitive even in post-stroke cohorts with dementia.[14] Thus the results of any such studies, though internally valid, might only obtain in a small number of patients that one would see in a clinic setting. To be conclusive, the studies will need to be long enough to overcome the effect of plateaus. Plateaus, coupled with variable presentations will decrease treatment effects while increasing the variance of estimates, and will reduce effect sizes. Lower effect sizes will mean that the studies will have to be large to yield statistically significant results. Low effect sizes also make clinical interpretation of the results more difficult.[15] In short, the variable nature of the phenomenon, and the varying ways that people have attempted to get to grips with it, can be expected to pose particular challenges for clinical trials in VaD.

Non-cholinergic drug studies

The current era in VaD pharmacotherapy has been defined by trials of cholinesterase inhibitors, which have, for better or for worse, largely adopted the trials design and instrumentation used for studying these compounds in Alzheimer's disease. Before reviewing these, however, we first turn to three non-cholinergic compounds, each of which has contributed to the modern understanding of VaD.

Nimodipine is a dihydropyridine calcium-antagonist that has been used in an early open-label study,[16] and more recently in the placebo-controlled Scandinavian Multi-Infarct Dementia trial.[17] The rationale for nimodipine was that it causes vasodilatation, without a so-called 'steal effect' (i.e. not inducing critical hypoperfusion), and reduces the influx of calcium ions into depolarised neurons. As calcium overload is a common mechanism of neuronal dysfunction and death, the result was that it would have a neuroprotective effect, in addition to whatever effect might arise form changes in cerebral blood flow.[18] The Scandinavian trial showed no beneficial effect of nimodipine in the primary analysis, despite showing fewer cerebrovascular and cardiac events in the actively treated group. A *post hoc* analysis, however, showed that patients with subcortical infarction showed benefit compared with those on placebo, although these results were not statistically significant.[19] The experience was encouraging enough to give rise to a trial that has recently been completed. Importantly for our understanding of VaD, the relative homogeneity of clinical and radiographic features of the patients with subcortical ischaemia has given rise to criteria for so-called 'subcortical vascular dementia'.[20] These criteria emphasise the executive dysfunction seen in such patients. As argued elsewhere,[21] and as reviewed in Chapter 21, an emphasis on executive dysfunction is likely to be sensitive, but not specific, as executive dysfunction is a core feature of any of the dementia syndromes. This has been the experience of the Canadian cohort study known as the Consortium to Investigate Vascular Impairment of Cognition.[5] That the problem is highlighted in the subcortical vascular dementia criteria reflects not just its prominence in that setting, but a general lack of clinically sensible instrumentation for executive dysfunction.[22,23] Most recently, as described below, nimodipine (plus aspirin) has been used in a comparison trial with rivastigmine.[24]

Propentofylline, a glial modulator, is no longer under study despite its observed beneficial effect on learning and memory. (Conflict of interest disclosure: KR acted as an investigator and paid consultant to Hoechst Marion Roussel (now merged as Aventis) on this compound.) The initial experience, using the 'multi-infarct dementia construct',[25,26] and otherwise largely unpublished results of European and Canadian double-blind, placebo-controlled, randomised, parallel group trials on the efficacy and safety of long-term treatment with propentofylline, showed benefits compared with placebo, in patients with mild-to-moderate VaD according to NINDS-AIREN criteria.[27] This 24-week study showed a significant symptomatic improvement and long-term efficacy in Alzheimer's Disease Assessment Scale–Cognitive Subscale (ADAS-Cog) and

Clinician's Interview-Based Impression of Charge, Plus Caregiver Input (CIBIC-Plus) up to 48 weeks. In addition, sustained treatment effects for at least 12 weeks after withdrawal were present, suggesting an effect on disease progression. The propentofylline experience in VaD also showed an important and persistent placebo response, presaging some of the cholinesterase inhibitor response.[28]

For a participant in the process, it is difficult to write entirely objectively about propentofylline. In short, the public record shows that while the Canadian/European studies showed benefit, the two American studies (in both Alzheimer's disease and VaD) did not, possibly due to a strong interaction with food and less well controlled food compliance in the negative studies.[29] With negative studies, the American regulatory authorities were understandably disinclined to support registration of propentofylline. The European regulatory process was then new, and seemed there and in Canada to possibly be susceptible to an early sense of 'buyer's remorse' over the registration of cholinesterase inhibitors. Whatever the actual rationale, some of the stated reasons for rejecting the drug reflect why the pharmaceutical industry might be sceptical of innovation in drug design. Two of the propentofylline studies used a design of a placebo-controlled delayed start (i.e. some actively treated patients were first assigned to placebo and started on treatment only months later) and early withdrawal design. Although such a design had been widely endorsed,[30,31] it resulted in four comparison groups (always on placebo; always on active treatment; first on placebo then on active treatment; first on active treatment, then on placebo). This proved to be difficult for regulators to follow, and was the subject of specific criticism. So too was their inclusion of patients with mixed dementia, even though such patients are, on a population basis, sure to be the most common patients encountered. Thus, despite placebo-controlled trials with up to 72 weeks of an active-treatment placebo comparison, showing treatment effects in favour of propentofylline, the drug is no longer manufactured or even under study.

One drug that has fared better than propentofylline, despite, at the outset, no evidently more unequivocal data, is memantine. Early studies with this compound enrolled patients with Alzheimer's disease and with vascular dementia.[32] Specific trials in mild to moderate vascular dementia showed that patients who received memantine 20 mg/day had less cognitive deterioration at 28 weeks as measured by the ADAS-Cog.[33–35] This effect was not clinically discernible, however, using the CIBC and CIBIC-Plus scales, respectively. This theme of apparent non-responsiveness of a clinical global measure in the face of apparently detectable change on the ADAS-Cog has been explored in Chapter 7 and will recur below as we consider the studies with cholinesterase inhibitors in VaD. Briefly, it is not yet clear whether the effect is truly not clinically detectable, or whether it requires clinicians to know what to look for in order to find it. The absence of systematic assessment of caregiver observations also means that any of their insights are relegated to the status of anecdote. Memantine has also been used in conjunction with a cholinesterase inhibitor, but not in patients with VaD.[36]

Other compounds studied in patients with VaD include anti-thrombotics, ergot alkaloids, nootropics, TRH-analogue, ginkgo biloba, plasma viscosity drugs, hyperbaric oxygen, antioxidants, serotonin and histamine receptor antagonists, vasoactive agents, xanthine derivates, and calcium antagonists. These now are chiefly of historical interest and are reviewed elsewhere.[37–39]

Cholinesterase inhibitors in vascular dementia

Three cholinesterase inhibitors (ChEIs; donepezil, galantamine, and rivastigmine) are currently used in the treatment of Alzheimer's disease. To better understand what evidence is available to support their use in *vascular* dementia, we searched the literature to identify systematic reviews and properly conducted randomised controlled trials (RCTs) that have evaluated

cholinesterase inhibitor use in this patient population.

The Cochrane Library was searched (on September 11 2004) via the internet (*www.update-software.com.ezproxy.library.dal.ca/projects/cochrane/*). The updated systematic reviews supported by the Cochrane Dementia and Cognitive Improvement Group include one completed systematic review evaluating donepezil in patients with vascular cognitive impairment.[40,41] This systematic review (which excluded open-label trials) included two randomised, double-blind, parallel-group controlled trials (the Donepezil 307 study,[42] and the Donepezil 308 study[43]). In total, 1219 people with mild to moderate cognitive decline due to probable or possible vascular dementia (according to the NINDS/AIREN criteria and the Hachinski Ischemia Scale (HIS)) were included in the meta-analysis. Donepezil, at doses of 5 or 10 mg a day was compared with placebo for 24 weeks. For each outcome measure, the mean change from baseline at weeks 12 and 24 was calculated using a last observation carried forward analysis. Data were included for subjects randomised to treatment who complied with treatment until the endpoint, and on the Intention-To-Treat – Last Observation Carried Forward (ITT-LOCF) population, who were randomised to treatment, received at least one dose of study medication and provided data at baseline and at least one post-baseline assessment.

The systematic review reported that the donepezil groups showed a statistically significantly better performance than the placebo groups on the cognitive subscale of the (ADAS-Cog). At the endpoint (ITT-LOCF), the weighted mean difference (WMD) between donepezil 5 mg/day and placebo was -1.66 (95% confidence interval (CI) -2.40 to -0.92, $P < 0.0001$). The WMD between donepezil 10 mg/day and placebo was -2.17 (95% CI -2.97 to -1.37, $P < 0.00001$).

In terms of assessment of global function, the seven-point CIBIC-Plus scale was dichotomised, with those showing no change or decline compared with those showing improvement. There was benefit associated with donepezil 5 mg/day compared with placebo at 24 weeks (odds ratio (OR) 1.56, 95% CI 1.15 to 2.11, $P = 0.004$), but not for donepezil 10 mg/day.

In their discussion, Malouf et al postulated that the classification of patients according to two diagnostic guidelines (NINDS-AIREN criteria and the HIS scale) probably resulted in the enrolment of patients with different vascular cognitive impairment aetiologies, and that the studies probably contained patients with mixed dementia.[41] The positive effect of donepezil might therefore be due to the medication effect on the Alzheimer's disease component. The review concluded by stating that:

> Evidence from the available studies supports the benefit of donepezil in improving cognition function, clinical global impression and activities of daily living in patients with probable or possible mild to moderate vascular cognitive impairment after 6 months' treatment. Extending studies for longer periods would be desirable to establish the efficacy of donepezil in patients with advanced stages of cognitive impairment. Moreover, there is an urgent need for establishing specific clinical diagnostic criteria and rating scales for vascular cognitive impairment.

The Cochrane Library also lists two 'protocols in development'; one for the use of galantamine,[44] and one for rivastigmine in patients with vascular cognitive impairment.[45] These two protocols have not yet collected or analysed data; the review process is ongoing.

A search was also undertaken of the Cochrane Library's database of randomised and controlled clinical trials using the terms: 'vascular dementia', 'vascular cognitive impairment', 'cholinesterase inhibitor(s)', 'donepezil (Aricept)', 'rivastigmine (Exelon)', and 'galanatamine (Reminyl)'. A total of 100 references were found; 58 of these were duplications or inappro-

priate, leaving 42 references linked to individual studies evaluating one of these three cholinesterase inhibitors.

Donepezil

The two randomised, placebo-controlled donezepil studies (307 and 308 noted in the systematic review) were identified.[42,43] One open-label, unpublished study was also listed; no data were available.

Rivastigmine

No RCTs were identified that evaluated the use of rivastigmine in an isolated population with vascular cognitive impairment. Two RCTs were found that examined cognitive outcomes in patients with Alzheimer's disease and vascular risk factors.[46,47] Erkinjuntti et al evaluated 6 and 12 mg of rivastigmine versus placebo in 725 patients with Alzheimer's disease and hypertension, and found that patients treated with rivastigmine had superior CIBIC-Plus scores; hypertensive patients receiving rivastigmine also showed improvements, suggesting that such benefits *may* be observed in those with vascular risk factors.[48] Kumar et al randomised patients aged 45–90 years with Alzheimer's disease and vascular risk factors to placebo ($n = 235$), low-dose rivastigmine (1–4 mg/day, $n = 233$), or high-dose rivastigmine (6–12 mg/day, $n = 231$) for 26 weeks.[46] The results indicated that patients randomised to rivistigmine had better scores on the CIBIC-Plus, and that those with vascular risk factors also experienced more benefit. Several references identified by the search strategy were linked to a small open-label pilot study (16 patients) comparing riviastigmine to aspirin for the treatment of symptoms specific to patients with subcortical vascular dementia.[47] The same research group also published an open-label 12-month study of rivastigmine in subcortical vascular dementia in 208 patients.[49] One reference to an as yet unpublished study was also noted,[50] as well as several references to abstracts that have not been otherwise published.

Galantamine

No RCTs were identified that evaluated the use of galanatamine in an isolated population with vascular cognitive impairment. Most reported studies assessing galantamine have used populations with probable vascular dementia (NINDS-AIRENS criteria), or possible Alzheimer's disease (NINCDS-ADRDA criteria). One multi-centre, double-blind trial randomly assigned such patients to galantamine 24 mg/day ($n = 396$) or placebo ($n = 196$) for a 6-month period.[51] Primary endpoints were cognition (ADAS-Cog) and global functioning (CIBIC-Plus). Galantamine showed greater efficacy than placebo on ADAS-Cog (galantamine change −1.7 (standard error (SE) 0.4) versus placebo 1.0 (SE 0.5); treatment effect 2.7 points; $P < 0.0001$) and CIBIC-Plus (213 (74%) versus 95 (59%) patients remained stable or improved, $P = 0.0001$). However, there was not a statistically significant improvement in the CIBIC-Plus in the subgroup ($n = 188$) diagnosed with vascular dementia ($P = 0.238$), although it was stated that the trial was not sufficiently powered to make this specific assessment. A subsequent open-label extension of this study (galantamine 24 mg/day for six months) was also identified.[52] Four- hundred and fifty-nine patients (mean age = 75.2 years) entered the open-label phase. Of these patients, 195 (42.5%) had a diagnosis (made at the time of initial enrolment in the RCT) of probable vascular dementia, and 238 (51.9%) had a diagnosis of Alzheimer's disease *with* cerebrovascular disease; the remainder had an inconclusive diagnosis. The primary efficacy endpoint was change in cognition, based on scores on the 11-item Alzheimer's Disease Assessment Scale–cognitive subscale (ADAS-Cog/11). At month 12 of the study, improvements from baseline (the start of the double-blind phase) in ADAS-Cog/11 scores were observed in both the group that received placebo during the double-blind phase (placebo/galantamine group: −0.3 points; 95% CI, −1.64 to 1.06), and the group that received galantamine during the double-blind phase (galantamine/galantamine group: −0.9 points; 95% CI, −1.73 to 0.03). Further interim analyses of this study group,[53] as well as *post hoc* subgroup analyses of this trial were identified.[54] Finally, several references to abstracts that have not been otherwise published were identified.

Lessons from clinical trails in patients with vascular dementia

Clearly, the nature of vascular dementia has proven to be an elusive target. As argued, this is in part due to the distribution of the disease (which favours VaD mixed with Alzheimer's disease versus pure VaD) and its clinical heterogeneity. It is also due to our evolving understanding of the disease and the changes in clinical criteria. Given that the clinical criteria that became dominant (the NINIDS-AIREN) defined a 'probable VaD' phenotype that was uncommon clinically, changes in the understanding of VaD were a particular challenge, given that it has taken a long time for patients to be recruited, and for the trials to be concluded.

Against this background, it is perhaps easy to understand why the regulatory interpretation of this disease entity has been such a delicate flower, and required so much care and attention, while still yielding incomplete results. Nevertheless, it is ironic to see the conceptually elegant European/Canadian studies of propentofylline rejected, and earlier studies discounted for relying on the Hachinski Ischemia Score, when later studies, with standard designs, inclusion/exclusion criteria that remain based on a multi-infarct model and outcome measures designed for Alzheimer's disease, effect sizes that are no larger, are rather more celebrated. The decline of the placebo-controlled withdrawal design also means that we have no way to assess the common clinical observation that, for some patients, drug treatment effects are most evident only when the drug is stopped.

Studies of the use of ChEIs in VaD have thus far yielded inconsistent results. Despite evidence for a cholinergic deficit in patients who have dementia without AD, the studies of ChEIs that most closely conform to the ChEI experience in AD are the studies that are most likely to have included patents with mixed AD/VaD. Still, from a regulatory perspective, it is troubling to understand the persisting controversy over whether the highly specific and poorly sensitive NINDS-AIREN criteria are actually defining VaD, or might still yield patients 'contaminated' by AD. Such a position is operationally indistinguishable from the position that there is no such thing as 'pure' VaD. While the latter might have a certain scientific support, much recent evidence is against it.[55,56] In any case, it was not the position taken by the regulatory authorities when trials in this area were first begun using ChEIs. How future regulators and future ventures will be affected by these variable interpretations is not clear, but they are unlikely to inspire enthusiasm for pure VaD studies. Given that most dementia affects people who are likely to have both vascular and neurodegenerative disorders,[6] and that the two appear to interact synergistically,[57,58] it might be as well to abandon studies in any but mixed dementia.

How should regulators deal with evolving information? The short answer would appear to be that they should do so judiciously. It is particularly important to separate opinion, even consensus opinion, from data. For example, the merit of rejecting studies that used the Hachinski Ischemia Score (which at least has been autopsy validated) to define VaD, in favour of consensus criteria that have no validation and that in any case are based on the same model of VaD as is the Ischemia Score, suggests that extra-scientific factors are at play. At the same time, a single piece of published evidence, unless it is the rare crucial experiment, should not too readily trump accumulated unpublished information. As always, judgement is difficult, but when it is so, an important guarantor of public confidence is transparency of the processes by which judgements were made. In this instance, the rest of the world would appear to have much to learn from the deliberations of the United States Food and Drug Administration, and their easily navigated website (*www.fda.gov*).

An exception to abandoning trials in probable VaD might be subcortical VaD, but it remains to be demonstrated that the apparently unique aspects of this disorder will give rise to a clinical radiographic profile that is both sensitive and specific.[59] Particular caution will be necessary, in that other studies of the neuropsychological profile of subcortical ischaemia do not show the predilection for impaired executive function.[60–63] Still, the entity has been studied in other trials,[24,64] and in other as yet unpublished subgroup analyses, with mixed support for the entity and for its treatment with a variety of compounds. Whether this is a specific effect remains to be determined.

Another target for a specifically vascular dementia might be post-stroke dementia. In addition to the methodological difficulties noted above (including those of diagnostic criteria) a pragmatic obstacle will be that, although there is a rich literature of stroke trials, there has been little work done on cognition in general, much less on some of the more subtle aspects of it which seem to be the best target for drug treatment effects. One area in which we might benefit from the lead of the stroke physicians, however, is in their careful control of vascular risk factors. The impressive degree of stabilisation seen in patients in the placebo arms of various studies,[42,43] probably argues for the importance of vascular risk factor control, although the stability of patients with VaD who are selected for clinical trials cannot be discounted.

Whichever areas within VaD are selected for study, the question of outcome measurement will be crucial. As in AD, the United States Food and Drug Administration requires two 'pre-specified co-primary outcome measures' being a test of cognitive function, and a clinical global assessment.[65] Whether the tests used thus far specifically evaluate executive function is debatable. If executive function is specifically impaired in VaD, this will diminish the sensitivity of the assessment to clinically important change. As argued in the chapter on global assessment and elsewhere,[66] clinical global measures require that clinicians know what to look for, and, in dementia, this is not as easy as looking for reversal of disease progression.

The challenges posed to investigators and regulators in testing drugs for use in patients with VaD well illustrate the many factors – including social ones, such as the extent to which we socially construct disease entities – that must be considered when interpreting 'the evidence'. Clearly, there is no alternative to this – no utopian world where what to do can be had unequivocally by virtue of a meta-analysis. For those embedded in the process, the best course would appear to be a transparent laying bare of the process by which scientific and regulatory decisions are made.

References

1. Rockwood K. Vascular cognitive impairment and vascular dementia. J Neurol Sci 2002; 203–204:23–7.

2. O'Brien JT, Erkinjuntti T, Reisberg B et al. Vascular cognitive impairment. Lancet Neurol 2003; 2(2):89–98.

3. Erkinjuntti T, Roman G, Gauthier S et al. Emerging therapies for vascular dementia and vascular cognitive impairment. Stroke 2004; 35(4):1010–7.

4. Erkinjuntti T, Hachinski VC. Rethinking vascular dementia. Cerebrovasc Dis 1993; 3:3–23.

5. Rockwood K, Davis H, MacKnight C et al. The Consortium to Investigate Vascular Impairment of Cognition: methods and first findings. Can J Neurol Sci 2003; 30(3):237–43.

6. Neuropathology Group. Medical Research Council Cognitive Function and Aging Study. Pathological correlates of late-onset dementia in a multicentre, community-based population in England and Wales. Neuropathology Group of the Medical Research Council Cognitive Function and Ageing Study (MRC CFAS). Lancet 2001; 357(9251): 169–75.

7. Bowler JV, Eliasziw M, Steenhuis R et al. Comparative evolution of Alzheimer disease, vascular dementia, and mixed dementia. Arch Neurol 1997; 54(6):697–703.

8. Desmond DW, Erkinjuntti T, Sano M et al. The cognitive syndrome of vascular dementia: implications for clinical trials. Alzheimer Dis Assoc Disord 1999; 13(Suppl 3):S21–9.

9. Wentzell C, Rockwood K, MacKnight C et al. Progression of impairment in patients with vascular cognitive impairment without dementia. Neurology 2001; 57:714–6.

10. Mungas D, Reed BR, Ellis WG, Jagust WJ. The effects of age on rate of progression of Alzheimer disease and dementia with associated cerebrovascular disease. Arch Neurol 2001; 58(8):1243–7.

11. Ballard C, Rowan E, Stephens S, Kalaria R, Kenny RA. Prospective follow-up study between 3 and 15 months after stroke: improvements and decline in cognitive function among dementia-free stroke survivors 75 years of age. Stroke 2003; 34:2440–4.

12. Zekry D, Duyckaerts C, Belmin J et al. Alzheimer's disease and brain infarcts in the elderly. Agreement with neuropathology. J Neurol 2002; 249(11): 1529–34.

13. Gold G, Bouras C, Canuto A et al. Clinicopathological validation study of four sets of clinical criteria for vascular dementia. Am J Psychiatry 2002; 159(1):82–7.

14. Tang WK, Chan SS, Chiu HF et al. Impact of applying NINDS-AIREN criteria of probable vascular dementia to clinical and radiological characteristics of a stroke cohort with dementia. Cerebrovasc Dis 2004; 18(2):98–103.

15. Rockwood K, MacKnight C. Assessing the clinical importance of statistically significant improvement in anti-dementia drug trials. Neuroepidemiology 2001; 20(2):51–6.

16. Pantoni L, Carosi M, Amigoni S et al. A preliminary open trial with nimodipine in patients with cognitive impairment and leukoaraiosis. Clin Neuropharmacol 1996; 19:497–506.

17. Pantoni L, Bianchi C, Beneke M et al. The Scandinavian multi-infarct dementia trial: a double-blind, placebo-controlled trial on nimodipine in multi-infarct dementia. J Neurol Sci 2000; 175: 116–23.

18. Kobayashi T, Mori Y. Ca^{2+} channel antagonists and neuroprotection from cerebral ischemia. Eur J Pharmacol 1998; 363(1):1–15.

19. Pantoni L, Rossi R, Inzitari D et al. Efficacy and safety of nimodipine in subcortical vascular dementia: a subgroup analysis of the Scandinavian multi-infarct dementia trial. J Neurol Sci 2000; 175:124–34.

20. Erkinjuntti T, Inzitari D, Pantoni L et al. Research criteria for subcortical vascular dementia in clinical trials. J Neural Tramsm Suppl. 2000; 59:23–30.

21. Voss SE, Bullock RA. Executive function: the core feature of dementia? Dement Geriatr Cogn Disord 2004; 18(2):207–16.

22. Royall DR, Lauterbach EC, Cummings JL et al. Executive control function: a review of its promise and challenges for clinical research. A report from the Committee on Research of the American Neuropsychiatric Association. J Neuropsychiatry Clin Neurosci 2002; 14(4):377–405.

23. Royall DR, Chiodo LK, Polk MJ. Executive dyscontrol in normal aging: normative data, factor structure, and clinical correlates. Curr Neurol Neurosci Rep 2003; 3(6):487–93.

24. Moretti R, Torre P, Antonello RM et al. Rivastigmine superior to aspirin plus nimodipine in subcortical vascular dementia: an open, 16-month, comparative study. Int J Clin Pract 2004; 58(4):346–53.

25. Mielke R, Moller HJ, Erkinjuntti T et al. Propentofylline in the treatment of vascular dementia and Alzheimer-type dementia: overview of phase I and phase II clinical trials. Alzheimer Dis Assoc Disord 1998; 12(Suppl 2):S29–35.

26. Rother M, Erkinjuntti T, Roessner M et al. Propentofylline in the treatment of Alzheimer's disease and vascular dementia: a review of phase III trials. Dement Geriatr Cogn Disord 1998; 9(Suppl 1):36–43.

27. Pischel T. Long-term-efficacy and safety of propentofylline in patients with vascular dementia. Results of a 12 months placebo-controlled trial. Neurobiol Aging 1998; 19(4S):s182 [Abstract].

28. Kittner B. Clinical trials of propentofylline in vascular dementia. European/Canadian Propentofylline Study Group. Alzheimer Dis Assoc Disord 1999; 3(Suppl 3):S166–71.

29. Rockwood K. Propentofylline. In: Qizilbash N, Schneider L, Chui H et al (eds). Evidence-Based Dementia: A Practical Guide to Diagnosis and Management. Oxford: Blackwell, 2002.

30. Bodick N, Forette F, Hadler D et al. Protocols to demonstrate slowing of Alzheimer disease progression. Position paper from the International Working Group on Harmonization of Dementia Drug Guidelines. The Disease Progression Sub-Group. Alzheimer Dis Assoc Disord 1997;11(Suppl 3):50–3.

31. Whitehouse PJ, Kittner B, Roessner M et al. Clinical trial designs for demonstrating disease-course-altering effects in dementia. Alzheimer Dis Assoc Disord 1998; 12(4):281–94.

32. Winblad B, Poritis N. Memantine in severe dementia: results of the 9M-Best Study (benefit and efficacy in severely demented patients during treatment with memantine). Int J Geriatr Psychiatry 1999; 14(2):135–46.

33. Wilcock G, Mobius HJ, Stoffler A for the MMM 500 group. A double-blind, placebo-controlled multicentre study of memantine in mild to moderate vascular dementia (MMM500). Int Clin Psychopharmacol 2002; 17(6):297–305.

34. Orgogozo JM, Rigaud AS, Stoffler A, Mobius HJ, Forette F. Efficacy and safety of memantine in patients with mild to moderate vascular dementia: a randomized, placebo-controlled trial (MMM 300). Stroke 2002l; 33(7):1834–9.

35. Rosen WG, Mohs RC, Davis KL. A new rating scale for Alzheimer's disease. Am J Psychiatry 1984; 141:1356–64.

36. Tariot PN, Farlow MR, Grossberg GT et al. Memantine treatment in patients with moderate to severe Alzheimer disease already receiving donepezil: a randomized controlled trial. JAMA 2004; 291(3):317–24.

37. Erkinjuntti T. Cerebrovascular dementia. Pathophysiology, diagnosis and treatment. CNS Drugs 1999; 12:35–48.

38. Erkinjuntti T, Rockwood K. Vascular Cognitive Impairment. Psychogeriatrics 2001; 1:27–38.

39. Inzitari D, Lamassa M, Pantoni L. Treatment of vascular dementias. In: Bowler JV, Hachinski V (eds). Vascular cognitive impairment preventable dementia. Oxford: Oxford University Press, 2003: 277–92.

40. Grimley Evans J, Birks J, Hermans D. Dementia and Cognitive Improvement Group. About the Cochrane Collaboration (Cochrane Review). In: The Cochrane Library, Issue 3, 2004. Oxford: Update Software.

41. Malouf R, Birks J. Donepezil for vascular cognitive impairment (Cochrane Review). In: The Cochrane Library, Issue 3, 2004. Oxford: Update Software.

42. Black S, Roman GC, Geldmacher DS et al. Donepezil 307 Vascular Dementia Study Group.

Efficacy and tolerability of donepezil in vascular dementia: positive results of a 24-week, multicenter, international, randomized, placebo-controlled clinical trial. Stroke 2003; 34:2323–30.

43. Wilkinson D, Doody R, Helme R et al. Donepezil 308 study. Donepezil in vascular dementia a randomized placebo controlled study. Neurology 2003; 61(4):479–86.

44. Craig D, Birks J. Galantamine for vascular cognitive impairment (Protocol for a Cochrane Review). In: The Cochrane Library, Issue 3, 2004. Oxford: Update Software.

45. Craig D, Birks J. Rivastigmine for vascular cognitive impairment (Protocol for a Cochrane Review). In: The Cochrane Library, Issue 3, 2004. Oxford: Update Software.

46. Kumar V, Anand R, Messina J, Hartman R, Veach J. An efficacy and safety analysis of Exelon in Alzheimer's disease patients with concurrent vascular risk factors. Eur J Neurol 2000; 7:159–69.

47. Moretti R, Torre P, Antonello RM et al. An open-label pilot study comparing rivastigmine and low-dose aspirin for the treatment of symptoms specific to patients with subcortical vascular dementia. Curr Ther Res Clin Exp 2002; 63:443–58.

48. Erkinjuntti T, Kurz A, Gauthier S et al. GAL-INT-6 Study Group. Efficacy of galantamine in probable vascular dementia and Alzheimer's disease combined with cerebrovascular disease: a randomized trial. Lancet 2002; 359:1283–90.

49. Moretti R, Torre P, Antonello RM, Cazzato G, Bava A. Rivastigmine in subcortical vascular dementia: a randomized, controlled, open 12-month study in 208 patients. Am J Alzheimers Dis Other Demen 2003; 18:265–72.

50. Haworth J. A 24-week prospective randomized multicentre double-blind placebo-controlled parallel-group comparison of the efficacy tolerability and safety of Exelon (Rivastigmine) capsules in patients with probable vascular dementia. National Research Register 2003.

51. Erkinjuntti T, Skoog I, Lane R, Andrews C. Rivastigmine in patients with Alzheimer's disease and concurrent hypertension. Int J Clin Pract 2002; 56:791–6.

52. Erkinjuntti T, Kurz A, Small GW et al. GAL-INT-6 Study Group. An open-label extension trial of galantamine in patients with probable vascular dementia and mixed dementia. Clin Ther 2003; 25:1765–82.

53. Kurz AF, Erkinjuntti T, Small GW, Lilienfeld S, Damaraju CR. Long-term safety and cognitive effects of galantamine in the treatment of probable vascular dementia or Alzheimer's disease with cerebrovascular disease. Eur J Neurol 2003; 10:633–40.

54. Small G, Erkinjuntti T, Kurz A, Lilienfeld S. Galantamine in the treatment of cognitive decline in patients with vascular dementia or Alzheimer's disease with cerebrovascular disease. CNS Drugs 2003; 17:905–14.

55. Di Carlo A, Baldereschi M, Amaducci L et al. Incidence of dementia, Alzheimer's disease, and vascular dementia in Italy. The ILSA Study. J Am Geriatr Soc 2002; 50(1):41–8.

56. Knopman DS, Parisi JE, Boeve BF et al. Vascular dementia in a population-based autopsy study. Arch Neurol 2003; 60(4):569–75.

57. Zekry D, Duyckaerts C, Belmin J et al. The vascular lesions in vascular and mixed dementia: the weight of functional neuroanatomy. Neurobiol Aging 2003; 24(2):213–9.

58. Riekse RG, Leverenz JB, McCormick W et al. Effect of vascular lesions on cognition in Alzheimer's disease: a community-based study. J Am Geriatr Soc 2004; 52(9):1442–8.

59. Rockwood K, Macknight C, Wentzel C et al. The diagnosis of 'mixed' dementia in the Consortium for the Investigation of Vascular Impairment of Cognition (CIVIC). Ann N Y Acad Sci 2000; 903:522–8.

60. Ingles JL, Wentzel C, Fisk JD, Rockwood K. Neuropsychological predictors of incident dementia in patients with vascular cognitive impairment, without dementia. Stroke 2002; 33(8):1999–2002.

61. Luis CA, Barker WW, Loewenstein DA et al. Conversion to dementia among two groups with cognitive impairment. A preliminary report. Dement Geriatr Cogn Disord 2004; 183:307–13.

62. Garrett KD, Browndyke JN, Whelihan W et al. The neuropsychological profile of vascular cognitive impairment – no dementia: comparisons to patients at risk for cerebrovascular disease and vascular dementia. Arch Clin Neuropsychol 2004; 9:745–57.

63. Jones S, Jonsson Laukka E, Small BJ, Fratiglioni L, Backman L. A preclinical phase in vascular dementia: cognitive impairment three years before diagnosis. Dement Geriatr Cogn Disord 2004; 18:233–9.

64. Moretti R, Torre P, Antonello RM, Cazzato G. Rivastigmine in subcortical vascular dementia: a comparison trial on efficacy and tolerability for 12 months follow-up. Eur J Neurol 2001; 8(4):361–2.

65. United States Food and Drug Administration. Peripheral and Central Nervous System Drugs Advisory Committee. Issues Paper on Vascular Dementia. February 6, 2001. *www.fda.gov/ohrms/dockets/ac/01/briefing/3724b2 01 VasDementia.doc* (accessed February 14 2005).

66. Rockwood K, Wallack M, TallisR. The treatment of Alzheimer's disease: success short of cure. Lancet Neurol 2003; 2:630–3.

16

Clinical Trials for Primary Prevention in Dementia

Ingmar Skoog and Deborah Gustafson

Introduction

Prevention and control of chronic diseases and the promotion of good health in human populations is an ultimate goal of epidemiological studies. The World Health Organisation reports a 223% expected increase from 1970 to 2025 in the number of adults age 60 years and above, so that by 2025, there will be 1.2 billion people in this age group worldwide. Of Western societies aged 65 years and older, the most rapidly growing group is age 85 years and older. Dementia disorders, such as Alzheimer's disease (AD) and vascular dementia, are chronic diseases of aging that may be prevented. The incidence of dementia is around 1% at age 70–74 years, and approaches 10% by age 85.[1] Its prevalence is 1% at age 70 years and approximately 30% at age 85 years.[2] These characteristics of dementia are of utmost importance as we witness an increasing lifespan and increasing prevalence of risk factors for dementia, such as hypertension, obesity and other vascular morbidities.

There are three major forms of prevention – primary, secondary and tertiary – the definitions of which depend on when an intervention is initiated in relationship to disease onset or its clinical symptoms (see Figure 16.1). The overall goal of primary prevention is to reduce the incidence of disease. This occurs by intervening before disease onset through promoting the initiation and maintenance of good health or removing potential causes of disease.[3,4] Thus, primary prevention trials of dementia may include non-demented individuals from the general population, individuals with potentially modifiable risk factors for dementia, e.g. hypertension or high cholesterol, or unmodifiable risk factors, such as family history or high age.

Secondary and tertiary prevention will only be discussed briefly in this chapter. The goal of secondary prevention is to prevent very early or preclinical forms of disease from progressing to more overt, manifest disease. Secondary prevention is best accomplished through early detection efforts followed by definitive treatments. In relationship to dementia, this would include interventions among those with mild cognitive impairment (MCI), memory complaints or a positive biological marker, to prevent further progression to dementia.[4] Tertiary prevention is designed to reduce disabilities and co-morbidities resulting from or accompanying disease, and to interfere with future disease progression. The goal of tertiary prevention is to reduce the burden of disease on society in terms of healthcare costs and numerous co-morbidities.

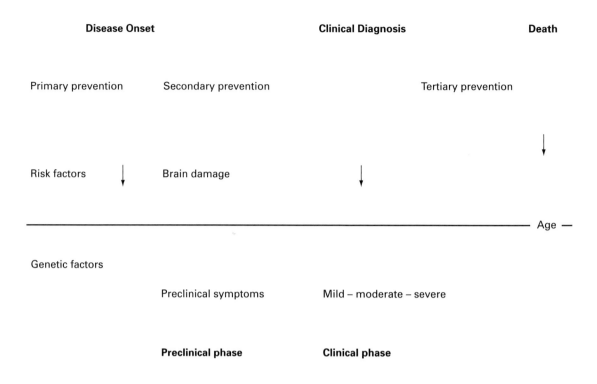

Figure 16.1

Placebo-controlled randomised trials

Placebo-controlled, randomised clinical trials are the gold standard for judging whether a specific factor or set of factors prevent disease or affect a disease outcome. Factors can include pharmaceutical agents, vitamins or nutraceuticals, or lifestyle habits or behaviours. Clinical trials are conducted in relationship to all forms of prevention. Primary prevention trials are extremely expensive, as they require large sample sizes and a sufficient follow-up period during which monitoring must occur and effects are observed.

Often, risk and protective factors for dementia have been identified from observational population studies and provide the basis for prevention trials. On the basis of observational studies, a number of risk factors that have been or may be targets of intervention studies in the future include hypertension, hypercholesterolaemia, hyperhomocysteinaemia, obesity, and the metabolic syndrome.

Observational studies have also identified potentially protective factors, such as pharmacological agents (e.g. anti-hypertensive agents, statins, postmenopausal hormone replacement therapy (HRT), non-steroidal anti-inflammatory drugs (NSAIDs), nutraceuticals, food components (e.g. antioxidants, B vitamins, soy) and physical activity.[5,6]

Data from placebo-controlled, randomised trials for dementia are now accumulating. Most primary prevention trials of dementia published thus far have been accomplished as part of large trials where dementia is actually a secondary outcome, and the primary outcome is reduction in a risk factor for dementia. These trials have focused on hypertension and hypercholesterolaemia. Three published primary prevention trials of dementia have dementia and cognitive function as primary outcome. These are the Women's Health Initiative Memory Study (WHIMS) HRT trials in which two forms of HRT were administered in an effort to reduce dementia and global cognitive impairment,[7-10] and a small soy trial.[11] The details of these trials are given in Tables 16.1–16.3. We await results from trials on NSAIDs, gingko balboa, antioxidants (vitamins E and C, beta-carotene) and B vitamins. Herein, are presented backgrounds, general outlines, methods and results of published placebo-controlled, randomised primary prevention trials for dementia, followed by a discussion about general methodological issues related to these trials.

Postmenopausal hormone replacement therapy (HRT)

HRT exists in many forms and has traditionally been taken for the treatment of menopausal symptoms, cardiovascular benefits and/or bone health. Many observational studies suggest a favourable effect of postmenopausal HRT on dementia and cognitive function.[5,6,12] In addition, data indicative of endogenous oestrogen exposure during premenopausal years, also suggest a protective effect of oestrogen.[13] However, data on oral contraceptive use, an exogenous premenopausal oestrogen and/or progestin exposure are lacking in the published literature. Experimental cell and animal studies suggest several possible mechanisms whereby oestrogens or other forms of postmenopausal HRT may exert a protective effect on cognition and dementia.[12] Most studies of oestrogen action in cell and animal models have used 17β-oestradiol, a different hormone from that used in human clinical studies. Nonetheless, resulting evidence has suggested various mechanisms of oestrogen action, including promotion of cholinergic and serotonergic activity, influences on amyloid metabolism, stabilisation of the microtubules, enhancement of synaptic plasticity, prevention of oxidative stress, inhibition of apoptosis, maintenance of neural circuitry, and enhancement of cerebral blood flow.[12]

Dietary forms of oestrogen are also interesting in relationship to this hypothesis. Soy, beans and peas (legumes), nuts and grains, and to a lesser extent, fruits and vegetables, contain phytoestrogens, compounds which enhance oestrogenicity in postmenopausal women. There has been much research on the beneficial effects of soy on menopausal symptoms, bone mineral density, and lipid levels. Phytoestrogens bind to the oestrogen receptor similarly to endogenous oestrogens and initiate transcription of oestrogen-responsive genes. Therefore the effects of plant-based oestrogen compounds on cognition are of interest.

So far, two primary prevention trials using postmenopausal HRT and one using soy protein have been published with dementia or cognitive function as endpoints (see Tables 16.1–16.3):

- Women's Health Initiative Memory Study (WHIMS)
  - WHIMS Estrogen + Progestin Trial (n = 4532 women) using conjugated equine oestrogen with medoxyprogesterone acetate (CEE + MPA) as active treatment. The trial was terminated in July 2002 due to significantly more heart disease, stroke, pulmonary emboli and breast cancer events among women taking HRT compared to those who were not[7,8]
  - WHIMS Estrogen-Alone Trial (n = 2947 women) using conjugated equine oestrogen (CEE) alone as active treatment. The Estrogen-Alone arm was terminated February 2004 because of excess risk of stroke among women on HRT[9,10]
- Soy Protein Supplement Trial[11] (n = 175 women) using soy protein containing 99 mg isoflavones (52 mg genistein, 41 mg daidzein, and 6 mg glycetein or total milk protein) as active treatment.

Both WHIMS trials reported an increased incidence of dementia and a smaller increase in the Modified Mini-Mental State Examination (3Ms) scores[14] in the group on active treatment compared to placebo, while the Soy Protein Supplement Trial showed no difference between the groups.

Hypertension

During the last decade, evidence has accumulated that hypertension may be a risk factor for cognitive

Table 16.1 Main characteristics of placebo-controlled randomised primary prevention trials

Study	Treatment	n	Ages (years) (mean)	Inclusion	Exclusion	Follow-up (mean years)
Hormone replacement therapy						
WHIMS Estrogen + Progestin Trial[7,8]	Conjugated equine oestrogen and medoxyprogesterone acetate	4532	65–79	Female volunteers Intact uterus No HRT last 3 months (washout period)	Dementia, breast cancer in the last 10 years, myocardial infarction (MI), stroke, or TIA last 6 months, chronic liver disease, severe hypertension, current use of oral corticosteroids	4.2
WHIMS Estrogen-Alone Trial[9,10]	Conjugated equine oestrogen	2947	65–79	Female volunteers Prior hysterectomy No HRT last 3 months (washout period)	See above	5.4
Soy Protein Supplement Trial[11]	Soy protein containing isoflavones	175	60–75 (67)	Female volunteers No HRT for last 6 months	Contraindications oestrogen	1
Antihypertensives						
SHEP[29,30]	Chlorthalidon (diuretic)	4736	60–94 (72)	SBP 160–219 mmHg DBP <90 mmHg	Major cardiovascular and other disorders	5
MRC Treatment Trial of Hypertension[31,32]	Atenolol (beta-blocker) hydrochlorthiazide (diuretic)	4396	65–74 (70)	SBP 160–209 mmHg DBP <115 mmHg	Using antihypertensive drugs or had indications for its use, MI or stroke in the last 3 months, diabetes	4.5
Syst-Eur[33–35]	Nitrendipine (calcium-channel blocker)	2418 (4695 main study)	>60 (70)	SBP 160–219 mmHg DBP <95 mmHg	Dementia Stroke or MI in the last year, severe concomitant disease	2

SCOPE[36]	Candesartan cilexetil (angiotensin II type 1 receptor blocker)	4937	70–89 (76.4)	SBP 160–179 mmHg DBP 90–99 mmHg	Dementia or treatment for dementia, MMSE <24 Stroke or MI in the last 6 months	3.7
PROGRESS[37,38]	Perindopril (angiotensin-converting enzyme inhibitor) indapamide (diuretic)	6105	(64)	History of prior stroke or TIA in the last 5 years	Subarachnoid haemorrhage	3.9
Statins						
PROSPER[48]	Pravastin	5804	70–82 (75)	Pre-existing vascular disease (coronary, cerebral, peripheral) or vascular risk factors (smoking, hypertension, diabetes, plasma total cholesterol 4.0–9.0 mmol/l)	Triglycerides less than 6.0 mmol/l MMSE <24	3.2
MRC/BHF Heart Protection Study[49]	Simvastatin	20536	40–80 (5806 >70)	Total cholesterol >3.5 mmol/l Coronary disease, occlusive disease of non-coronary arteries, diabetes mellitus, treated hypertension in males >age 65 years	Indications for statin treatment, chronic liver or renal disease, abnormal liver or renal function, severe stroke, dementia, psychiatric disorder, child-bearing potential, life-threating condition	5
Antioxidants						
MRC/BHF Heart Protection Study[65]	Antioxidant vitamin supplementation (600 mg vitamin E, 250 mg vitamin C and 20 mg beta-carotene)	20536	40–80			5

Table 16.2 Main outcomes in placebo-controlled randomised primary prevention trials

Study	Main outcome	Secondary outcome	Cognitive variables	Test repeated	Cognitive endpoints
Hormone replacement therapy					
WHIMS Estrogen + Progestin Trial[7,8]	Dementia, cognitive function, stroke		Modified Mini Mental State Examination (3MSE) CERAD in those that screened positive	Annually	Dementia, rates of change in 3MS
WHIMS Estrogen-Alone Trial[9,10]	Dementia, cognitive function, stroke		See above	Annually	See above
Soy Protein Supplement Trial[11]			MMSE, Rey Auditory Verbal Learning Test Doors Test (visual memory), Digit Span test, Boston naming test, digit symbol substitution test, Trail-making Test	Baseline and last exam	Change in score for each cognitive test
Antihypertensives					
SHEP[29,30]	Fatal and non-fatal stroke	Cardiovascular and coronary morbidity and mortality, total mortality	Short-CARE, CES-D In subsample: Digit symbol substitution Addition test Finding A's test Boston naming test Letter sets test Delayed recognition test	Every 6 months Annually	Dementia, change in cognitive scores

					Rate of change
MRC Treatment Trial of Hypertension[31,32]	Mortality and morbidity due to stroke and cardiovascular disease. Total mortality	Treatment response in men and women	Paired associate learning (PALT) Trail-making test (TMT) A	At entry, 1, 9, 21 and 54 months	
Syst-Eur[33-35]	Fatal and non-fatal stroke	All death MI, congestive heart failure,	MMSE	Annually	Dementia, mean change in MMSE
SCOPE[36]	Major cardiovascular events (cardiovascular death, non-fatal stroke, non-fatal MI)	Cardiovascular death Non-fatal stroke Fatal stroke MI, cognitive endpoints	MMSE	Every 6 month	Dementia, 'significant cognitive decline' (decline of ≥3 points in MMSE), mean change in MMSE
PROGRESS[37,38]	Fatal and non-fatal stroke	Other major vacular events Dementia, cognitive decline	MMSE Screen for evaluation: MMSE ≤25, decline in MMSE of 3 points, investigator suspecting dementia	At baseline, 6 months, 12 months and then annually	Dementia 'significant cognitive decline' (decline of ≥4 in MMSE) mean change in MMSE
Statins PROSPER[48]	Combined endpoint of definite or suspect death from coronary heart disease, non-fatal myocardial infarction, and fatal or non-fatal stroke	TIA, disability and cognitive function	MMSE, picture-word learning test, Stroop colour word test, letter digit coding test	Annually	Difference between the last on-treatment and the second baseline measurements

Table 16.2 continued

Study	Main outcome	Secondary outcome	Cognitive variables	Test repeated	Cognitive endpoints
MRC/BHF Heart Protection Study[49]	Mortality (total, vascular, coronary) Events (coronary, stroke, revascularisation)		Modified Telephone Interview for Cognitive Status (TICS-m)	At last examination	TICS-m score below 22 (out of maximum 39)
Antioxidants					
MRC/BHF Heart Protection Study[65]			Modified Telephone Interview for Cognitive Status (TICS-m)	At last examination	TICS-m score below 22 (out of maximum 39)

Table 16.3 Main results of placebo-controlled randomised primary prevention trials

Study	n dementia	Non-cognitive endpoint	Cognitive endpoint
Hormone replacement therapy			
WHIMS Estrogen + Progestin Trial[7,8]	61	**Significant increase in:** stroke (1.8% vs 1.3%)	**Placebo significantly:** larger increase in 3MS (+0.15 vs +0.21/year) less incident dementia (1.8% vs 0.9%) **No significant difference in:** MCI (2.5% vs 2.4%)
WHIMS Estrogen-Alone Trial[9,10]	47		**Placebo significantly:** larger increase in 3MS (+1.07 vs +1.43) **No significant difference in:** incident dementia (1.9% vs 1.3%) MCI (5.2% vs 3.9%)
Soy Protein Supplement Trial[11]	–		**No significant difference in:** change in any cognitive tests
Antihypertensives			
SHEP[29,30]	81	**Significant reduction in:** BP (11–14/3–4 mmHg stroke (5.2% vs 8.2%) coronary heart disease (5.9% vs 6.7%) all cardiovascular (12.2% vs 17.5%) **No significant difference in:** total mortality (9.0% vs 10.2%)	**No significant difference in:** change in s-CARE score (−0.11 vs −0.06) change in any cognitive tests incidence of dementia (1.6% vs 1.9%)

Table 16.3 continued

Study	n dementia	Non-cognitive endpoint	Cognitive endpoint
MRC Treatment Trial of hypertension[31,32]	–	**Significant reduction in:** SBP (diuretic −34 mmHg, beta-blocker −31 mmHg, placebo −16 mmHg) stroke (4.6% vs 6.1%) coronary events (5.9% vs 7.2%) all cardiovascular endpoints (11.8% vs 14.0%) **No significant difference in:** total mortality (13.8% vs 14.2%)	**No significant difference in:** deterioration in PALT score (diuretic −0.31, beta-blocker −0.33, placebo −0.30) improvement in TMT A score (diuretic −2.73, beta-blocker −2.08, placebo −3.01)
Syst-Eur[33,35]	32	**Significant reduction in:** BP (10.1/4.5 mmHg) all stroke (2.0% vs 3.4%) non-fatal stroke (1.4% vs 2.5%) all cardiac endpoints (3.7% vs 5.0%) all vascular endpoints (5.7% vs 8.1%) **No significant difference in:** myocardial infarction (1.4% vs 2.0%) TIA (1.1% vs 1.3%)	**Significant reduction in:** dementia (0.4% vs 0.8%) **No significant difference in:** change in MMSE (+0.08 vs +0.01)
SCOPE[36]	119	**Significant reduction in:** BP (3.2/1.6 mmHg) non-fatal stroke (2.7% vs 3.8%) **No significant difference in:** major cardiovascular events (9.8% vs 10.8%) all stroke (3.6% vs 4.7%) fatal stroke (1.0% vs 1.1%) all MI (2.8% vs 2.6%)	**No significant difference in:** dementia (2.5% vs 2.3%) significant cognitive decline (4.7% vs 5.2%) mean change in MMSE (−0.49 vs −0.64)

Trial	n	Vascular/other outcomes	Cognitive outcomes
PROGRESS[37,38]	410	**Significant reduction in:** BP (9/4 mmHg) stroke (10% vs 14%) non-fatal stroke (9.0% vs 12.4%) coronary events (3.8% vs 5.0%) total vascular events (15.0% vs 19.8%) **No significant difference in:** total mortality (10.0% vs 10.4%)	**Significant reduction in:** significant cognitive decline (9.1% vs 11.0%) mean change in MMSE (–0.05 vs –0.24) **No significant difference in:** dementia (6.3% vs 7.1%)
Statins PROSPER[48]	–	**Significant reduction in:** LDL cholesterol combined vascular endpoints (14.1% vs 16.2%) total coronary heart (10.1% vs 12.2%) TIA (2.7% vs 3.5%) **No difference in:** all stroke (4.7% vs 4.5%)	**No significant difference in:** change from 2nd baseline to last on treatment measure for any test
MRC/BHF Heart Protection Study[49]	62	**Significant reduction in:** total death (12.9% vs 14.7%) vascular death (7.6% vs 9.1%) coronary death (5.7% vs 6.9%) non-fatal MI (3.5% vs 5.6%) any coronary event (8.7% vs 11.8%) first stroke (4.3% vs 5.7%) ischaemic stroke (2.8% vs 4.0%) TIA (2.0% vs 2.4%) any major vascular event (19.8% vs 25.2%)	**No significant difference in:** dementia (0.3% vs 0.3%) cognitive impairment (23.7% vs 24.2%)
Antioxidants MRC/BHF Heart Protection Study[65]	62	See above	**No significant difference in:** dementia (0.3% vs 0.3%) cognitive impairment (23.4% vs 24.4%)

decline and dementia, independent of the presence of cerebrovascular disease. Hypertension is currently defined as systolic blood pressure (SBP) above 140 mmHg and/or diastolic blood pressure (DBP) above 90 mmHg. The threshold for hypertension has decreased during recent years, thus most trials have used a systolic blood pressure (SBP) above 160 mmHg as cut-off. Hypertension is a risk factor for stroke, ischaemic white matter lesions, silent infarcts, general atherosclerosis, myocardial infarction and cardiovascular morbidity and mortality.[15] This risk increases with increasing blood pressure also within normal blood pressure ranges.[16] Treatment of hypertension has been proven to reduce cardiovascular risk substantially, but a large proportion of people with hypertension in the general population are not diagnosed or treated.

Several longitudinal studies suggest an association between AD and previous hypertension,[17–22] while some do not,[23–25] or report associations with vascular dementia.[23,24] Some observational studies also suggest that the use of anti-hypertensive drugs may reduce the incidence of AD and dementia,[26–28] while others have been less conclusive, with non-significant odds ratios below 1.00 for risk of AD.[29] It thus appears that treatment of hypertension might be one tool for prevention of dementia/cognitive decline.

Thus far, five hypertension trials with dementia or cognitive function as secondary endpoints have been published (see Tables 16.1–16.3):

- the Systolic Hypertension in the Elderly Program (SHEP) ($n = 4736$) using chlorthalidon, a diuretic, as active treatment[30,31]
- Medical Research Council's (MRC) Treatment Trial of hypertension ($n = 4396$) using atenolol, a beta-blocker, or hydrochlorthiazide, a diuretic, as active treatments[32,33]
- the Systolic Hypertension in Europe Study (Syst-Eur) ($n = 2418$) using nitrendipine, a calcium-channel blocker, as active treatment[34–36]
- the Study on Cognition and Prognosis in the Elderly (SCOPE) ($n = 4937$) using candesartan

cilexetil, an angiotensin II type 1 (AT1) receptor blocker, as active treatment[37]
- Perindopril Protection against Recurrent Stroke Study (PROGRESS) ($n = 6105$) using perindopril, an angiotensin-converting enzyme inhibitor, and indapamide, a diuretic, as active treatment.[38,39]

These hypertension trials with dementia or cognitive function as secondary endpoints observed significant reductions in cardiovascular outcomes (see Table 16.3), but only Syst-Eur reported a significantly reduced incidence of dementia in the treatment group.[34] Interestingly, this effect remained during an open-label follow-up period.[35] Regarding other cognitive outcomes, PROGRESS reported a significant 19% decreased incidence of 'significant cognitive decline',[38] while SCOPE reported a non-significant 11% reduction.[37] PROGRESS reported less decline in Mini-Mental State Examination (MMSE) in the active treatment group than in the placebo group,[40] while MRC, SHEP, SCOPE and Syst-Eur observed no differences between the groups in mean change in different cognitive tests (see Table 16.2 and 16.3). No study reported a higher risk for dementia or cognitive decline in the active treatment group.

Some of these studies reported subanalyses. PROGRESS reported that dementia with recurrent stroke was reduced by 34%.[38] SCOPE found that mean MMSE declined significantly less in the treatment group than in controls among those with mild cognitive dysfunction (defined as 28 points or less).[41] In SHEP, active treatment reduced incidence of dementia if drop-outs were assigned a prevalence of 20–30% of dementia.[42]

Statins

Two Finnish population studies reported that individuals with AD had high cholesterol levels 15–30 years before disease onset,[18,43] and cholesterol concentrations in the brain cortex are increased in AD.[44] Statins are used to lower cholesterol in humans. Two case record-based studies reported a reduced frequency of dementia and AD in statin-treated individuals,[45,46] and the longitudinal population study of the Canadian Study of Health and Aging,[47] reported that the incidence of AD was reduced in individuals using statins. Several experimental studies support these observations. Simvastatin and lovastatin reduced intracellular and extracellular levels of β-amyloid-42 and β-amyloid-40 in primary cultures of hippocampal neurons and mixed cortical neurons.[48] One explanation may come from a report that lowering cellular cholesterol with lovastatin reduced the conversion of Amyloid Precursor Protein (APP) to β-amyloid in cell culture.[49] So far, two randomised controlled trials with dementia or cognitive function as secondary endpoints and with statin as active treatment have been published:

- pravastatin in elderly individuals at risk of vascular disease (PROSPER) ($n = 5804$) using pravastin 40 mg per day, as active treatment[50]
- MRC/British Heart Foundation (BHF) Heart Protection Study ($n = 20\ 536$), using simvastatin 40 mg per day as active treatment.[51]

These studies found significant reductions of cardiovascular morbidity (see Table 16.3), but cognitive function declined at the same rate in the treatment and in the placebo group in PROSPER, irrespective of which test was used, and the percentage of participants classified as cognitively impaired or demented in MRC/BHF Heart Protection Study, was similar in the simvastatin group and the placebo group. Furthermore, the lack of difference was also found in different age groups and among those with previous stroke.

Non-steroidal anti-inflammatory drugs (NSAIDs)

Several observational studies report that anti-inflammatory agents may prevent dementia.[5,6] Inflammation and inflammatory processes may play an important role in the pathogenesis of AD.[52] Inflammatory proteins are found in AD lesions and in brain regions affected by AD. Damaged neurons and neurites and amyloid beta peptide deposits, the major constituent of senile plaques, and neurofibrillary tangles may provide stimuli for inflammation, which may lead to further neuronal damage.[53] Trials in individuals with manifest AD show no effect on cognitive decline in demented patients.[6,54] However, observational studies suggest that NSAIDs lose their protective effect about two years before the onset of the full dementia syndrome.[55] Therefore, NSAIDs may only be effective in long-term prevention trials, not in trials conducted in patients with incipient or manifest AD. So far, no prevention study has been reported regarding NSAIDs, but one large trial is ongoing, the Alzheimer's Disease Anti-inflammatory Prevention Trial (ADAPT), comprising 2625 cognitively normal individuals aged 70 years and above, with a family history of AD to be followed for 7 years.[56]

Antioxidants

Oxidative stress with the formation of free radicals has been suggested to be involved in the aetiology of AD.[57,58] β-Amyloid induces oxidative stress in neurons,[59,60] and endothelial cells,[61] possibly by activating the receptor for advanced glycation end-products.[60] One common explanation for the risk factors for Alzheimer's disease is the formation of free oxygen radicals.[62] Findings that dietary antioxidants may be protective for cognitive impairment and incident AD may support this possibility.[63,64] A variety of nutritional supplements and dietary components have exhibited antioxidant properties including vitamins E, C and A, superoxide dismutase, glutathione, urea, selenium and the carotenoids.[6] Vitamin E is incorporated into cell membranes and performs a local free radical scavenging function. Animal studies show that vitamin E may reduce A-beta levels, particularly when administered at the early stages of oxidative and inflammatory processes.[65] One randomised, double-blind, placebo-controlled trial in patients with manifest AD, suggested that selegiline, a monoamine oxidase B inhibitor with antioxidant properties, and vitamin E significantly reduced the risk of reaching a primary outcome (institutionalisation, loss of basic ADLs, severe dementia).[66] So far, one primary prevention trial using antioxidant vitamin supplementation with cognitive function/dementia as secondary outcomes has been published:

- MRC/BHF Heart Protection Study ($n = 20\,536$), using antioxidant vitamin supplementation (600 mg vitamin E, 250 mg vitamin C and 20 mg beta-carotene) daily as active treatment.[67] The percentages of participants classified as demented or cognitively impaired were similar for those on antioxidant vitamin supplementation and those in the placebo group.

Gingko

Gingko has long been touted as a potential beneficial agent in relationship to the brain. A large prevention trial is underway and awaited with interest.

General methodological considerations

Despite the seemingly negative findings from these trials that may give rise to pessimism regarding possibilities for dementia prevention, there are several methodological factors that might explain some of these results. These factors will be discussed below.

Factors related to outcome

Cognitive endpoints in the primary prevention trials have been dementia, significant cognitive decline and change in cognitive function based on longitudinal performance on certain tests (see Table 16.2).

Dementia

In large trials involving several thousands of patients, where the main outcome is related to cardiovascular diseases, it may be difficult to make valid diagnoses of dementia, which requires comprehensive evaluations. Methods for detection of dementia vary between studies. Generally patients screened positive for dementia (based on cut-off scores or change in scores on different tests), or patients where the investigators suspected dementia, have been referred to a more formal investigation performed by local specialists. In some studies (PROGRESS, SCOPE, Syst-Eur, WHIMS) final diagnoses were decided centrally by an endpoint committee, based on all collected information.

SHEP used the Comprehensive Assessment and Referral Evaluation (short-CARE) questionnaire for screening of dementia.[30,68] Patients with two values above 3 were referred for clinical evaluations. Out of 192 who screened positive, 165 were referred for evaluation, and 81 of those were found to have dementia.

In PROGRESS, individuals with a MMSE score below 26, those who declined with 3 or more points on MMSE and those where the investigators had the impression that the patients became demented were referred to a local specialist for further investigations.[38] Dementia was finally diagnosed according to the Diagnostic and Statistical Manual of Mental Disorders, Fourth Revision (DSM-IV) by a 'dementia committee'. Among all participants, 1580 were screened positive, and the local specialists diagnosed 358 of these as demented. Finally, the dementia committee reclassified 32 of those as non-demented and 84 who were classified as non-demented were reclassified as demented by the dementia committee.

In Syst-Eur, those who reached a MMSE score below 24 and those who had reports on dementia were diagnosed according to the Diagnostic and Statistical Manual of Mental Disorders, Third Revision Revised (DSM-III-R) by the investigators.[34] Cases of dementia were validated by a central review board.

In SCOPE,[37] those who declined with 4 or more points on the MMSE, or where the investigators reported a dementia onset, were rated by the investigators with a checklist based on th Tenth Revision of the International Classification of Diseases (ICD-10) criteria.[37] All information was then evaluated centrally by two dementia experts for a final diagnosis. One-hundred and forty-four cases were reported by the investigators but only 71% were confirmed by the event committee.

In WHIMS, women with <8 years of education who scored 72 or lower on 3MSE, or who had >9 years of education and scored 76 or lower (3MSE cutpoints for those going on to more intensive examinations were set higher as the study progressed) underwent a more extensive neuropsychological battery and clinical examination by a local specialist, which included the Consortium to Establish a Registry for Alzheimer's Disease (CERAD)[69] neuropsychological battery and other standardised interviews. Finally, a central adjudication committee reviewed all data and made a final diagnosis. Overall, 12.6% of those on CEE, 11.6% of those on placebo in the oestrogen-alone trial, 7.9% of those on CEE + MPA and 6.1% of those on placebo in the oestrogen plus progestin trial were referred for further evaluation. Thus, only around one-tenth of those referred were finally diagnosed with dementia. The agreement between local specialists and the central adjunction committee was 75%.

Thus, the diagnoses made by local specialists were not always confirmed by the endpoint committee. This illustrates the problems of making diagnoses when many different specialists are involved, and maybe also the difficulties that may arise with collecting sufficient information for the endpoint committees. It is also not clear how many cases of dementia are missed by the screening procedure.

Cognitive function

Several studies use change in cognitive function, based on performance in psychometric tests from baseline to last examination, as outcome. If active treatment is related to less decline in cognitive function, it may indicate that treatment has an effect on very early stages of dementia. The most commonly used test has been the MMSE, which is a test of global cognitive function.[40] This test has a maximum score of 30, and a score below 24 is believed to indicate cognitive impairment compat-

ible with dementia. Tests of cognitive function are influenced by several factors, as outlined below.

Practice and learning effect

Repeated administration of psychometric instruments may result in higher scores at retesting, the so-called practice or learning effect.[70] This may be the result of memorising items or developing better strategies in performing the tests, which may be one reason why several trials (e.g. MRC trial in hypertension, SYST-EUR, PROSPER and WHIMS) reported improvement in scores on some cognitive tests during the study period. In SHEP, a training effect was evident as a progressive improvement in the short-CARE score throughout year 1 in both study groups.[71] Thereafter, cognitive scores deteriorated, more in the placebo than in the active treatment group. When follow-up scores were adjusted based upon baseline data, the difference between the study groups reached statistical significance, but the apparent overall trend towards deterioration in cognitive score was no longer observed.[71] Trials using changes in cognitive test scores as endpoint should be designed appropriately to estimate and minimise the consequences of a training effect in follow-up data. In the PROSPER trial, the second cognitive testing was used as a baseline in the analyses, to minimise the influence of the learning effect. Other strategies may include increasing the interval between the testings.

Ceiling effect

Ceiling effect refers to the fact that a test may not be able to detect changes in cognitive function if most individuals score close to the maximum score.[70] Around 70% of participants in trials using the MMSE had scores of 29 or 30, and the mean 3MS score was 95 out of a maximum of 100 in the WHIMS. Thus, even if subjects with a maximum score of 30 decline in intellectual function, it may not be detected by MMSE, and improvements in scores are not possible in those with maximum scores, which may affect the results, as scores have often improved in the trials done so far.

Sensitivity to change

Most studies used the MMSE to test cognitive function. This test may not be sensitive enough to detect changes over time in unselected populations. Furthermore, it measures global cognitive function, but it might be better to examine specific cognitive functions to detect changes over time. Areas of cognitive function that tend to change with time include memory, attention and cognitive speed.[70] The SHEP, MRC Treatment Trial of Hypertension, PROSPER and the Soy Protein Supplement Trial used tests which should measure several of these areas (see Table 16.2), but no difference in change of test score was detected between the active treatment group and placebo in these studies. In a substudy of SCOPE, the investigators in Newcastle reported that the group taking the anti-hypertensive candesartan showed less decline than the control group in attention and episodic memory with a similar trend for speed of cognition.[72] This indicates that SCOPE might have had a treatment effect if more sensitive tests had been used.

Rare outcomes

Clinical trials should not be conducted without sample size and power calculations. As may be seen in Table 16.3, the proportion developing dementia in the different trials has generally been very low, ranging from 0.3% to 2.5% in most studies, except PROGRESS where it was 6.3% in the active treatment group and 7.1% in the placebo group. The high number of demented in PROGRESS might be due to the fact that this study only included patients with a previous stroke or transient ischaemic attack (TIA). In WHIMS, which reported an increased incidence of dementia in the actively treated group, 1.8% of those on active treatment and 1.1% on

placebo in the combined analyses developed dementia. This means that 98.2% of those on HRT did not develop dementia, despite the alarming media reports.

Factors related to the intervention sample
Age

Age is the strongest predictor of dementia, with an incidence increasing from approximately 1% at age 70–74 years to 10% at age 85 years, and a prevalence increasing from 1% at age 70 years to approximately 30% at age 85 years. Therefore, the age of participants when an intervention is initiated, and the age at which events are evaluated influences the number of outcomes observed, and subsequently the interpretation of the results. Trials to date have generally recruited relatively young participants where dementia risk is low (see Table 16.1). Therefore, relatively small numbers of persons with dementia have been observed in several of these studies.

The roles and relative influence of preventable risk factors in relationship to dementia may also differ depending on whether they occur in early, mid, or late life. For example, if there is a critical time during the perimenopausal period when oestrogen levels are declining at the fastest rate and when HRT is most beneficial related to dementia prevention, WHIMS did not target such a time period. Thus, even if risk factors for dementia are treated, it might be too late to treat them in those age groups in which trials have been conducted so far.

Healthy volunteer effect

It is well-known that individuals who take part in studies are healthier than the general population. This may have more than one effect on observations from trials for dementia prevention. The incidence of dementia may not be as high as in the general population, since these healthy volunteers practise lifestyle and healthcare habits that reduce their risk. On the other hand, since healthy volunteers theoretically live longer, they are potentially more at risk for dementia over a long time period since they live to the high ages where dementia is most likely to occur.

High-risk groups

In some intervention trials, participants are selected based on a risk factor, such as hypertension, hypercholesterolaemia or stroke. While this may assist in the accrual of more outcomes, generalisability is limited to individuals harbouring these characteristics.

Cognitive function at baseline

Most population studies report that individuals with high cognitive test scores at baseline, experience less decline in test scores and have a lower incidence of dementia during follow-up than those with lower scores at baseline.[73] Both SYST-EUR and SCOPE reported mean baseline MMSE scores of 28.5 (maximum score 30) and almost 70% of the participants had a score of 29 or 30. PROSPER reported a mean MMSE score of 28.0.[70] WHIMS participants had a mean score of 95.1 (maximum score 100) on the 3MS. This suggests that participants in these trials had a rather low short-term risk of developing dementia. Interestingly, WHIMS women receiving HRT and with lower baseline cognitive function, were at higher risk for dementia. Perhaps this is indicative of acceleration of latent disease, similar to what has been suggested for the observations relating HRT to breast cancer in the Women's Health Initiative. The generally high cognitive performance in participants of the published trials may also be positive, as it suggests that the great majority of participants did not have incipient dementia.

Sex

Since female sex is strongly associated with risk for dementia in very high ages, since men do not live as long as women, and since risk factors for dementia may vary by sex, the relative inclusion or exclusion of women and men in trials for dementia prevention is critical to interpretation of results. Results from trials in women, for example, cannot be generalised as relevant to men.

Factors related to study design
Time of follow-up

It is likely that prevention trials need lengthy periods of follow-up to be able to detect a difference between placebo and active treatment. Length of follow-up in the large trials performed has been between 2 and 5 years (see Table 16.1). Thus, all trials have measured the short-term effects of treatment. In contrast, studies reporting associations between hypertension and AD reported that follow-ups had to be at least 5 years to detect differences between AD and controls.[17–21] Regarding cholesterol, high cholesterol measured in midlife is associated with AD in late life, but there is no short-term association.[19,43] Thus, length of follow-up might have been too short to detect an effect on the incidence of dementia in the trials conducted so far. In contrast, the only trial reporting that treatment reduced the incidence of dementia was SYST-EUR, which had the shortest time of follow-up (2 years).[34]

Timing

The roles and relative influence of risk factors in relationship to dementia may differ, depending on their timing in relationship to disease onset. Perhaps the biggest lesson to be learned from clinical trials of preventive agents in dementia is how to address the question of timing. When is a potentially preventive agent maximally effective in retarding the onset of dementia? And, when is it appropriate to intervene for any given individual? Trials of HRT in women illustrate this dilemma. As mentioned before, WHIMS suggested an unexpected, albeit small, increased risk of dementia among women taking HRT. However, what appears to be bad in the short-term, may be beneficial in the long-term. For example, the Cache County study reported that HRT (any type) increased AD risk in current users who had used HRT for 0–10 years (i.e. a short time before disease onset), but those who had a longer exposure and former users with more than 3 years' exposure had a decreased risk of dementia.[74] This indicates that both the timing in relation to dementia onset, the duration of HRT use, and the age when the compound is used is of importance. Interestingly, the negative short-term effect of HRT in WHIMS was especially evident in individuals with lower cognitive function at baseline, indicating that HRT may have a negative effect once the disease process has started.

Another example is blood pressure, where cross-sectional studies,[75–77] and studies with short follow-ups,[21,78] consistently report associations between low blood pressure and prevalent or incident AD, and that blood pressure decreases in the years preceding AD onset.[17] Thus, those enrolled in hypertension trials, i.e. those with high blood pressure, might in fact have a decreased short-term risk for dementia, and those who develop dementia in these trials may have other characteristics than dementias in general. Similar temporal relationships have been reported in relationship to cholesterol,[19,43] and being overweight.[79] Lack of effect in the years before dementia onset has also been suggested for NSAIDs.[55] Thus, the timing of potential risk or benefit factors in relation to dementia onset needs always to be considered.

Selective attrition/missing data

Selective attrition may affect the results of trials if missed assessments differ between treatment and placebo groups. An analysis of SHEP data revealed that the placebo group had more missed assessments regarding cognitive function than the treatment group.[40] Furthermore, those with missed cognitive assessments had more cardiovascular events during the study period. In a *post hoc* analysis of SHEP data, an assumption was made that 20–30% of those with missed assessments were cognitively impaired. With this assumption, active treatment reduced the risk of cognitive impairment.[42] It is yet not clear whether selective attrition influenced the results in other trials.

Factors related to the drugs
Type of drug

There is a possibility that some anti-hypertensive drugs may have effects on cognitive function beyond that of lowering blood pressure. This may also partly explain different results in trials. Calcium-channel blockers have been suggested to have neuroprotective properties.[34] This might be a reason why SYST-EUR, despite its short time of follow-up, was the only hypertension trial that found an effect in the treatment group compared to controls. Also, other anti-hypertensive agents may have an effect on cognition beyond that of blood pressure lowering.

Angiotensin II type 1 (AT1) receptor blockers and angiotensin-converting enzyme inhibitors may affect cognitive function by a direct effect on the renin–angiotensin system in the brain. Elevated levels of angiotensin II have been shown to impair learning, especially acquisition and recall of newly learned material, and other memory functions,[80] and angiotensin II receptor antagonists have been shown to improve cognitive function in mice.[81] Finally, anti-hypertensive treatment may reverse the changes in the small vessels of the brain, independently of the blood pressure reduction, which may also have an effect on cognitive function.[82]

In the WHIMS, two forms of HRT were used, which although widely prescribed, are not the only types of HRT on the market, nor are they the hormonal preparations used in animal studies of brain health. In addition, many forms of HRT administration are available, aside from oral, including patches, implants, injections and creams; and different regimens and doses may be used during a woman's course of HRT, such as adding progestin only during the first 14 days of the month or during days 15–25. The WHIMS regimen was daily intake of the same preparation over an approximate 4-year period. The advent of many new and different types of HRT, including SERMS (selective oestrogen receptor modulators), will cause this field to be continually filled with questions in the future. The WHIMS experience may also illustrate that a focus entirely placed on oestrogen and, to some extent, progesterone, is misplaced. Oestrogens and androgens, such as testosterone, and other sex hormones, such as progesterone, act in concert with one another, and are linked to the functioning of the hypothalamic-pituitary-gonadal axis in general. The role of many of the key players in this axis in relationship to dementia aetiology, is a relatively unexplored area.[83]

Drug–drug interactions

Intervention studies among the elderly must consider the potential interactions between the intervention agent and other drugs a participant may be taking. Interactions may influence interpretation of study results, or interfere with the proposed mechanism of action.

Previous drug treatment

Prior to initiation of a clinical trial, a washout period is often required, during which time the participant ceases to engage in behaviours that may interfere with the study intervention. For example, the women about to participate in WHIMS, who were already taking HRT as part of their daily regimen (21–22%), were asked to refrain from their usual preparation for 3 months prior to initiation of the trial. In the case of vitamin supplementation, a similar strategy may be used depending on the vitamin, in addition to requesting a refrain from taking any other daily vitamin supplements during the course of the study, so as not to contaminate the study results. In the case of vitamin E and other fat-soluble vitamins, the effects of a vitamin supplement intervention among those already consuming high levels of vitamin E or beta-carotene supplements, may be attenuated because of the body's long-term storage of these particular compounds. Also in the hypertension trials, a large proportion of the subjects had used different types of anti-hypertensive agents during different times of follow-up before the trials started. Previous drug use may thus dilute the short-term effects of the study drugs on the risk for dementia and cognitive decline.

Changes in treatment guidelines

During clinical trials, especially those that last over a long period of time, changes in treatment guidelines that involve an endpoint under study, can dramatically affect the validity of a trial, and perhaps render it obsolete. One problem in SCOPE was that treatment guidelines for hypertension changed during the course of the study, and it became necessary to make a protocol amendment that the use of open-label active anti-hypertensive therapy to control blood pressure was tolerated.[37] This resulted in 49% of the patients in the group taking candesartan and 66% of those in the control group receiving treatment with other anti-hypertensive agents. Therefore, both treatment regimens effectively lowered blood pressure, and the difference in blood pressure reductions between the treatment groups was only 3.2/1.6 mmHg.

Regarding hypertension, it is unlikely that it will be possible to perform placebo-controlled studies with the outcome of dementia or cognitive function in the future. Another example of the impact of changing treatment guidelines comes from WHIMS, where women were discouraged from using HRT due to higher breast cancer risks prior to data on cognition and dementia becoming available.

Conclusions and direction to the future

Primary prevention trials published so far have not given conclusive answers to whether dementia might be prevented, and they have only reported on the short-term effect of intervention. We have reviewed a number of factors that need to be considered in future prevention trials, including length of follow-up, timing in relation to dementia onset, use of outcomes that are sensitive to change, learning effects, and the selection of participants at risk for dementia.

References

1. Jorm A, Jolley D. The incidence of dementia: a meta-analysis. Neurology 1998; 51:728–33.

2. Skoog I, Nilsson L, Palmertz B, Andreasson LA, Svanborg A. A population-based study of dementia in 85-year-olds. N Engl J Med 1993; 328:153–8.

3. Last J. A Dictionary of Epidemiology (2nd edn). New York: Oxford University Press; 1988.

4. Skoog I. Possibilities for secondary prevention of Alzheimer's disease. In: Mayeux R, Christen Y (eds). The Epidemiology of Alzheimer's Disease: From Gene to Prevention. Berlin: Springer Verlag; 1999: 121–34.

5. Veld BA, Launer L, Breteler MMB et al. Pharmacological agents associated with a preventive effect on Alzheimer's disease: A review of the epidemiological evidence. Epidem Rev 2002; 24:248–68.

6. Zandi P, Breitner J, Anthony J. Is pharmacological prevention of Alzheimer's a realistic goal? Expert Opin Pharmacother 2002; 3:365–80.

7. Shumaker S, Legault C, Rapp S et al. Estrogen plus progestin and the incidence of dementia and mild cognitive impairment in postmenopausal women. The Women's Health Initiative Memory Study: a randomised controlled trial. JAMA 2003; 289:2651–62.

8. Rapp S, Espeland M, Shumaker S et al. Effect of estrogen plus progestin on global cognitive function in postmenopausal women. The Women's Health Initiative Memory Study: a randomized controlled trial. JAMA 2003; 289:2663–72.

9. Shumaker S, Legault C, Kuller L et al. Conjugated equine estrogens and incidence of probable dementia and mild cognitive impairment in postmenopausal women. Women's Health Initiative Memory Study. JAMA 2004; 291:2947–58.

10. Espeland M, Rapp S, Shumaker S et al. Conjugated equine estrogens and global cognitive function in postmenopausal women. Women's Health Initiative Memory Study. JAMA 2004; 291:2959–68.

11. Kreijkamp-Kaspers S, Kok L, Grobbee D et al. Effect of soy protein containing isoflavones on cognitive function, bone mineral density, and plasma lipids in postmenopausal women. A randomized controlled trial. JAMA 2004; 292:65–74.

12. Skoog I, Gustafson D. HRT and dementia. J Epidemiol Biostat 1999;b4(3):227–51; discussion 252.

13. McLay R, Maki P, Lyketsos C. Nulliparity and late menopause are associated with decreased cognitive decline. J Neuropsychiatry Clin Neurosci 2003;b15:161–7.

14. Teng EL, Chui HC. The Modified Mini-Mental State (3Ms) examination. J Clin Psychiatry 1987; 48:314–18.

15. Skoog I, Gustafson D. Hypertension, hypertension-clustering factors and Alzheimer's disease. Neurol Res 2003; 25:675–80.

16. Kannel WB. Risk stratification in hypertension: new insights from the Framingham Study. Am J Hypertens 2000; 13:3S–10S.

17. Skoog I, Lernfelt B, Landahl S et al. 15-year longitudinal study of blood pressure and dementia. Lancet 1996; 347:1141–5.

18. Launer LJ, Ross GW, Petrovitch H et al. Midlife blood pressure and dementia: the Honolulu-Asia aging study. Neurobiol Aging 2000; 21:49–55.

19. Kivipelto M, Helkala E-L, Laakso M et al. Midlife vascular risk factors and Alzheimer's disease in later life: longitudinal, population based study. Br Med J 2001; 322:1447–51.

20. Qiu C, von Strauss E, Fastbom J et al. Low blood pressure and risk of dementia in the Kungsholmen project: a 6-year follow-up study. Arch Neurol 2003; 60(2):223–8.

21. Ruitenberg A. Vascular factors in dementia. Observations in the Rotterdam Study. Rotterdam: Erasmus University; 2000.

22. Wu C, Zhou D, Wen C et al. Relationship between blood pressure and Alzheimer's disease in Linxian County, China. Life Sci 2003; 72:1125–33.

23. Posner HB, Tang MX, Luchsinger J et al. The relationship of hypertension in the elderly to AD, vascular dementia, and cognitive function. Neurology 2002; 58:1175–81.

24. Yoshitake T, Kiyohara Y, Kato I et al. Incidence and risk factors of vascular dementia and Alzheimer's disease in a defined elderly Japanese population: the Hisayama Study. Neurology 1995; 45:1161–8.

25. Morris MC, Scherr PA, Hebert LE et al. Association of incident Alzheimer disease and blood pressure measured from 13 years before to 2 years after diagnosis in a large community study. Arch Neurol 2001; 58:1640–6.

26. Guo Z, Fratiglioni L, Zhu L et al. Occurrence and progression of dementia in a community population aged 75 years and older: relationship of anti-

hypertensive medication use. Arch Neurol 1999; 56:991–6.

27. Skoog I, Lernfelt B, Landahl S. High blood pressure and dementia. Lancet 1996; 348:65–6.

28. Guo Z, Fratiglioni L, Viitanen M et al. Apolipoprotein E genotypes and the incidence of Alzheimer's disease among persons aged 75 years and older: variation by use of antihypertensive medication? Am J Epidemiol 2001; 153:225–31.

29. Richards SS, Emsley CL, Roberts J et al. The association between vascular risk factor-mediating medications and cognition and dementia diagnosis in a community-based sample of African-Americans. J Am Geriatr Soc 2000; 48:1035–41.

30. SHEP Cooperative Research Group. Prevention of stroke by antihypertensive drug treatment in older persons with isolated systolic hypertension: Final results of the systolic hypertension in the elderly program. JAMA 1991; 265:3255–64.

31. Applegate WB, Pressel S, Wittes J et al. Impact of the treatment of isolated systolic hypertension on behavioral variables. Results from the Systolic Hypertension in the Elderly Program. Arch Intern Med 1994; 154:2154–60.

32. Prince MJ, Bird AS, Blizard RA, Mann AH. Is the cognitive function of older patients affected by antihypertensive treatment? Results from 54 months of the Medical Research Council's trial of hypertension in older adults. BMJ 1996; 312:801–5.

33. Peart S, Brennan P, Broughton P et al. Medical Research Council trial of treatment of hypertension in older adults: principal results. BMJ 1992; 304:405–12.

34. Forette F, Seux M-L, Staessen J et al. Prevention of dementia in randomised double-blind placebo-controlled Systolic Hypertension in Europe (syst-Eur) trial. Lancet 1998; 352:1347–51.

35. Forette F, Seux ML, Staessen JA et al. The prevention of dementia with antihypertensive treatment: new evidence from the Systolic Hypertension in Europe (Syst-Eur) study. Arch Intern Med 2002; 162:2046–52.

36. Staessen JA, Fagard R, Thijs L et al. Randomised double-blind comparison of placebo and active treatment for older patients with isolated systolic hypertension. Lancet 1997; 350:757–64.

37. Lithell H, Hansson L, Skoog I et al. The Study on Cognition and Prognosis in the Elderly (SCOPE):

principal results of a randomized double-blind intervention trial. J Hypertension 2003; 21:875–86.

38. Tzourio C, Anderson C, Chapman N et al. Effects of blood pressure lowering with perindopril and indapamide therapy on dementia and cognitive decline in patients with cerebrovascular disease. Arch Intern Med 2003; 163:1069–75.

39. PROGRESS Collaborative Group. Randomised trial of a perindopril-based blood-pressure-lowering regimen among 6105 individuals with previous stroke or transient ischemic attack. Lancet 2001; 358:1033–41.

40. Folstein M, Folstein S, McHugh P. 'Mini-mental state': a practical method for grading the cognitive state of patients for the clinician. J Psychiatr Res 1975; 12:189–98.

41. Skoog I, Lithell H, Hansson L et al for the SCOPE Study Group. Effect of baseline cognitive function on cognitive and cardiovascular outcomes in the Study on Cognition and Prognosis in the Elderly (SCOPE) – a randomized double-blind trial. Am J Hypertension (in press).

42. Di Bari M, Pahor M, Franse LV et al. Dementia and disability outcomes in large hypertension trials: lessons learned from the systolic hypertension in the elderly program (SHEP) trial. Am J Epidemiol 2001; 153:72–8.

43. Notkola I-L, Sulkava R, Pekkanen J et al. Serum total cholesterol, apolipoprotein E e4 allele, and Alzheimer's disease. Neuroepidemiology 1998; 17:14–20.

44. Sparks D. Coronary artery disease, hypertension, ApoE, and cholesterol: a link to Alzheimer's disease? Ann N Y Acad Sci 1997; 826:128–46.

45. Jick H, Zornberg GL, Jick SS et al. Statins and the risk of dementia. Lancet 2000; 356:1627–31.

46. Wolozin B, Kellman W, Ruosseau P et al. Decreased prevalence of Alzheimer disease associated with 3-hydroxy-3-methyglutaryl coenzyme A reductase inhibitors. Arch Neurol 2000; 57:1439–43.

47. Rockwood K, Kirkland S, Hogan DB et al. Use of lipid-lowering agents, indication bias, and the risk of dementia in community-dwelling elderly people. Arch Neurol 2002; 59:223–7.

48. Fassbender K, Simons M, Bergmann C et al. Simvastatin strongly reduces levels of Alzheimer's disease beta -amyloid peptides Abeta 42 and Abeta

40 in vitro and in vivo. Proc Natl Acad Sci U S A 2001; 10:10.

49. Simons M, Keller P, De Strooper B et al. Cholesterol depletion inhibits the generation of beta-amyloid in hippocampal neurons. Proc Natl Acad Sci U S A 1998; 95:6460–4.

50. Shepherd J, Blauw G, Murphy M et al. Pravastatin in elderly individuals at risk of vascular disease (PROSPER): a randomised controlled trial. Lancet 2002; 360:1623–30.

51. Heart Protection Study Collaborative, Group. MRC/BHF Heart Protection Study of cholesterol lowering with simvastatin in 20 536 high-risk individuals: a randomised placebo-controlled trial. Lancet 2002; 360:7–22.

52. van Gool W, Aisen P, Eikelenboom P. Anti-inflammatory therapy in Alzheimer's disease: is hope still alive? J Neurol 2003; 250:788–92.

53. Akiyama H, Barger S, Barnum S et al. Inflammation and Alzheimer's disease. Neurobiol Aging 2000; 21:383–421.

54. Aisen P, Schafer K, Grundman M et al. Effects of rofecoxib or naproxen vs placebo on alzheimer disease progression. A randomized controlled trial. JAMA 2003; 289:2819–26.

55. Breitner J. NSAIDs and Alzheimer's disease: how far to generalize from trials? Lancet Neurology 2003; 2:527.

56. Martin B, Meinert C, Breitner J. Double placebo design in a prevention trial for Alzheimer's disease. Control Clin Trials 2002; 23:93–9.

57. Lethem R, Orrell M. Antioxidants and dementia. Lancet 1997; 349:1189–90.

58. Smith M, Sayre L, Monnier V, Perry G. Radical Ageing in Alzheimer's disease. Trends Neurosci 1995; 18:172–6.

59. El Khoury J, Hickman S, Thomas C et al. Scavenger receptor-mediated adhesion of microglia to b-amyloid fibrils. Nature 1996; 382:716–19.

60. Yan S, Chen X, Fu J et al. RAGE and amyloid-b peptide neurotoxicity in Alzheimer's disease. Nature 1996; 382:685–91.

61. Thomas T, Thomas G, McLendon C et al. b-Amyloid-mediated vasoactivity and vascular endothelial damage. Nature 1996; 380:168–71.

62. Henderson A. The risk factors for Alzheimer's disease: a review and a hypothesis. Acta Psychiatr Scand 1988; 78:257–75.

63. Warsama Jama J, Launer L, Witteman J et al. Dietary antioxidants and cognitive function in a population-based sample of older persons. The Rotterdam Study. Am J Epidemiol 1996; 144:275–80.

64. Zandi P, Anthony J, Khachaturian A et al. Reduced risk of Alzheimer disease in users of antioxidant vitamin supplements: the Cache County Study. Arch Neurol 2004; 61:82–8.

65. Sung S, Yao Y, Uryu K et al. Early vitamin E supplementation in young but not aged mice reduces Abeta levels and amyloid deposition in a transgenic model of Alzheimer's disease. FASEB J 2004; 18:323–5.

66. Sano M, Ernesto C, Thomas R et al. A controlled trial of selegiline, alpha-tocopherol, or both as treatment for Alzheimer's disease. N Engl J Med 1997; 336:1216–22.

67. Heart Protection Study Collaborative Group. MRC/BHF Heart Protection Study of antioxidant vitamin supplementation in 20 536 high-risk individuals: a randomised placebo-controlled trial. Lancet 2002; 360:23–33.

68. Gurland B, Golden RR, Teresi JA, Challop J. The SHORT-CARE: an efficient instrument for the assessment of depression, dementia and disability. J Gerontol 1984; 39:166–9.

69. Morris JC, Heyman A, Mohs RC et al. The Consortium to Establish a Registry for Alzheimer's Disease (CERAD). Part I. Clinical and neuropsychological assessment of Alzheimer's disease. Neurology 1989; 39:1159–65.

70. Houx P, Shepherd J, Blauw G-J et al. Testing cognitive function in elderly populations: the PROSPER study. J Neurol Neurosurg Psychiatry 2002; 73:385–9.

71. Di Bari M, Pahor M, Barnard M et al. Evaluation and correction for a 'training effect' in the cognitive assessment of older adults. Neuroepidemiology 2002; 21:87–92.

72. Saxby BK, Harrington F, McKeith IG et al. The effect of candesartan on cognitive function in older adults with mild hypertension [abstract]. First Congress of the International Society for Vascular Behavioural and Cognitive Disorders (Vas-Cog), Göteborg, Sweden; 2003.

73. Aevarsson O, Skoog I. A longitudinal population study of the mini-mental state examination in the very old: relation to dementia and education. Dement Geriatr Cogn Disord 2000; 11:166–75.

74. Breitner J, Zandi P. Effects of estrogen plus pro-gestin on risk of dementia. JAMA 2003; 290:1706.

75. Guo Z, Viitanen M, Fratiglioni L, Winblad B. Low blood pressure and dementia in elderly people: the Kungsholmen project. Br Med J 1996; 312:805–8.

76. Skoog I, Andreasson LA, Landahl S, Lernfelt B. A population-based study on blood pressure and brain atrophy in 85-year-olds. Hypertension 1998; 32:404–9.

77. Hogan D, Ebly E, Rockwood K. Weight, blood pres-sure, osmolarity, and glucose levels across various stages of Alzheimer's disease and vascular dementia. Dement Geriatr Cogn Disord 1997; 8:147–51.

78. Guo Z, Viitanen M, Winblad B, Fratiglioni L. Low blood pressure and incidence of dementia in a very old sample: dependent on initial cognition. J Am Geriatr Soc 1999; 47:723–6.

79. Gustafson D, Rothenberg E, Blennow K et al. An 18-year follow-up of overweight and risk of Alzheimer disease. Arch Internal Med 2003; 163:1524–8.

80. Wright J, Harding J. Brain angiotensin receptor subtypes in the control of physiological and behav-ioral responses. Neurosci Biobehav Res 1994; 18:21–53.

81. Barnes N, Champaneria S, Costall B et al. Cognitive enhancing actions of DUP 753 detected in a mouse habituation paradigm. Neuroreport 1990; 1:239–42.

82. Struijker Boudier H. Vascular growth and hyperten-sion. In: Swales J (ed). Textbook of Hypertension. Oxford: Blackwell Scientific Publications; 1994.

83. Casadesus G, Zhu X, Atwood C et al. Beyond estro-gen: targeting gonadotropin hormones in the treat-ment of Alzheimer's disease. Curr Drug Targets CNS Neurol Disord 2004; 3:281–5.

17

Clinical Trials for Memantine

Michael Borrie and Matthew Smith

Introduction

What is memantine, its mechanism of action and its place in anti-dementia therapy? Which clinical trial statistical evidence is significant for which people with dementia? Does this evidence meet clinical relevance criteria that are meaningful to clinicians and also to patients and their families? Monotherapy, with cholinesterase inhibitors in clinical practice, is well established. Is there sufficient evidence to herald a new era, for sufferers of dementia, of *bona fide* combination therapy? This chapter will examine these questions in the light of the published literature on memantine up to the end of 2004.

Glutamate and NMDA receptors

The body of evidence supporting the concept that overstimulation of neuron membrane receptors by excitatory amino acids injures neurons continues to grow. L-glutamate and aspartate are two of these excitatory neurotransmitters with glutamate being most prevalent.[1] Excessive activation of glutamate receptors, so-called 'excitotoxicity', has been implicated as a final mechanism in a number of acute and chronic neurological disorders resulting in death of neurons in the cortex and hippocampus.[1,2] Disturbed glutamate neurotrans-

mission and sustained changes in intracellular calcium ions have been hypothesised as a cause for neural degeneration as the final event in the pathogenesis pathway in Alzheimer's disease (AD).[3] Glutamate receptors are subtyped as either ionotropic, coupled directly to membrane ion channels, or metabotropic, modulating intracellular second messengers.[1] The ionotropic receptors are further divided into three types according to their selective agonists N-methyl-D-aspartate (NMDA), alpha-amino-3-hydroxy 5-methyl-4-isoxazolepropionate (AMPA) and kainate.[1] The NMDA receptor is activated by glutamate and is involved in learning and memory.[4,5] Overstimulation of NMDA receptor-activated channels may allow excessive entry of calcium into the neuron.

Sustained elevated levels of glutamate, possibly in combination with increased sensitivity of glutamate receptors causing prolonged influx of calcium, may result in neurodegeneration.[3,6,7]

Overstimulation of NMDA receptors produces pathologically enhanced synaptic noise. It is also postulated that selective blockade of pathological activation of the NMDA receptor, while still allowing normal physiological activity to occur, could improve memory-related symptoms in AD.[8] Noncompetitive NMDA receptor antagonists have been suggested as possible therapeutic agents in AD.[9] Several NMDA receptor antagonists have been tested but not all have been effective as they competed with glutamate to bind to the NMDA receptor or blocked the NMDA receptor channel.[10] Of these, memantine prolongs duration of learning and synaptic plasticity *in vivo* and memory in rats.[11,12] Recently memantine has also been shown to inhibit and reverse hyperphosphorylation of tau/neurofibrillary degeneration in rat brain.[13]

Properties of memantine

Memantine is a low to moderate affinity non-competitive NMDA receptor antagonist with strong voltage dependency and rapid blocking/unblocking kinetics.

Following single oral doses, memantine is completely absorbed with nearly 100% bio-availability. Time to maximum plasma concentration is three to seven hours with approximately 45% bound to plasma proteins.[14] It is distributed extensively through the body and rapidly crosses the blood–brain barrier. It does not inhibit cytochrome P450 pathways, has a half-life of up to 100 hours and is excreted largely unchanged in the urine.[14] Trials in Germany of memantine, given intravenously in patients with Parkinson's disease, were reported first in 1977.[15] Memantine was subsequently tested to treat detrusor instability and testing for dementia began in the mid-1980s.[14]

Clinical trial evidence

Evaluation of the randomised controlled trial (RCT) published data on memantine has been reported most recently in August 2004 as part of a Cochrane systematic review.[16] Nine RCTs, three phase II and six phase III studies, using memantine in dementia have been published (one as a poster) in the last 14 years.[17–25] The first three phase II studies, all of 6 weeks' duration, were conducted in Germany and each had a total of less than 90 subjects. In each of these three studies, the subjects had either AD or vascular dementia or mixed dementia (AD and VaD).[16]

Moderate to severe AD and VaD

In the first phase III trial of 12 weeks' duration, 49% of the subjects had a clinical diagnosis of AD, 51% had VaD and were recruited from nursing homes in Latvia.[20] The subjects had severe dementia with a mean Mini Mental State Examination (MMSE) score of 6.3 (2.7 baseline), and were relatively young (compared to the other trials) with a mean age of 68.4 years. Subjects were randomised to memantine 5 mg per day for one week followed by 10 mg per day for 11 weeks or placebo. Because of the severity of dementia, rather than a cognitive scale, a functional measure, the Behaviour Rating Scale for Geriatric Patients (BGP),[26] as rated by the nursing staff, and the clinical global impression of change (CGI-C),[27] rated by physicians, were chosen for the primary outcome measures. Using the CGI, benefit was in favour of memantine, compared to placebo (60/82 compared with 38/84; odds ratio (OR) 3.30, 95% confidence interval (CI) 1.72 to 6.33, $P = 0.0003$). Function evaluated with the BGP and a Wilcoxon stratified test used by the authors, found a significant difference in favour of memantine. However, there was no statistically significant difference when using the OR analysis as reported in the Cochrane review.

Moderate to severe AD

The most convincing data to date for the use of memantine in people with dementia are in moderate to severe AD, at 20 mg per day. This is the indication for which memantine is presently licensed in the European Union (2002), in the United States (October 2003) and Canada (December 2004). The supporting evidence comes from two phase III trials over 24 to 28 weeks.[23,24] The Reisberg trial evaluated memantine monotherapy, and the Tariot trial combination therapy with a stable dose of donepezil. On the combined Clinicians Interview-Based Impression of Change scale (CIBIC-Plus),[28] and using an intention to treat analysis, there was a significant difference in favour of memantine weighted mean difference (WMD -0.27, 95% CI -0.43 to -0.10,

$P = 0.002$).[16] In these two trials, cognition was measured using the Severe Impairment Battery (SIB). Using a combined intention to treat (ITT) analysis, there was a statistically significant result in favour of memantine (WMD 4.13, 95% CI 2.51 to 5.74, $P < 0.00001$).[16] Both studies used the AD Cooperative Study Activities of Daily Living Inventory (ADCS-ADL) and both found significant results in favour of memantine (WMD 1.70, 95% CI 0.63 to 2.76, $P = 0.002$).[16]

Combined overall mood and behaviour symptom/symptoms score measured by the Neuropsychiatric Inventory (NPI),[29] was lower in subjects on memantine (WMD -3.64, 95% CI -5.9 to -1.38, $P = 0.002$). The combined drop-out rate for the two trials was significantly lower for those on memantine compared to placebo (59/329 versus 93/327, WMD 0.55, 95% CI 0.30 to 0.79, $P = 0.002$). Although there was no difference between groups in the number of subjects suffering at least one adverse event, agitation was less likely to occur on memantine 42/338 versus 64/327 on placebo, OR 0.59 (95% CI 0.39 to 0.91). On the other hand, confusion as an adverse event, which was uncommon, occurred more frequently in those on memantine 16/202 versus placebo 4/201, OR 4.24 (95% CI 1.49 to 12.90).

Mild to moderate AD

The efficacy of memantine monotherapy 20 mg a day over 24 weeks has also been evaluated in subjects with mild to moderate AD. This RCT has been presented in poster format at least four international conferences in 2004 and a journal publication is anticipated. The CIBIC-Plus showed a significant difference in favour of memantine (WMD -0.30, 95% CI -0.31 to -0.29, $P < 0.00001$). Consistent with RCTs using cholinesterase inhibitor monotherapy, cognition in these subjects was measured using the Alzheimer's Disease Assessment Scale cognitive subscale (ADAS-Cog).[30] On the ITT analysis, a change from baseline was statistically significant in favour of memantine (WMD -1.81, 95% CI -1.92 to -1.70, $P < 0.00001$).

Unlike each of the cholinesterase inhibitor (ChEI) RCTs in mild to moderate AD, there was no significant change in function using the ADCS-ADL. There was a significant difference in disturbances of neuropsychiatric symptoms measured with the NPI in favour of memantine (WMD -3.50, 95% CI -3.74 to -3.26, $P < 0.00001$). There are two completed, but unpublished, trials of memantine in mild to moderate AD referenced in the Cochrane Review as Anonymous 2003B and Anonymous 2004.[15] Both were multi-centre RCTs of memantine. The first examined combination therapy with a stable dose of a cholinesterase inhibitor and the second evaluated memantine as a monotherapy at 20 mg a day. The reported preliminary analyses of both of these studies have not demonstrated a statistically significant difference on cognition, function or global measures in this population.[16] Further information may be available if, and when, these trials are published in full.

Mild to moderate VaD

Two studies of very similar design have been published assessing the efficacy and safety of memantine in patients with mild to moderate VaD.[21,22] Combined, they form the largest group of subjects assessed with memantine for one 'type' of dementia (900 subjects). Using the CIBIC-Plus, at 28 weeks there was no difference between memantine and placebo (WMD -0.29, 95% CI -0.66 to 0.08, $P = 0.13$). Cognition was assessed using the ADAS-Cog, and the scores at 28 weeks were statistically improved relative to placebo (WMD -2.19, 95% CI -3.16 to -1.21, $P < 0.0001$). Both studies used the Nurses Observational Scale for Geriatric Patients that measures behaviour (NOSGER).[31] On this secondary outcome measure, there was no difference between the memantine and placebo groups (WMD -0.92, 95% CI -2.90 to 1.05, $P = 0.4$). Likewise, on the Gottfries–Brane–Steen Scale,[32] that measures motor performance, intellectual and emotional capacity and six dementia symptoms, there was no difference between the two groups (WMD 1.81, 95% CI -4.21 to 0.58, $P = 0.14$).

Meta-analyses

Pooled data of the five trials of 6 months' duration in patients with AD and VaD provide more robust examination.[16] In each of the four domains, memantine was favoured over placebo:

- *global impression*: (SMD −0.74, 95% CI −0.86 to −0.61)
- *cognition*: (SMD −0.33, 95% CI −0.42 to −0.23)
- *behaviour*: (SMD −0.22, 95% CI −0.32 to −0.11)
- *function*: (SMD −0.11, 95% CI −0.21 to −0.01).

Safety

Memantine is well tolerated and, overall, is safe when compared to placebo. Using the pooled 6-month data, those on memantine were less likely to become agitated (OR 0.65, 95% CI 0.48 to 0.89, $P = 0.007$). Although in individual trials confusion was significantly more likely on memantine, the pooled data showed a non-significant trend only (OR 1.35, 95% CI 0.80 to 2.05, $P = 0.15$). Combining all six phase III studies, those on memantine were more likely to continue treatment compared to placebo (OR 0.84, 95% CI 0.68 to −1.05, $P = 0.12$). Somnolence was significantly more common in the one mild to moderate AD study, and constipation more common in one of the VaD studies.[22,33]

Cost-effectiveness

The cost-effectiveness of memantine has been published from the moderate to severe AD monotherapy trial.[34] Benefits attributed to memantine were less care time (45.8 hours per month 95% CI 10.37 to 81.27, $P = 0.017$), and an average monthly cost to society of US$1090 (1999 dollar; $P = 0.01$).

Two secondary cost-effectiveness extrapolations explored beyond 6 months using an Amarkov model have been published.[35,36] One includes assumptions from additional epidemiological data from the UK and Denmark (dependency data) and quality-adjusted life years (QALYs) and extrapolates cost-effectiveness of memantine over 2 years.[35] The other includes assumptions from an epidemiological study in Finland and extrapolates resource use cost and patient outcome over 5 years. Both models were conservative in estimating that memantine would only be effective for one year. Both studies, which had pharmaceutical involvement, concluded that the cost of memantine was offset by increased independence and delayed institutionalisation.

A look beyond statistical significance: interpreting clinical meaningfulness

What is the place of memantine in dementia therapy? The six published trials of 12–28 weeks' duration vary in the type and severity of dementia and use different outcome measures, depending on the severity of dementia.[20–25] Direct comparison between them or with published trials of one of the four ChEIs in a similar population has to be considered with caution.

Numbers needed to treat

One approach that sheds light on the clinical relevance of different drugs is a numbers needed to treat (NNT) analysis. It is the average number of patients needed to be treated with a drug to achieve an improvement in an outcome compared to placebo or other drug for a treatment period.[37] Since the NNT is only a point estimate, the 95% CIs provide an indication that the true value falls within a specific range. It is only appropriate to compare NNTs for different interventions for the same condition and the same outcome.[37] The smaller the NNT, the more effective the treatment is.

The NNT analysis in controlled trials of ChEIs in dementia has been published,[38] and recently updated to include data on galantamine,[39,40] and two of the studies with memantine.[19,22]

Table 17.1 adds to the calculations made by Livingston and Katona,[39] and shows NNT values of 6–8 for moderate–severe AD on global, cognitive, and functional outcome scales. For severe dementia, the NNT are somewhat lower, with scores of

Table 17.1 Numbers needed to treat (NNT) in dementia trials with memantine

Study/indication	Scale	Criteria	Analysis	NNT (95% CI)
Reisberg et al 2003[23]/ moderate–severe AD	CIBIC-Plus	Stabilisation or improvement	ITT-LOCF	6 (4–26)
Reisberg et al 2003[23]/ moderate–severe AD	SIB	Stabilisation or improvement	ITT-LOCF	7 (4–74)
Reisberg et al 2003[23]/ moderate–severe AD	ADCS-ADLsev	Stabilisation or improvement	ITT-LOCF	8 (4–98)
Winblad and Poritis 1999[20]/ severe dementia	BGP-D	Improvement ≥15%	ITT-worst rank	4 (2–10)
Winblad and Poritis 1999[20]/ severe dementia	CGI-C	Final score <4	ITT-worst rank	3 (2–8)
Orgogozo et al 2002[21]/ mild–moderate VaD	CGI-C	Stabilisation or improvement	PP-completers	5 (3–13)
Orgogozo et al 2002[21]/ mild–moderate VaD	CGI-C caregiver	Stabilisation or improvement	PP-completers	10 (5 – no upper limit)

3 and 4 on tests of behaviour and global ratings, respectively. Orgogozo's study on patients with mild–moderate VaD provided NNT of 5 for global rating from a physician.[21] Interestingly, when caregivers completed the same scale, the NNT increased to 10. This discrepancy between physician and caregiver ratings of disease progression casts some doubt on the validity of the CGI-C as a valid tool to monitor the progression of VaD. To provide some comparison, the NNT to prevent decline in global rating (CIBIC-Plus) with donepezil from the first published trial in moderate–severe AD was 5 (95% Cl, 4–11).[41]

Treatment effect size

A second way to understand the benefit of drugs in dementia is to calculate the effect size. The clinical implications of effect size in AD have been discussed by Rockwood and MacKnight,[42] and also in Chapter 1 of this book.

Treatment effect size can be calculated in a variety of ways. We have chosen to use Cohen's d statistic to allow for comparison with Livingston and Katona.[39]

If one accepts that an effect size of at least 0.20 is clinically detectable, 0.50 has a moderate clinical effect and greater than 0.80 represents a clinically large effect, then memantine has at least a clinically detectable effect for the treatment of dementia.

For the sake of comparison from one of the first published trials on donepezil the treatment effect size for donepezil on cognition using the ADAS-Cog was 0.25.[43] These numbers are also within the range reported for all the ChEIs.[44]

Convergence of measures

As noted in Chapter 1, convergence of measures will be defined as occurring when more than one outcome measure is significantly affected by the

Table 17.2 Treatment effect size in dementia trials with memantine

Study/indication	Analysis type	Test	Mean treatment difference (P-value)	Treatment effect size (d)
Peskind et al 2004[25]/ mild–moderate AD	ITT	ADAS-cog	−1.9 (<0.00001)	−0.24
		CIBIC-Plus	−0.3 (<0.00001)	−0.29
		NPI	−3.5 (<0.00001)	−0.21
Reisberg et al 2003[23]/ moderate–severe AD	ITT-LOCF	SIB	6.1 (<0.001)	0.49
		CIBIC-Plus	−0.3 (0.06)	−0.27
		ADCS-ADLsev	2.1 (0.02)	0.32
		FAST	−0.4 (0.02)	−0.30
Tariot et al 2004[24]/ moderate–severe AD	ITT-LOCF	SIB	3.4 (<0.001)	0.36
		CIBIC-Plus	−0.25 (0.03)	−0.24
		ADCS-ADLsev	1.4 (0.03)	0.20
		NPI	−3.8 (0.002)	−0.28
Wilcock 2002[46] mild–moderate VaD	ITT-LOCF	ADAS-cog	1.75 (0.04)	0.24
Orgogozo et al 2002[21]/ mild–moderate VaD	ITT-OC	ADAS-cog	−2.83 (0.0016)	−0.48
		MMSE	1.23 (0.0121)	0.33
Winblad and Poritis 1999[20]/ severe dementia	ITT-WCR	BGP care dependence	−2 (0.016)	−0.17
		CGI-C	−0.38 (<0.001)	−0.55

ITT, intention to treat; LOCF, last observation carried forward; OC, observed cases; WCR – worst case replacement

treatment. Table 17.3 demonstrates the occurrences of convergence by disease type and severity.

The data on patients with mild to moderate AD show convergence occurring across three domains. A replication study is necessary to fully understand the clinical meaningfulness of these findings. The data from the two studies on patients with moderate to severe AD is fairly robust, with convergence across three domains and complete replication between studies. There was no convergence achieved in the two studies on patients with mild to moderate VaD, which further cast doubts on the usefulness of memantine in this population. There was some convergence in Winblad's study in people with severe dementia, with the global and functional ratings achieving significance.[20] Caution should be used in interpreting these results as there may be considerable expected overlap between these two rating scales. To some degree, they may be measuring the same domain.

Conclusion

The clinical trial data make sense to researchers and clinicians. However, to our patients and informed caregivers, the relevance to their personal circumstance is often lost. What is missing

Table 17.3 Convergence of outcome measures in dementia trials with memantine

Study/indication	Convergence	Measures
Peskind et al 2004[25]/ mild–moderate AD	Cognition, global rating, behaviour	ADAS-cog, CIBIC-Plus, NPI
Reisberg et al 2003[23]/ moderate–severe AD	Cognition, global rating, function	SIB, CIBIC-Plus, ADCS-ADLsev
Tariot et al 2004[24]/ moderate–severe AD	Cognition, global rating, function	SIB, CIBIC-Plus, ADCS-ADLsev
Mild–moderate VaD[22,46]	None	
Winblad and Poritis 1999[20]/ severe dementia	Global rating, function	CGI-C, BGP care dependence

from the memantine studies and the ChEI trials alike are measures that can capture improvement in quality of life or positive changes that are relevant to each unique individual mind that is being destroyed. Specific quality of life measures for use in dementia trials are only just now being to be explored. Individualised measures that are responsive to change have been tried in some dementia studies and are promising. They have not reached the mainstream of dementia clinical trial design and are not required by regulatory agencies. Combination drug therapies for symptomatic treatment for at least moderate to severe AD is here.

Several drugs with potential to be disease modifying, based on preclinical data, and each with different mechanisms of action, are presently in phase II and phase III trials. Treating AD as a chronic illness, with drug combinations and non-drug management, much like congestive heart failure, ischaemic heart disease or acquired immune deficiency, rather than as a terminal illness, is becoming a reality. Measuring small additive clinically relevant effects that capture quality of life or individualised improvement will be needed and expected.

References

1. Lipton SA, Rosenberg PA. Mechanisms of disease: excitatory amino acids as a final common pathway for neurologic disorders. NEJM 1994; 613–22.

2. Olney JW. Excitotoxic amino acids and neuropsychiatric disorders. Ann Rev Pharmacol Toxicol 1990; 30:47–71.

3. Cacabelos R, Takeda M, Winblad B. The glutamatergic system and neurodegeneration in dementia: preventive strategies in Alzheimer's disease. Int J Geriatr Psychiatry 1999; 14:3–47.

4. Collingridge GL, Blis TVP. NMDA receptors – their role in long-term potentiation. Trends Neurosci 1987; 10:288–93.

5. Danysz W, Zajaczkowski W, Parsons CG. Modulation of learning processes by ionotropic glutamate receptor ligands. Behav Pharmacol 1995; 6:455–74.

6. Lancelot E, Beal MF. Glutamate toxicity in chronic neurodegenerative disease. Prog Brain Res 1998; 116:331–47.

7. Greenamyre JT, Young AB. Excitatory amino acids and Alzheimer's disease. Neurobiol Aging 1989; 10(5):593–602.

8. Parsons CG, Danysz W, Quack G. Memantine is a clinically well tolerated N-methyl-D-aspartate (NMDA) receptor antagonist – a review of the preclinical data. Neuropharmacology 1999; 38:735–67.

9. Muller WE, Mutschler E, Riederer P. Noncompetitive NMDA receptor antagonists with fast open-channel blocking kinetics and strong voltage-dependency as potential therapeutic agents for Alzheimer's dementia. Pharmacopsychiatry 1995; 28(4):113–24.

10. Albensi BC, Igoechi C, Janigro D et al. Why do many NMDA antagonists fail, while others are safe and effective at blocking excitotoxicity associated with dementia and acute injury? Am J Alz Oth Dementias 2004; 19:269–74.

11. Zajaczkowski W, Quack G, Danysz W. Infusion of (=) −MK-801 and memantine – contrasting effects on radial maze learning in rats with entorhinal cortex lesion. Eur J Pharmacol 1996; 296(3):239–46.

12. Barnes CA, Danysz W, Parsons CG. Effects of the uncompetitive NMDA receptor antagonist memantine on hippocampal long-term potentiation, short-term exploratory modulation and spatial memory in awake, freely moving rats. Eur J Neurosci 1996; 8(3):565–71.

13. Li L, Sengupta A, Haque N et al. Memantine inhibits and reverses the Alzheimer type abnormal hyperphosphorylation of tau and associated neurodegeneration. FEBS Lett 2004; 566:261–9.

14. Mobius HJ, Stoffler A, Graham SM. Memantine hydrochloride pharmacological and clinical profile: drugs today 2004; 40:685–95.

15. Fischer PA, Jacobi P, Schneider E, Schonberger B. Effects of intravenous administration of memantine in parkinsonian patients. Arzneimittelforschung 1977; 27(7):1487–9.

16. Areosa Sastre A, McShane R, Sherriff F. Memantine for dementia (Cochrane Review). In: The Cochrane Library, Issue 4, 2004. Oxford: Update Software.

17. Ditzler K. Efficacy and tolerability of memantine in patients with dementia syndrome. A double-blind, placebo controlled trial. Arzneimittelforschung 1991; 41(8):773–80.

18. Gortelmeyer R, Erbler H. Memantine in the treatment of mild to moderate dementia syndrome. A double-blind placebo-controlled study. Arzneimittelforschung 1992; 42(7):904–13.

19. Pantev M, Ritter R, Gortelmeyer R. Clinical and behavioural evaluation in long-term care patients with mild to moderate dementia under memantine treatment. Zeitschrift fuer Gerontopsychologie und Psychiatrie 1993; 6(2):103–17.

20. Winblad B, Poritis N. Memantine in severe dementia: results of the M-Best Study (Benefit and Efficacy in Severely Demented Patients During Treatment with Memantine). Int J Geriatr Psychiatry 1999; 14:135–46.

21. Orgogozo JM, Rigaud AS, Stoffler A et al. Efficacy and safety of memantine in patients with mild to moderate vascular dementia: a randomized, placebo-controlled trial (MMM 300). Stroke 2002; 33:1834–9.

22. Wilcock G, Mobius HJ, Stoffler A. A double-blind, placebo-controlled multicentre study of memantine in mild to moderate vascular dementia (MMM500). Int Clinical Psychopharmacol 2002; 17:297–305.

23. Reisberg B, Doody R, Stoffler A et al. Memantine in moderate-to-severe Alzheimer's Disease. NEJM 2003; 348:1333–41.

24. Tariot PN, Farlow MR, Grossberg GT et al . Memantine treatment with moderate to severe Alzheimer's disease already receiving donepezil. A randomized control trial. JAMA; 2004:317–24.

25. Peskind E, Potkin S, Pomara N et al. Memantine monotherapy is effective and safe for the treatment of mild to moderate alzheimer's disease: a randomized control trial. Poster presented at: The 17th Annual Meeting of the American Association for Geriatric Psychiatry, Baltimore, MD; February 21–24, 2004.

26. van der Kam P, Hoeksma BH. ADL and behaviour rating scales for the evaluation of nurses' workload in psychogeriatric nursing homes [De bruikbaarheid van BOP en SIVIS voor het schatten van de werklast in het psychogeriatrisch verpleeghuis]. Tijdschrift voor Gerontologie en Geriatrie 1989; 20:159–66.

27. Guy W. CGI: Clinical Global Impressions. In: Guy W (ed). ECDEU Assessment Manual for Psychopharmacology. Rev Edition. Rockville: National Institutes of Health; 1976: 217–20.

28. Reisberg B, Schneider L, Doody R et al. Clinical global measures of dementia: Position Paper from the International Working Group on Harmonization of Dementia Drug Guidelines. Alzheimer Dis Assoc Disord 1997; 11:(Suppl 3):8–18.

29. Cummings JL, Mega M, Gray K et al. The Neuro-Psychiatric Inventory comprehensive assessment of psycho-pathology in dementia. Neurology 1994; 44:2308–14.

30. Rosen WG, Mohs RC, Davis KL. A new rating scale for Alzheimer's disease. Am J Psychiatry 1984; 11:1356–64.

31. Spiegel R, Brunner C, Ermini-Funfschilling D et al. New behavourial assessment scale for geriatric out and inpatients: The NOSGER (Nurses Observational Scale for Geriatric Patients) J Am Geriatr Soc 1991; 39:339–47.

32. Gottfries CG, Brane G, Steen G. A new rating scale for dementia syndromes. Gerontology 1982: 28 Suppl 2:20–31.

33. Peskind E, Potkin S, Pomara N et al. Memantine monotherapy is effective and safe for the treatment of mild to moderate Alzheimer's disease: a randomized control trial. Poster presented at: The 17th Annual Meeting of the American Association for Geriatric Psychiatry, Baltimore, MD; February 21–24, 2004.

34. Wimo A, Winblad B, Stoffler A et al. Resource utilization and cost analysis of Memantine in patients with moderately severe to severe Alzheimer's disease. Pharmacoeconom 2003; 21(5):327–40.

35. Jones RW, McCrone P, Guilhaume C. Cost effectiveness of memantine in Alzheimer's disease: an analysis based on a probabilistic Markov model from a UK perspective. Drugs Aging 2004; 21:607–20.

36. Francois C, Sintonen H, Sulkava R et al. Cost effectiveness of Memantine in moderately severe to severe Alzheimer's disease. A Markov model in Finland. Clin Drug Invest 2004; 24(7):373–84.

37. McQuay HJ, Moore RA. Using numerical results from systematic reviews in clinical practice. Ann Intern Med 1997; 126:712–20.

38. Livingston G, Katona C. How useful are cholinesterase inhibitors in the treatment of Alzheimer's disease? A number needed to treat analysis. Int J Geriatric Psychiatry 2000; 15(3):203–7.

39. Livingston G, Katona C. The place of memantine in the treatment of Alzheimer's disease: a number needed to treat analysis. Int J Geriatr Psychiatry 2004; 19:919–25.

40. Wilcock GK, Lilienfeld S, Gaens E. Efficacy and safety of galantamine in patients with mild to moderate Alzheimer's disease: multicentre randomized controlled trial. Galantamine International-1 Study Group. BMJ 2000; 321:1445–9.

41. Feldman H, Gauthier S, Hecker J, Vellas B, Subbiah P, Whalen E; Donepezil MSAD Study Investigators Group. A 24-week, randomized, double-blind study of donepezil in moderate to severe Alzheimer's disease. Neurology 2001; 57(4):613–20.

42. Rockwood K, MacKnight C. Assessing the clinical importance of statistically significant improvement in anti-dementia drug trials. Neuroepidemiology 2001; 20:51–6.

43. Rogers SL, Farlow MR, Doody RS, Mohs R, Friedhoff LT. A 24-week, double-blind, placebo-controlled trial of donepezil in patients with Alzheimer's disease. Donepezil Study Group. Neurology 1998; 50(1):136–45.

44. Rockwood K. Size of the treatment effect on cognition of cholinesterase inhibition in Alzheimer's disease. J Neurol Neurosurg Psychiatry 2004; 75:677–85.

45. Wilcock G, Mobius HJ, Stoffler A. A double-blind, placebo-controlled multicentre study of memantine in mild to moderate vascular dementia (MMM500). Int Clinical Psychopharmacol 2002; 17:297–305.

18

Clinical Trials for Psychotropic Agents in Alzheimer's Disease

David M Blass and Peter V Rabins

Introduction

Psychotropic medications are frequently prescribed for patients with Alzheimer's disease (AD). They target a variety of neuropsychiatric symptoms and syndromes that include depression, psychosis, physical aggression, other behavioural disturbances and sleep disorders. These symptoms, often referred to as 'non-cognitive symptoms,' are highly prevalent in demented patients, with 61% of demented patients demonstrating at least one symptom over any one-month period.[1] Once present, these symptoms tend to persist over time.[2,3] In a one-year longitudinal study of 181 patients with AD, recurrence rates for depression, psychosis and agitation were 85%, 95% and 93%, respectively.[3] These symptoms are associated with increased caregiver burden which in turn is associated with nursing home placement.[4] Numerous clinical trials assessing the efficacy of psychotropic medications have been conducted. This chapter discusses the conceptual and methodological challenges that are inherent in defining and classifying neuropsychiatric syndromes in AD, identifying appropriate treatment outcomes, and designing meaningful clinical trials that can inform routine clinical decision making.

The terms 'neuropsychiatric', 'non-cognitive' and 'behavioural' will be used throughout this chapter to refer to symptoms that are not *traditionally* considered 'cognitive'. Each phrase has a different emphasis and no single term encompasses all of these symptoms. *In toto*, these terms refer to abnormalities in the realms of emotion (mood and affect), perceptual experience (hallucinations and illusions), belief (delusions), initiative (apathy and disinhibition), motor activity (over and under activity), specific actions (hitting) and drive behaviours (sleep, sexual behaviour and eating).[5] We will use the phrases 'non-cognitive' and 'neuropsychiatric' to refer to the class of symptoms as a whole because they have the fewest implications regarding the aetiology and genesis of the symptoms discussed in this chapter.

Syndrome definition

Medical knowledge typically progresses through a series of stages: definition of a clinical syndrome, identification of pathophysiology, and ultimate discovery of underlying aetiology.[6] This simple but powerful algorithm leads not only to greater scientific understanding, but lays the groundwork for the development of rational therapeutics that target not only symptoms but more fundamental targets in the chain of pathophysiology. The initial step in clinical description is the appreciation that symptoms, often superficially unrelated, often cluster in recognisable syndromes. Recognition of this allows for the definition of populations for epidemiological investigations of prevalence, incidence and risk factors, the characterisation of natural history and the development of empirical treatment, even before the elucidation of pathophysiology and aetiology is complete.

Progress in psychiatry has been greatest in the realm of syndrome description. Validation of a psychiatric syndrome, even before a causal mechanism has been discovered, occurs through seeking concurrent sources of validity. These include demonstrating that the syndrome occurs in an idiopathic form as well as being present in other brain disorders, the occurrence of familial and genetic clustering, a predictable course, biomarkers such as neuroimaging abnormalities, and a predictable response to treatment. Universally accepted disorders that are syndromic in nature include bipolar disorder, major depression, schizophrenia and autism. Although correlation can be made between dysfunction of certain brain regions and abnormalities of psychological experience, true explanation is lacking. This is true of both normal psychological experience and its pathological counterpart. This fact is an outgrowth of the mind–brain mystery and partially explains what distinguishes neuroscience from other areas of medical research in which a direct causal path can be discerned linking molecular or cellular abnormalities to organ dysfunction.

Alzheimer's disease

The clinical syndrome of AD, defined by its core features of cognitive and functional decline, slow progression, and lack of focal neurological abnormalities has been well characterised clinically and can be diagnosed with fairly high accuracy in life. Significant progress has also been made in discovering the underlying pathophysiology of the disease, and in familial cases numerous genetic aetiologies have been found.

A major advance in the care of patients with AD has been the appreciation of the significance of neuropsychiatric symptoms, both in terms of prevalence and caregiver burden.[7] In keeping with the developmental schema of syndrome, pathology and aetiology outlined earlier, the widespread recognition of symptom prevalence has stimulated attempts to define specific syndromes of neuropsychiatric disturbances that may be unique to AD.[8] To date, two syndromes have been proposed: depression of AD (DAD) and psychosis of AD (PAD).[9–11]

The next section will briefly summarise the current literature regarding pharmacological treatment of neuropsychiatric disturbances in AD, and will be divided into two sections: depression and psychosis and behavioural disturbances. The remainder of the chapter will focus on methodological issues relevant to design of treatment trials for these disturbances.

Depression

An association between some dementing illnesses and depression has been long recognised. Emil Kraepelin noted that depression was common in patients with Huntington's disease and stroke 100 years ago,[12] and in the 1960s, Martin Roth suggested that cerebrovascular disease might be aetiologic for late-life depression.[13] Understanding of the precise relationship between depression and dementia has evolved. During the 1970s, depression was highlighted as a reversible cause of cognitive impairment in some patients. Long-term follow-up studies by Alexopoulos and Reynolds in the mid-1980s however, demonstrated that more than 50% of older individuals presenting with depression and mild cognitive impairment went on to develop irreversible dementia when followed for 5 years.[14,15] Evidence on whether early life affective disorder predisposes patients to develop dementia is mixed, and no consensus exists about this possibility.[16]

In AD specifically, prevalence rates for depression appear to be approximately 20%.[1,16] Some of these patients have lifelong affective disorder, while others appear to have depression as an early symptom of AD or an emergent symptom during the course of the dementia.

This high prevalence of depressive symptoms in patients with Alzheimer's disease has led to clinical investigations into the response of these symptoms to standard depression therapies. Results of clinical trials have been mixed. An early antidepressant study by Reiffler demonstrated equal responses to placebo and imipramine.[17] However, because imipramine has significant anti-cholinergic properties and because the study used a forced titration, fixed dose strategy, it does not meet today's standards of clinical trial design. There have been other negative trials as well.[18] Several more recent randomised double-blind studies have shown benefit from tricyclics, selective serotonin reuptake inhibitors (SSRIs) and monamine oxidase (MAO) inhibitors,[19–21] but the total number of patients studied is still small. A number of studies have compared two medications with each other without including a placebo control group.[22,23]

Symptom scales used in these trials have included the Hamilton Depression Rating Scale,[24] the Cornell Scale for Depression in Dementia,[25] the Montgomery–Asberg Depression Rating Scale,[26] and the Clinical Global Impression Improvement Scale.[27]

Psychosis and behavioural disturbances

The original patient reported by Alois Alzheimer presented with the delusion that her husband was having an affair.[28] Thus, the association between psychosis and AD has been known since the first report of the disease's neuropathology. However, it is only in the past 30 years that the high prevalence of such symptoms in AD has been noted, and only in recent years that the morbidity caused by these symptoms has been widely appreciated as causing significant morbidity in the patient, family members and friends.

A specific syndrome, psychosis of AD, has been narrowed from the original conception of Behavioural and Psychological Symptoms of Dementia

(BPSD).[29] Psychotic symptoms are generally defined as hallucinations and delusions, and these are commonly accompanied by behavioural symptoms such as aggression (targeted behaviour that has the potential to harm others), agitation (untargeted overactive behaviour), wandering, and yelling, to name a few. In many clinical trials, behaviour symptoms as well as psychosis have often been grouped together, a methodological problem addressed later in this chapter. A number of classes of medications have been studied for use in treatment of these symptoms.

Anti-psychotics

A number of randomised, placebo-controlled clinical trials have established that anti-psychotic medications are effective in treating psychotic and behavioural symptoms of AD. These include trials of risperidone,[30–32] olanzapine,[33,34] and a number of other agents including thiothixene, thioridazine, haloperidol and others.[35] Outcome measures used in these studies included the Neuropsychiatric Inventory (NPI) regular and nursing home versions, total scores as well as subscores (agitation, delusions and hallucinations items, for example),[36] the Behavioral Pathology in Alzheimer's Disease scale (BEHAVE-AD),[37] and the Cohen–Mansfield Agitation Inventory (CMAI).[38] As reviewed by Schneider more than 10 years ago, 18% more trial subjects with agitation who received active drug responded compared to those who received placebo.[39] This is an effect size that is similar to that observed in more recent trials.

Anti-convulsants

A number of randomised, placebo-controlled clinical trials have suggested that anti-convulsant medications are effective in treating psychotic and behavioural symptoms of AD. These include trials of carbamazepine,[40,41] and divalproex sodium,[42] although some studies of divalproex sodium have had negative results.[43] Outcome measures used in these studies included the Brief Psychiatric Rating Scale (BPRS) and the Social Dysfunction and Aggression Scale-9 (SDAS-9).[44,45]

Other agents

A number of psychotropic agents including cholinesterase inhibitors, SSRIs, trazodone, benzodiazepines, beta-blockers and buspirone may be useful in the treatment of psychosis and behavioural disturbance in AD but only have case series or small clinical trials supporting their use.[46] A meta-analysis of cholinesterase inhibitor trials found only a modest benefit in neuropsychiatric symptoms (as measured by the NPI and the ADAS-noncog).[47] However, these trials of cholinesterase inhibitors were not designed with psychosis and behavioural disturbance as primary outcome measures, and therefore many patients in the trials did not have these symptoms. Randomised, placebo-controlled trials of the SSRI citalopram have demonstrated benefits over placebo in both outpatients and inpatients.[48,49] One randomised, placebo-controlled trial of trazodone failed to show any benefit over placebo.[50] Older trials of benzodiazepines indicate that they may be effective in the treatment of agitation in patients with dementia.[51] Their use has generally been limited due to concerns about their deleterious effects on cognition and their potential to worsen agitation by leading to disinhibition. We agree with the practice of limiting their widespread and long-term use, but believe that further study is warranted.

Specific behaviours

A number of specific behaviours or behavioural syndromes lend themselves particularly well to individual study, but in general, there are few controlled studies. One example is hypersexuality, a symptom that develops in many patients with dementia in the absence of a prior history of this symptom. Case reports suggest that anti-androgen or other agents may be beneficial for this symptom but there are no controlled trials to date.[52] Hyperorality, a symptom commonly seen early in the course of frontotemporal dementia and somewhat later in the course of AD, causes significant morbidity for patients. This symptom can also be conceptualised and measured independently from other symptoms. There are no controlled studies of this symptom either. Calling out, repetitive self-mutilation, sleep disturbance and aggression are other examples.[53]

Methodological issues

Syndrome definition – revisited

Much progress has been made in defining neuropsychiatric syndromes that occur during the course of AD, particularly with respect to depression.[11] Although the syndrome of DAD has been clearly defined and is now accepted by the Food and Drug Administration (FDA), a number of questions still remain. The first is the relationship between the numerous patients who may have several symptoms of depression but who do not meet the criteria for the full syndrome. This 'threshold problem' also exists for major depression in non-demented patients.[54] Although the diagnosis of major depression is treated as a categorical variable (present or absent), in fact symptom distribution (number of symptoms, duration and severity) falls along a continuum.

Another challenging issue is whether patients with a single symptom from the major depression syndrome are best conceptualised as being at the low extreme end of the major depression spectrum, or as having a distinct neuropsychiatric symptom, unrelated to the syndrome of major depression. One example of this is apathy. Although there is evidence to suggest that apathy and major depression can be reliably distinguished from one another in patients with dementia,[55] their relationship in terms of underlying physiology is not yet fully clear. Other examples include poor appetite, insomnia, mood lability and the tearfulness of pseudobulbar palsy.

With respect to the two syndromes PAD and behavioural disturbances (including the specific behaviours hitting, calling out, motor overactivity or threatening actions), further refinement is necessary for several reasons. First, despite the frequent co-occurrence of hallucinations/delusion with these other behaviours,[56] the two are phenomenologically distinct sets of symptoms. Therefore, combining them is problematic both in terms of defining populations for clinical trials, and in terms of defining outcome measures (e.g. total NPI score). To a certain extent this was addressed by removing PAD from the BPSD category but studies continue to be published that combine the two.

Second, the individual constructs of PAD and behavioural disturbance are themselves still problematic. For example, although the linking of delusions and hallucinations within the category of PAD is in keeping with the linkage of these two symptoms in other psychiatric syndromes such as schizophrenia and psychotic affective disorder, there are few data by which this pairing can be validated in AD. Some data suggest minimal association in the same patient. A recent latent class analysis of neuropsychiatric symptoms in a population-based sample of 198 AD patients from the Cache County Study of Memory in Aging, for example, found that 42% of patients with hallucinations did not have delusions and 67% of patients with delusions did not have hallucinations.[57] Moreover, while the hallucinations (typically visual or auditory) seen in AD are phenomenologically similar to those seen in other psychiatric conditions, (although differences exist in specific symptom prevalence in PAD versus schizophrenia[10,58]), the delusions that are described in AD tend to be different. For example, delusions in dementia tend to be simple persecutory delusions often involving elementary misbeliefs,[59] in contrast to the elaborate delusions often seen in schizophrenia or psychotic affective disorder. Complicating this issue even more is the fact that the delusions in AD can sometimes be considered to be a direct consequence of the cognitive impairment. Thus, the symptom of persistently misidentifying a relative as a stranger or imposter, while often classified as a delusion (a fixed, false, idiosyncratic belief), may be more appropriately classified as an agnosia (a misperception due to impairment of the cortical structures underlying recognition of the familiar), especially if the

patient is misidentifying other objects as well. To date there is no consensus on the issue; therefore, classifying an AD patient as psychotic or agnosic lacks the reliability necessary as the first step in classification. This calls into question the unity of the category 'psychosis' and its similarity to psychosis in other conditions.

Likewise, the breadth of the category 'behavioural disturbance' limits its utility and constrains further investigation into specific behavioural symptoms or syndromes with unique underlying pathologies. While a behaviour may be equally disturbing to caregivers without regard to its aetiology, rational therapeutics will be more likely to develop if behaviours can be classified into subsyndromes that can be validated through linkage to dysfunction in specific brain regions or genetic polymorphisms.[60] The approach of equating the behaviour of an agnosic patient who hits and kicks during personal care, with an aggressive patient who strikes at anyone in his vicinity or with a hypersexual patient who persistently speaks to and touches other patients and staff members inappropriately, does not advance our understanding of the pathogenesis of these activities. This is the approach taken, however, in studies where the primary outcome or severity measure, regardless of specific behaviour abnormality, is a general measure such as the distress evoked by the behaviour. The heterogeneity of behavioural disturbances suggests that aetiology will be best understood through careful description and linkage to underlying biological substrates. Certain individual behavioural symptoms suggest a unique pathophysiology. This is the case for symptoms such as hypersexuality that can be seen in patients with other psychiatric conditions but without cognitive impairment.

Further work towards subsyndrome definition will build upon previously identified clusters of symptoms such as sleep disturbance with wandering,[61] or aggression with depression[62] or delusions.[63] For example, a recent proposal based on results of a latent class analysis suggested three subsets of AD patients with regard to neuropsychiatric symptomatology: one group with no symptoms or a single symptom, a second with pre- dominantly affective disturbances (e.g. depression, anxiety, irritability), and a third with primarily psychotic symptomatology. This proposal takes a statistical approach to the question of syndrome definition.[11] Ultimate validation will depend upon linkage to specific biological substrates.

Clinical trial design

Clinical trial design for neuropsychiatric symptoms of AD is still evolving. In this section we review and comment upon a number of aspects of trial design. Table 18.1 summarises this discussion. The goals outlined below apply not only to the design of individual trials but to the process of drug development overall. It may not be possible to achieve all of these goals with one study alone, and multiple studies of the same medication may be required.

Inclusion/exclusion criteria

Inclusion and exclusion criteria should be chosen to maximise the relevance and generalisability of study results. This concern affects the definition of both the patient population and the syndrome to be studied. The minimum criterion should be that the symptom induces morbidity. Morbidity can be experienced by the patient, the caregiver whether a family member, friend or professional, another individual such as another patient in the facility, the facility itself or society. The presence of a symptom or syndrome in the absence of morbidity does not in and of itself generate a need for treatment. A study showing a treatment effect in that case may not be clinically relevant.

Patients with dementia typically have significant medical co-morbidity. For this reason, clinical trials in dementia must be as inclusive as possible of patients with medical co-morbidity in order to demonstrate efficacy and safety in the patient population in which the medication will ultimately be

---

Table 18.1 Suggestions to improve clinical trial design for treatment studies of neuropsychiatric symptoms in dementia

Goal	Method
Maximise generalisability	Include subjects with significant medical co-morbidity Do not use placebo run-in prior to randomisation Allow concomitant medications Study neuropsychiatric symptoms along the full range of cognitive impairment (multiple studies may be required)
Demonstrate clinical effectiveness	Use placebo control group Do not use placebo run-in prior to randomisation Allow concomitant medications Use syndrome-specific outcome measures Use dementia-specific rating scales Randomise group assignment with the neuropsychiatric symptom as the primary inclusion criterion
Assure clinical relevance	Include patients with clinically meaningful symptom severity Use clinically relevant outcome measures. Consider using individually tailored outcome measures in addition to standardised ones

used. Large studies should be stratified by medical co-morbidity in order to ensure group equivalence and allow for prospective subgroup comparisons.

Neuropsychiatric symptoms are prevalent and cause morbidity at all stages of dementia. Given the neuropathological changes in the brain as dementia progresses, clinical efficacy demonstrated in one stage of the illness cannot be assumed to hold true during a different stage. Therefore it is necessary to study treatments of neuropsychiatric symptoms in dementia patients at all stages. It may not be possible to accomplish this goal in a single trial, and therefore multiple trials may be necessary.

Placebo control

In our opinion, the use of placebo controls in clinical trials for neuropsychiatric symptoms in AD is still a necessity and is ethically justifiable for a number of reasons. First, the response to placebo in the already published trials is quite high, and numerous placebo control studies fail to demonstrate benefit of drug over placebo. This is probably due to a significant waxing and waning of symptomatology over any given time interval and to variability in observer ratings. As a result, treatment efficacy must initially be established against the placebo standard. Second, there is not yet a standard of care for pharmacological intervention for these symptoms, and many patients respond to non-pharmacological treatments. Many medications with demonstrated efficacy in certain clinical trials have also had negative trials, suggesting that there are still many unanswered questions about the proper indication and use of these agents. Therefore, withholding an active pharmacological treatment from a patient, particularly in a trial of limited duration, is ethically justifiable. Moreover, given the lack of proven effective treatments for these conditions, a continued requirement for placebo-controlled studies will not simply produce redundant treatments but may actually advance clinical practice beyond its current capabilities. This is in contrast to a growing body of opinion challenging the use of placebo-controlled trials for

agents treating cognitive impairment in AD.[64] Finally, head-to-head comparisons of two agents should not be performed until the efficacy of both agents has been established in placebo-controlled trials.

Use of a placebo washout period to screen out 'placebo responders' prior to randomisation is a practice commonly used in drug trials to reduce the placebo response and facilitate detection of the treatment effects of the active compound. We find little scientific justification for this practice because it limits application of the trial to the clinical setting, and may overstate the benefit of the study drug. In clinical practice there is no *a priori* method of distinguishing between placebo responders and non-responders. For trials to accurately inform clinical decision making, they must demonstrate superiority over placebo under the conditions in which they will in fact be prescribed.

Concomitant medications

Clinical drug trials attempt to limit concomitant medications as a matter of routine trial design. This is done to simplify interpretation of results by eliminating possible confounds and by limiting group heterogeneity. However, this practice is a significant deviation from the way in which medications are prescribed in actual practice, particularly for neuropsychiatric symptoms in which the use of more than one pharmaceutical agent is sometimes necessary. Application of study results to typical clinical practice may thereby be limited. Moreover, requiring patients to be on no concomitant medications may potentially bias the study sample away from patients who are unable or unwilling (for reasons of symptom severity) to be tapered off other medications. The net result may be that participants in clinical trials may represent only specific subgroups of dementia patients as a whole.[65]

An alternative to a washout of concomitant medications is to allow patients to remain on their current medications but to require stable doses for a particular period of time prior to study entry. This may allow a broader range of patients to participate in the trial as well as prevent baseline ratings from becoming artificially increased by the effects of drug withdrawal. Moreover, it may enhance applicability of results, strengthening the case for superiority over placebo in typical clinical settings. One drawback to this approach is a heightened potential for side-effects due to drug–drug interactions that might be mistakenly attributed to the new study drug alone. Additionally, there is the risk of mistakenly attributing therapeutic efficacy to the new study drug alone, when in fact it is due to the unique combination of the study drug and one of the concomitant medications. Given these confounds, we recommend that new drugs be studied preliminarily in the most 'pristine' situation and that potential confounds such as other drugs be eliminated as much as possible. We recommend that subsequent more definitive studies allow subjects to remain on concomitant medications but would restrict dose changes during the study period. Use of a placebo should be encouraged but not required if the drug under study has established efficacy in identical circumstances.

Outcome measures

With further refinement of neuropsychiatric syndrome definition in AD, greater specificity in outcome measures will be possible. Until that time, consideration should be given to reducing reliance on neuropsychiatric scale 'total' scores and increasing the reliance on specific subscales that most closely resemble the target syndrome of interest. Most scales commonly used in clinical research studies of dementia-related neuropsychiatric disturbances have subscales that relate to particular clusters of symptoms.[66] For example, the BEHAVE-AD has seven symptom-clusters – delusions, hallucinations, activity, aggression, diurnal rhythm, affective and anxiety symptoms – while the CMAI comprises physical, verbal and total aggression scores, and physical, verbal and total non-aggression subscores. Ideally, the specific behavioural syndrome of interest and appropriately matched subscale would be chosen prospec-

tively in order to avoid *post hoc* subgroup definitions after randomisation has occurred.[67]

Numerous studies in which cognition was the primary outcome variable report neuropsychiatric symptom response as a secondary outcome. It is important that conclusions about medication efficacy for neuropsychiatric symptoms are not primarily drawn from these findings, as group equivalence cannot be assumed if randomisation and group composition are not primarily driven by the presence of these symptoms. Despite this caveat, there are numerous outcome measures that could be used as secondary outcomes to help characterise the significance of the effects on the primary outcome measure. Examples include patient quality-of-life,[68] caregiver wellbeing and nursing home placement.

Outcome measures that are specifically tailored to the symptoms of the individual patient and whose content is driven to a great extent by the priorities assigned by family members and caregivers are beginning to be studied. One method using this approach, called goal attainment scaling (GAS) was found to be sensitive to the effects of treatment. Improvement with GAS did not fully correlate with change on standard outcome measures such as the Mini-Mental Status Examination or the Alzheimer's Disease Assessment Scale cognitive subscale (ADAS-Cog).[69] Adding this type of outcome measure to standardised ones would enhance the clinical relevance of studies. Demonstrating that a medication can improve patient performance in a way defined *a priori* as important to the patient and/or family would put into perspective the improvements on standardised measures of behaviour, mood or cognition.

Because dementia impairs memory, abstraction and other cognitive abilities, the choice of scales to measure outcome is more limited than in traditional trials. For example, the use of self-report scales such as the Hamilton Depression and Anxiety scales is possible only in studies of individuals with relatively intact cognitive capacity. Some scales have been adapted for use in persons with dementia; for example, the Cornell Scale for

Depression in Dementia (CSDD) is an adaptation of the Hamilton Depression Rating Scale. Other scales have been developed specifically for use in persons with dementia. For example, the NPI, was developed as a scale inclusive of multiple psychiatric domains, such as depression, psychosis and apathy, and also includes scales that assess motor behaviour such as heightened motor activity. The CMAI assesses aggression, breaking this set of behaviours into physical and verbal aggression. Given the limited capacity of many persons with dementia to reliably report symptoms that might date back hours, days or weeks, we recommend the use of informant-rated scales developed specifically for persons with dementia.

Non-pharmacological interventions

In clinical practice, pharmacological and non-pharmacological interventions are often used together rather than alone or sequentially.[70] In contrast, controlled clinical trials have generally studied these interventions independently. Although more complicated to perform, large trials that combine or compare both methods of intervention will most closely resemble actual clinical practice, and therefore have the potential to be the most informative.[71] These more complex trials should follow pilot studies that establish disease or syndrome-specific therapeutic efficacy of drug alone under controlled trial circumstances.

Conclusions

Non-cognitive neuropsychiatric symptoms are highly prevalent in AD and cause significant morbidity for both the patient and caregivers. Clinical trials of medications should ideally target well-defined syndromes that have been identified using epidemiological investigation and ultimately validated by other scientific methods such as neuroimaging, neuropathology or genetics. The outcome measures used in trials should reflect these syndromes. Definitive clinical trials should be designed to maximise applicability of results rather than focus primarily on the detection of a drug effect. Ideally, clinical trials should include non-pharmacological interventions to both maximise potential benefits to patients and identify and define the proper role of pharmacotherapy.

References

1. Lyketsos CG, Steinberg M, Tschanz JT et al. Mental and behavioral disturbances in dementia: findings from the Cache County Study on Memory in Aging. Am J Psychiatry 2000; 157:708–14.
2. Keene J, Hope T, Fairburn CG et al. Natural history of aggressive behavior in dementia. Int J Geriatr Psychiatry 1999; 14:541–8.
3. Levy ML, Cummings JL, Fairbanks LA et al. Longitudinal assessment of symptoms of depression, agitation, and psychosis in 181 patients with Alzheimer's disease. Am J Psychiatry 1996; 153:1438–43.
4. Hebert R, Dubois MF, Wolfson C, Chambers L, Cohen C. Factors associated with long-term institutionalization of older people with dementia: data from the Canadian Study of Health and Aging. J Gerontol A Biol Sci Med Sci 2001; 56:M693–9.
5. Rabins PV. Noncognitive symptoms in Alzheimer Disease. In: Terry RD, Katzman R, Bick KL (eds). Alzheimer Disease. New York: Raven Press; 1994: 419–29.
6. McHugh PR, Slavney PR. The Perspectives of Psychiatry (2nd edn). Baltimore: The Johns Hopkins University Press; 1998.
7. Rabins PV, Mace NL, Lucas MJ. The impact of dementia on the family. JAMA 1982; 248:333–5.
8. Laughren T. A regulatory perspective on psychiatric syndromes in Alzheimer disease. Am J Geriatr Psychiatry 2001; 9:340–5.
9. Olin JT, Schneider LS, Katz IR et al. Provisional criteria for depression of Alzheimer disease. Am J Geriatr Psychiatry 2002; 10:125–8.
10. Jeste DV, Finkel SI. Psychosis of Alzheimer's disease and related dementias. Diagnostic criteria for a distinct syndrome. Am J Geriatr Psychiatry 2000; 8:29–34.
11. Lyketsos CG, Breitner JCS, Rabins PV. An evidence-based proposal for the classification of neuropsychiatric disturbance in Alzheimer's disease. Int J Geriatr Psychiatry 2001; 16:1037–42.

12. Diefendorf AR. Clinical Psychiatry – A Textbook for Students and Physicians. Abstracted and adapted from: Kraeplin E Lehrbuch der Psychiatrie (7th edn). New York: The Macmillan Company; 1915.

13. Kay DW, Roth M, Hopkins B. Affective disorders arising in the senium. I. Their association with organic cerebral degeneration. J Ment Sci 1955; 101:302–16.

14. Alexopoulos GS, Meyers BS, Young RC, Mattis S, and Kakuma T. The course of geriatric depression with 'reversible dementia': a controlled study. Am J Psychiatry 1993; 150:1693–9.

15. Reynolds CF, Kupfer DJ, Hoch CC et al. 2-year follow-up of elderly patients with mixed depression and dementia. J Am Geriatr Soc 1986; 34:793–9.

16. Lee HB, Lyketsos CG. Depression in Alzheimer's disease: heterogeneity and related issues. Biol Psychiatry 2003; 54:353–62.

17. Reifler BV, Teri L, Raskind M et al. Double-blind trial of imipramine in Alzheimer's disease patients with and without depression. Am J Psychiatry 1989; 146:45–9.

18. Petracca GM, Chemerinski E, Starkstein SE. A double-blind, placebo-controlled study of fluoxetine in depressed patients with Alzheimer's disease. Int Psychogeriatr 2001; 13:233–40.

19. Petracca G, Teson A, Chemerinski E, Leiguarda R, Starkstein SE A double-blind placebo-controlled study of clomipramine in depressed patients with Alzheimer's disease. J Neuropsychiatry Clin Neurosci 1996; 8:270–5.

20. Lyketsos C, Delcampo L, Steinberg M et al. Treating depression in Alzheimer disease: efficacy and safety of sertraline therapy, and the benefits of depression reduction: The DIADS. Arch Gen Psychiatry 2003; 60:737–46.

21. Roth M, Mountjoy CQ, Amrein R. Moclobemide in elderly patients with cognitive decline and depression: an international double-blind, placebo-controlled trial. Br J Psychiatry 1996; 168:149–57.

22. Katona CL, Hunter BN, Bray J. A double-blind comparison of the efficacy and safety of paroxetine and imipramine in the treatment of depression with dementia. Int J Geriatr Psychiatry 1998; 13:100–8.

23. Taragano FE, Lyketsos CG, Mangone CA, Allegri RF, Comesana-Diaz E. A double-blind, randomized, fixed-dose trial of fluoxetine vs. amitriptyline in the treatment of major depression complicating Alzheimer's disease. Psychosomatics 1997; 38:246–52.

24. Hamilton M. A rating scale for depression. J Neurol Neurosurg Psychiatry 1960; 23:56–62.

25. Alexopoulos GS, Abrams RC, Young RC, Shamoian CA. Cornell Scale for Depression in Dementia. Biol Psychiatry 1988; 23:271–84.

26. Montgomery S, Åsberg M. A new depression scale designed to be sensitive to change. Br J Psychiatry 1979; 134: 382–9.

27. Guy W. ECDEU Assessment Manual for Psychopharmacology (Revised). NIMH, Rockville, MD: US Department of Health, Education and Welfare; 1967: 217–22.

28. Alzheimer A. A characteristic disease of the cerebral cortex. Translated in: Bick K, Amaducci L, Pepeu G (eds). The Early Story of Alzheimer's Disease. New York: Raven Press; 1987: 1–3.

29. Finkel SI, Costa e Silva J, Cohen G, Miller S, Sartorius N. Behavioral and psychological signs and symptoms of dementia: a consensus statement on current knowledge and implications for research and treatment. Int Psychogeriatr 1996; 8(Suppl 3):497–500.

30. Brodaty H, Ames D, Snowdon J et al. A randomized placebo-controlled trial of risperidone for the treatment of aggression, agitation, and psychosis of dementia. J Clin Psychiatry 2003; 64:134–43.

31. Katz IR, Jeste DV, Mintzer JE et al. Comparison of risperidone and placebo for psychosis and behavioral disturbances associated with dementia: a randomized, double blind trial. Risperidone Study Group. J Clin Psychiatry 1999; 60:107–15.

32. De Deyn PP, Rabheru K, Rasmussen A et al. A randomized trial of risperidone, placebo, and haloperidol for behavioral symptoms of dementia. Neurology 1999; 53:946–55.

33. Street JS, Clark WS, Gannon KS et al. Olanzapine treatment of psychotic and behavioral symptoms in patients with Alzheimer disease in nursing care facilities: a double-blind, randomized, placebo-controlled trial. The HGEU Study Group. Arch Gen Psychiatry 2000; 57:968–76.

34. Meehan KM, Wang H, David SR et al. Comparison of rapidly acting intramuscular olanzapine, lorazepam, and placebo: a double blind, random-

ized study in acutely agitated patients with dementia. Neuropsychopharmacology 2002; 26:494–504.

35. Snowden M, Sato K, Roy-Byrne P. Assessment and treatment of nursing home residents with depression or behavioral symptoms associated with dementia: a review of the literature. J Am Geriatr Soc 2003; 51:1305–17.

36. Cummings JL, Mega M, Gray K et al. The Neuropsychiatric Inventory: comprehensive assessment of psychopathology in dementia. Neurology 1994; 44:2308–14.

37. Reisberg B, Auer SR, Monteiro IM. Behavioral pathology in Alzheimer's disease (BEHAVE-AD) rating scale. Int Psychogeriatr 1996; 8(Suppl 3):301–8; discussion 51–4.

38. Cohen-Mansfield J, Billig N. Agitated behaviors in the elderly. I. A conceptual review. J Am Geriatr Soc 1986; 34:711–21.

39. Schneider LS, Pollock VE, Lyness SA. A metaanalysis of controlled trials of neuroleptic treatment in dementia. J Am Geriatr Soc 1990; 38:553–63.

40. Tariot PN, Erb R, Podgorski CA et al. Efficacy and tolerability of carbamazepine for agitation and aggression in dementia. Am J Psychiatry 1998; 155:54–61.

41. Olin JT, Fox LS, Pawluczyk S, Taggart NA, Schneider LS. A pilot randomized trial of carbamazepine for behavioral symptoms in treatment-resistant outpatients with Alzheimer disease. Am J Geriatr Psychiatry 2001; 9:400–405.

42. Porsteinsson AP, Tariot PN, Erb R et al. Placebo-controlled study of divalproex sodium for agitation in dementia. Am J Geriatr Psychiatry 2001; 9:58–66.

43. Sival RC, Haffmans PM, Jansen PA, Duursma SA, Eikelenboom P. Sodium valproate in the treatment of aggressive behavior in patients with dementia – a randomized placebo controlled clinical trial. Int J Geriatr Psychiatry 2002; 17:579–85.

44. Overall JE, Gorham DR. Introduction to the Brief Psychiatric Rating Scale (BPRS): recent developments in ascertainment and scaling. Psychopharmacol Bull 1988; 24:97–8.

45. Wistedt B, Rasmussen A, Pedersen L et al. The development of an observer-scale for measuring social dysfunction and aggression. Pharmacopsychiatry 1990; 23:249–52.

46. Rabins PV, Blacker D, Bland W et al. Practice guideline for the treatment of patients with Alzheimer's disease and other dementias of late life. American Psychiatric Association. Am J Psychiatry 1997; 154(5 Suppl):1–39.

47. Trinh N-H, Hoblyn J, Mohanty S, Yaffe K. Efficacy of cholinesterase inhibitors in the treatment of neuropsychiatric symptoms and functional impairment in Alzheimer disease: a meta-analysis. JAMA 2003; 289:210–16.

48. Nyth AL, Gottfries CG. The clinical efficacy of citalopram in treatment of emotional disturbances in dementia disorders: a Nordic multicentre study. Br J Psychiatry 1990; 157:894–901.

49. Pollock BG, Mulsant BH, Rosen J. Comparison of citalopram, perphenazine and placebo for the acute treatment of psychosis and behavioral disturbances in hospitalized, demented patients. Am J Psychiatry 2002; 159:460–5.

50. Teri L, Logsdon RG, Peskind et al. Treatment of agitation in AD: a randomized, placebo-controlled clinical trial: Alzheimer's Disease Cooperative Study. Neurology 2000; 55:1271–1278; correction: 2001; 56:426.

51. Stern RG, Duffelmeyer ME, Zemishlani Z, Davidson M. The use of benzodiazepines in the management of behavioral symptoms in demented patients. Psychiatr Clin North Am 1991; 14:375–84.

52. Cooper AJ. Medroxyprogesterone acetate (MPA) treatment of sexual acting out in men suffering from dementia. J Clin Psychiatry 1987; 48:368–70.

53. Rabins PV, Lyketsos CG, Steele CD. Practical Dementia Care. New York: Oxford University Press; 1999.

54. Kendler KS, Gardner CO Jr. Boundaries of major depression: an evaluation of DSM-IV criteria. Am J Psychiatry 1998; 155:172–7.

55. Marin RS, Firinciogullari S, Biedrzycki RC. Group differences in the relationship between apathy and depression. J Nerv Ment Dis 1994; 182:235–9.

56. Hwang JP, Tsai SJ, Yang CH, Liu KM, Lirng JF. Persecutory delusions in dementia. J Clin Psychiatry 1999; 60:550–3.

57. Lyketsos CG, Sheppard JME, Steinberg M et al. Neuropsychiatric disturbance in Alzheimer's disease clusters into three groups: the Cache County study. Int J Geriatr Psychiatry 2001; 16:1043–53.

58. Leroi I, Voulgari A, Breitner JCS, Lyketsos CG. The epidemiology of psychosis in dementia. Am J Geriatr Psychiatry 2003; 11:83–91.

59. Cummings JL, Miller B, Hill MA, Neshkes R. Neuropsychiatric aspects of multi-infarct dementia and dementia of the Alzheimer type. Arch Neurol 1987; 44:389–93.

60. Sweet RA, Pollock BG, Sukonick DL et al. The 5-HTTPR polymorphism confers liability to a combined phenotype of psychotic and aggressive behavior in Alzheimer disease. Int Psychogeriatr 2001; 13:401–409.

61. Klein DA, Steinberg M, Galik E et al. Wandering behavior in community-residing persons with dementia. Int J Geriatr Psychiatry 1999; 14:272–9.

62. Lyketsos CG, Steele C, Galik E et al. Physical aggression in dementia patients and its relationship to depression. Am J Psychiatry 1999; 156:66–71.

63. Flynn FG, Cummings JL, Gombein J. Delusions in dementia syndromes: investigation of behavioral and neuropsychological correlates. J Neuropsychiatry Clin Neurosci 1991; 3:364–70.

64. Knopman D, Kahn J, Miles S. Clinical research designs for emerging treatments for Alzheimer disease: moving beyond placebo-controlled trials. Arch Neurol 1998; 55:1425–9.

65. Schneider LS, Olin JT, Lyness SA, Chui HC. Eligibility of Alzheimer's disease clinic patients for clinical trials. J Am Geriatr Soc 1997; 45:923–8.

66. De Deyn PP, Wirshing WC. Scales to assess efficacy and safety of pharmacologic agents in the treatment of behavioral and psychological symptoms of dementia. J Clin Psychiatry 2001;62(Suppl 21):19–22.

67. Finkel SI, Mintzer JE, Dysken M et al. A randomized, placebo-controlled study of the efficacy and safety of sertraline in the treatment of the behavioral manifestations of Alzheimer's disease in outpatients treated with donepezil. Int J Geriatr Psychiatry 2004; 19:9–18.

68. Rabins PV, Kasper JD. Measuring quality of life in dementia: conceptual and practical issues. Alzheim Dis Assoc Disord 1997; 11(Suppl 6): 100–104.

69. Rockwood K, Graham JE, Fay S; ACADIE Investigators. Goal setting and attainment in Alzheimer's disease patients treated with donepezil. J Neurol Neurosurg Psychiatry 2002; 73:500–507.

70. Blass DM, Steinberg M, Leroi I, Lyketsos CG. Successful multimodality treatment of severe behavioral disturbance in a patient with advanced Huntington's disease. Am J Psychiatry 2001; 158:1966–72.

71. Opie J, Doyle C, O'Connor DW. Challenging behaviors in nursing home residents with dementia: a randomized controlled trial of multidisciplinary interventions. Int J Geriatr Psychiatry 2002; 17:6–13.

19

Clinical Trials for Psychosocial Interventions Aimed at Caregivers of People with Dementia

Georgina Charlesworth and Stanton Newman

Introduction

The costs of family caregiving include high levels of stress and depression,[1] compromised physical health and premature mortality,[2] increased financial burden and social isolation.[3] Caregivers (CGs) for people with dementia (PwD) have higher stress than other caregivers,[4] and psychosocial interventions for this population have received considerable attention. For example, a recent systematic review identified 43 distinct quantitative intervention studies published between 1996 and 2001, and cited nine previous reviews published between 1989 and 2001.[5] Reviews have generally included a 'mixed bag' of interventions, and have highlighted the methodological limitations of many trials. The quality of research has increased over the years,[6] yet studies highlight the need to increase the efficacy of interventions, as caregivers of PwD benefit less from interventions than other caregiver populations on measures of burden, depression, subjective well-being and knowledge.[7]

In this chapter we: describe psychosocial interventions that have been evaluated for use with family caregivers of PwD, and consider their impact, summarise the lessons learned on evaluation methodologies, and outline the developing areas of caregiver research including use of 'theory-based' interventions and improved treatment integrity.

Psychosocial interventions for caregivers

Types of intervention

CG interventions are diverse, ranging from one-off education sessions and telephone helplines to comprehensive, intensive and long-term care packages delivered by multi-disciplinary health and social care teams. The heterogeneity of intervention aims, content, duration and intensity has presented a challenge to reviewers in how to categorise studies and synthesise findings. This has resulted in diverse organisational frameworks in CG literature. For example, in reviews of interventions predominantly but not exclusively for caregivers of people with dementia, Knight et al categorised interventions into 'respite', 'group' and 'individual psychological',[8] and Acton and Winter used the categories of 'education', 'support group and education', 'counselling', 'respite', 'case management' and 'multi-component'.[9] In reviews of intervention studies focused specifically on carers of people with dementia, Pusey and Richards created the four categories of 'technology-based', 'group-based', 'individually-based' and 'service configuration' studies;[10] Cooke et al used a coding system to identify 'social' or 'psychological' elements of interventions;[11] and Schulz et al considered studies in four domains' outcome measures, namely 'symptomatology', 'quality of life', 'social significance' and 'social validity'.[5]

In their review of nursing research, Acton and Winter noted a lack of individualisation in matching caregiver need to intervention strategy.[9] One framework for considering caregiver need is the developmental model proposed by Aneshensel and colleagues which acknowledges the changing nature of caregiver needs over the course of the 'caregiving career' from the point of initial engagement in the caregiving role, through the long enactment phase before finally disengaging from caregiving and re-engaging with life beyond.[12] Aneshensel et al's work has many similarities with Rolland's writing on family responses to gradually progressive and incapacitating illness.[13,14] He highlighted the differing needs of families during 'crisis', 'chronic' and' terminal' time phases of a chronic illness. Both Aneshensel and Rolland recommend different interventions at different stages of the illness and caregiver development. For example, in Rolland's crisis phase, which includes the symptomatic period before actual diagnosis and the initial period of readjustment and coping after diagnosis, facilitation of adjustment and provision of information about the illness and available services is generally called for. The chronic phase, also referred to as the 'long haul' or 'day-to-day living with chronic illness', draws heavily on, and often exceeds, the family's resources, necessitating engagement with services for provision of support and respite. Caregivers may need to develop new skills to cope with the demands on their time and emotions, and to understand changing behaviours and characteristics of the person with dementia. Carers in this phase are at risk of losing both their autonomy and their social supports. The last phase or 'terminal' period includes any pre-terminal stage in which the inevitability of death becomes apparent and dominates family life, and also encompasses the periods of mourning, resolution of loss and resumption of 'normal' family life beyond the loss. Interventions at this stage may most appropriately be emotional support to facilitate mourning, and tangible support for, and respite from, palliative care.

For the purposes of this chapter, we shall consider interventions within the categories 'information/education', 'skills', 'social support' and 'respite' interventions, reflecting the typical interventions for caregivers of a person with a chronic, deteriorating illness. However, in acknowledgement of the chronicity and complexity of the caregiving task, research into multi-component interventions has become more common, and is therefore also considered.[15]

Information/education

Acton and Winter define education as 'those interventions designed to provide specific information about the disease process, disruptive behaviour and caregiving skills'.[9] In line with this definition, educational programmes typically include information on: medical aspects of dementia; treatments and services, benefits and legal advice. Acton and Winter identified 23 studies in which education interventions were tested, but only nine of these measured knowledge.[9] Where knowledge measures have been used, educational programmes show significant benefit. For example, in Cooke et al, of 16 studies that included knowledge of illness as an outcome measure, 11 showed improvements in knowledge.[11] However, of these 11, only three also reported improvements in carer wellbeing or burden.

There are, however, individual differences to carers' desire for knowledge, with some carers wanting 'anything and everything on dementia' as soon as possible,[16] and others preferring to learn information on a 'need to know' basis. The latter may particularly apply to anxious carers for whom greater knowledge has been associated with heightened anxiety.[17] Furthermore, it has been suggested that carers with high levels of distress may be unable to acquire new knowledge until their psychological distress has reduced.

Skills

Skills learning programmes for carers have often been based on cognitive behavioural models, and include stress, time and behaviour management, assertiveness and communication skills. Skills interventions are focused on the activities of caregiving and caregivers' appraisal of care-related tasks or behaviours of the person with dementia. In a 14-session individual programme, based on a successful intervention for family carers of people with schizophrenia, Marriott and colleagues included three sessions on education about dementia, followed by six sessions on stress management and five sessions on coping skills.[18] Other skills training programmes which have been developed and refined over many years include: the Palo Alto 'Coping with caring' group programmes including interventions for feelings of stress, loss, depression and anger,[19] Bob Knight's four-element programme of stress level monitoring, relaxation training, scheduling relaxing events and cognitive restructuring,[20] and Levine and Gendron's programme of assertion, problem solving and cognitive restructuring.[21]

In Cooke et al's review,[11] three coding elements were included for cognitive components, namely: 'cognitive problem solving', defined as any intervention that identified strategies to overcome or cope with problems of caregiving, e.g. dealing with behavioural problems of the care recipient such as wandering; 'cognitive therapy', defined as an intervention to alter an individual's cognition – ranging from suggestions that other beliefs may be more appropriate to cognitive restructuring; and, 'cognitive skills', defined as the teaching of specific skills such as distraction, imagery or attention refocusing. In controlled studies, 7 of 13 (54%) interventions that had one or more cognitive components showed improvements in psychological wellbeing, and 4 of 8 (50%) showed a reduction in caregiver burden. However, all except one intervention also included social components, such as social support, social skills training or social activities, making a clear attribution to the efficacy of cognitive components difficult.

Social support

Social support has been variously defined and covers a range of concepts including structural and functional aspects.[22] Social support interventions focus on the role of caregiving rather than the activity. Peer support has been encouraged among carers of people with dementia in line with evidence on people's tendency to realign associations in social networks, to reduce contact with associates with whom they have become less similar, and to intensify existing relationships (or develop new ones) with others to whom they have become more similar.[23] Typically the evaluation of social support interventions has focused on satisfaction and the size of networks.

Respite

Respite is a widely provided intervention in dementia care in which the care recipient is temporarily supported in a hospital/nursing home environment or at home. Briggs and Askham summarise the aims of respite as being: to reduce the objective burden of care for the primary caregiver, thus exerting a positive effect on caregiver psychological and physical wellbeing, and to help delay long-term institutionalisation of the care-recipient.[24] Some reviewers have excluded respite intervention studies on the grounds that they are 'patient focused'.[6] Others consider respite a legitimate 'carer intervention'. Where respite has been the subject of review, evidence for benefits for the carer are equivocal. McNally and coworkers found that the impact on physical health had rarely been studied, and that psychological wellbeing and burden returned to baseline levels after around 1 week.[25] Compared to other interventions (home social services, day centre, expert centres and group living) respite in hospital produced the least reduction in burden for caregivers.[26]

There is some evidence that those carers who re-establish social relationships have better outcome,[27] yet carers often use respite time to 'catch up' on household chores and maintenance that cannot be completed while the care recipient is in the home.[24] This leaves little time for the carers to rest, make social contacts, or re-engage with pleasurable activities.

Multi-component programmes

Multi-component programmes include those described by Brodaty and Gresham, Mittelman and colleagues, and Moniz-Cook et al.[28–31]

Brodaty and Gresham's intervention comprised a 10-day intensive residential programme for people with dementia and their carers in which groups of up to four family carers took part in didactic education, discussion and social activities, while care recipients took part in sessions in memory retraining, reminiscence therapy, environmental reality orientation and general ward activities. Session topics for carers, delivered by professionals from within the multi-disciplinary team, were: reducing caregiver distress, combating isolation, guilt and separation, new ways of thinking (assertiveness, re-roleing, relaxation and stress management), new coping skills (communication, reality orientation, therapeutic use of activities, reminiscence, coping with physical frailty, fitness, diet, organising the day and home), medical aspects of dementia, using community services, planning for the future and coping with problem behaviours.[32] After the residential course, fortnightly telephone conference calls were arranged to link carers from a training cohort and the programme coordinator. The frequency of calls gradually decreased over 12 months to 4- and then 6-weekly, with two of the final telephone conferences being arranged by the coordinator but held in her absence to encourage cohorts to become self-supporting.

Mittelman et al's three-component programme consisted of: two individual and four family counselling sessions in the first 4 months after study enrolment; weekly support groups from 4 months post-enrolment; and continuous availability of counsellors to caregivers and families to help them deal with crises and with the changing nature and severity of the care recipients' symptoms. The initial counselling sessions were task oriented, promoting communication among family members,

teaching techniques for problem solving and management of challenging behaviours and improving emotional and instrumental support for the primary caregiver.[30]

The Moniz-Cook et al intervention was provided by a clinical psychologist in the context of a primary care memory clinic to both the carer and the person with dementia.[31] Interventions included: post-diagnostic counselling including structured written and verbal information on diagnosis, crisis prevention, individualised memory rehabilitation for preservation of abilities and skills, strategies for maintaining social activity, and sharing responsibility with family and social networks.

All three multi-component interventions showed a significant benefit compared to control conditions on both carer wellbeing and time to institutionalisation for the PwD. All three provided fairly intensive and individualised interventions, encouraged communication between family members and social support networks and family sessions, facilitated peer support and provided extensive follow-up by the intervention coordinator.

Impact of interventions

The impact of CG interventions has generally been considered in terms of statistically significant between-group differences, or within-group change, on psychometric measures with carers (burden, mood state, knowledge), or on the impact of the status of the person with dementia (institutionalisation, death). More recently, the level of behavioural disturbance in the person with dementia has also been measured.

Given the variety of outcome measures used, reviewers have needed to find ways of comparing studies that use different measures. In quantitative syntheses that use meta-analysis, the comparator has been effect size. Knight and colleagues,[8] using Hunter and Schmidt's procedures for estimating effect size,[33] calculated a mean effect size of 0.15 for respite; 0.15 for burden and 0.31 for emotional dysphoria for group interventions; and 0.41 for burden and 0.58 for emotional dysphoria for individual interventions (0 = no effect, 0.3 = small, 0.5 = moderate, 1 = strong). A decade later,[6] Brodaty and colleagues concluded that interventions producing the greatest effect were multi-component and comprehensive interventions, especially those involving the PwD. Unsuccessful interventions included short educational programmes (beyond enhancement of knowledge), support groups alone, single interviews and brief interventions or courses that were not supplemented with long-term contact.

A common assumption among early CG researchers was that carer-focused interventions would have little or no impact on the person with dementia, but may influence the length of time the caregiver feels able to continue caring at home. Retention of the care recipient in the community rather than admission to residential/nursing care has frequently been cited as an outcome of interest, even though it is not necessarily universally desirable. Although it is generally argued that a person with dementia is likely to feel more safe and secure in the familiar surroundings of their own home, the care recipient may at times be better served by residential or nursing care than by an emotionally and physically exhausted caregiver.

Where outcome measures on the PwD are available, it has been demonstrated that CG interventions can also show a measurable impact on the PwD. For example, the intervention by Marriott and colleagues demonstrated not only a reduction in caregiver stress, but also a smaller increase in behavioural disturbance in the person with dementia.[18] This finding would be predicted by systemic models, where one individual's appraisal of their own coping-efficacy is thought to influence their behaviour towards others, and in turn influence the behaviours of others. It would seem appropriate now for studies of caregiver interventions to include methods for monitoring their impact on the PwD.

Quality of existing research

Problems with methodological quality have been extensively covered in CG intervention reviews,

with recurring themes across the decades.[8,34] We summarise here some commonly raised issues, namely sample size, attrition, recruitment and selection, measurement sensitivity, 'dosage', follow-up and analysis. We also consider issues of study identification through search strategies, and the rating of methodological quality – two issues that have come to the fore with the development of the 'evidence based healthcare' movement. As will be seen, even information derived from the most systematic approach remains open to interpretation, and small differences in review methodology can result in large discrepancies in results.

Study identification/search strategy

Table 19.1 shows characteristics of four recent systematic reviews of intervention studies in which the target population was exclusively carers of community-dwelling people with dementia.[6,7,10,11] The table includes a description of search strategies, the inclusion and exclusion criteria for intervention type and trial design, and outcomes of interest. All four reviews included only quantitative studies published in the English language. The number of studies selected for inclusion ranged from 30 to 45, with the proportion of randomised controlled trials ranging from 35% to 70%. Schulz et al reviewed a significantly different set of publications due to their focus on more recent dates of publication (1996–2001), and by including respite studies and evaluations of cognitive enhancing medication.[7] Figure 19.1 shows the number of studies identified and included in the Brodaty et al, Cooke et al and Pusey and Richards reviews, including predominantly studies published between 1985 and 1999.[6,10,11] Between March 1999 and December 2000 additional studies became available to reviewers with Brodaty et al identifying ten studies that had not been published in time for the Pusey and Richards search. A second reason for discrepancy between reviews is the inclusion criteria for study design. Cooke et al included uncontrolled trials (n = 11),[11] whereas Brodaty et al and Pusey and Richards limited the studies included to controlled trials.[6,10]

Table 19.1 Search strategies and inclusion/exclusion criteria for trial type and study design used in systematic reviews of psychosocial interventions for family carers of community dwelling people with dementia, and number of studies identified

Search strategies	Type of intervention	Trial designs and outcome measures	n studies
Brodaty et al 2003[6] Medline (1985–December 2000), PsycInfo (1984–December 2000), Ageline (1985–December 2000), CINAHL (1985–2000) Cochrane Library 1998 Issue 3, EBM Reviews Best Evidence (1991–December 2000); EBM Reviews – Cochrane (4th Quarter 2000); EMBASE (1988–2000 Wk51) *Search terms:* random allocation; control group, dementia, Alzheimer's disease combined with keywords caregiver, carer and intervention types	*Inclusion:* self-help groups, support groups, education, training, skills training, counselling, psychotherapy	*Designs:* controlled studies – randomised or quasi-experimental trials Excluded from meta-analysis if: sample size <5 in treatment or control; insufficient outcome data to calculate effect size or nursing home delay; extreme values *Outcome measures:* psychological morbidity, burden, knowledge, other	45 met search criteria; 30 met criteria for inclusion in meta-analysis (21 RCTs)
Pusey and Richards 2001[10] Medline (1969–March 1999) Embase (1980–March 1999); CINAHL (1983–March 1999) PsycLit (1967–March 1999); Cochrane; CRIB; HMIC; ISI Science and Social Science Citation Indices; Age Info; NRR; Health CD; SIGLE	*Inclusion:* interpersonal interventions concerned with provision of information, education or emotional support together with individual psychological intervention addressing a specific health and social care outcome	*Designs:* Randomised controlled or controlled trials *Outcome measures:* psychological health, physical health, quality of life (including perception of burden)	30 (18 RCTs)

Study and search details	Inclusion/Exclusion	Designs/Outcome measures	Number
Cooke et al., 2001[11] PsycLit (1970–June 2000), Medline (1966–October 2000) ISI Science and Social Science Indices, EMBASE (1980–September 2000) Cochrane 2000 Issue 3) *Search items:* dement* or Alzheimer*; carer*; caregiver* or supporter* and trial*; intervention* or program*	*Inclusion:* focus on improving caregivers' psychological wellbeing and/or social wellbeing directly. Techniques designed to utilise cognitive, behavioural or social mechanisms of action *Exclusion:* respite; interventions directed at the care-recipient or at caregiver practical skills	*Designs:* controlled and uncontrolled studies *Outcome measures:* psychological; burden; social; knowledge of illness	40 (14 RCTs)
Schulz et al 2002[5] 1996–2001: Medline, PsycINFO, CINAHL MeSH: caregivers and either dementia or Alzheimer's disease *Exclusion:* dissertation abstracts; citations mapping to MESH acquired immune deficiency syndrome, infant care, neoplasms, child care	*Inclusion:* interventions with caregivers and/or care recipients (including cognition-enhancing medications and respite)	*Designs:* use of comparative statistics evaluating between- and/or within-group differences *Outcome measures:* 'clinically relevant outcomes' categorised as symptoms, quality of life, social significance, social validity	43 (27 RCTs)

CINAHL, Cumulative Index to Nursing and Allied Health; CRIB, Current Research in Britain; HMIC, Health Management Information Consortium; NRR, National Research Register; SIGLE, System for Information on Grey Literature

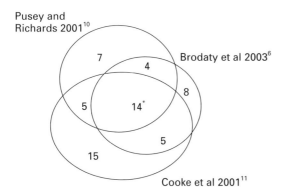

Pusey and Richards 2001[10]

Brodaty et al 2003[6]

Cooke et al 2001[11]

Figure 19.1 Venn diagram illustrating the numbers of studies unique to each review or in common with other reviews.[6,10,11]
* Hinchcliffe et al was classified as being included in Pusey and Richard's review.

Other discrepancies are likely to be due to the extent to which grey literature was surveyed, and to differences in search terms.

Methodology of quality ratings

In terms of quality ratings, a comparison of Brodaty et al and Pusey and Richards demonstrates that differences in rating methodologies result in considerable discrepancy in quality ratings, and as can be seen from Figure 19.2.[6,10] The correlation coefficient for the ratings was 0.49. Brodaty et al rated studies on an 11-point scoring system based on the Cochrane Collaboration Guidelines for examining quality of methodological design and analysis.[6,35] A single point was awarded for each of the following features: controlled trials (or those using a comparison group), random allocation, standardised diagnostic criteria, objective outcomes, well-validated and reliable measures, accounting for all subjects, consideration of statistical significance, adjustment for multiple comparisons, evidence of sufficient power, blind ratings and follow-up assessment at 6 months or beyond. In contrast, Pusey and Richards used an eight-point grading system in which random allocation earned four points, sufficient power gave three points (if no power calculation, a sam-

ple of $>50 = 2$ and $<50 = 1$), and blinding of outcome measures $= 1$.[10] The majority of discrepancies, including the most marked difference seen on ratings for Moniz-Cook et al,[31] were explained by the differing value placed on randomised controlled trial methodology.

Sample sizes

The numbers recruited into studies have generally been small and therefore often without the power to detect a significant group difference. Cooke et al calculated that for alpha of 0.05 and 80% power, only one of 21 controlled studies using between-group analysis would have been able to detect a small effect size (0.2), and five each could have detected a medium (0.5) and large (0.8) effect size.[11] Similarly, Schulz et al calculated that only one of the 43 distinct studies in their review would have been able to detect a small effect size, 14 medium and 13 large.[5] That is 10 of 21 (48%),[11] and 15 of 43 (35%) studies were too small to detect even a large effect.[5] As highlighted by Brodaty et al,[6] some caregiver interventions can have effect sizes greater than 1, hence some smaller trials have demonstrated clinically significant results. However, by underpowering studies and failing to capture small effect sizes, the overall picture of the literature may appear less positive than it should have if studies had adequate sample sizes.

Attrition

The 'flow' of participants through the recruitment, intake and follow-up assessments has generally been poorly described, as evidenced by the discrepancies in reporting of sample size and drop-out by different review groups. For example, 10 of the 29 studies reviewed in at least two of the systematic reviews had summarised different sample sizes,[6,10,11] and six had discrepancies for attrition, presumably due to differences in definition of drop-out. Differences may arise through participants being lost to the study between recruitment and randomisation, or pre-, during or post-intervention. Guidelines for reporting of participant flow through trials, as detailed by the

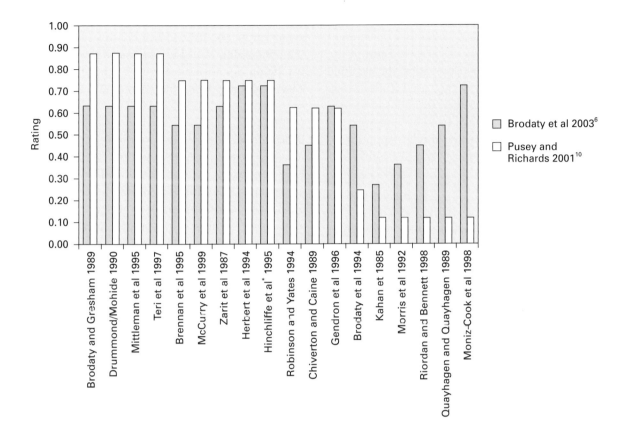

Figure 19.2 Comparison of quality ratings for studies reviewed by both Brodaty et al 2003 and Pusey and Richards 2001.
Quality scores were generated using Pusey and Richard's scoring methodology. Each bar represents a quality score as a proportion of the maximum possible within each review. See source reviews for full references to studies.
*Hinchcliffe et al was reviewed by Pusey and Richards, but was missing from the published table of studies.

Consolidated Standards of Reporting Trials (CONSORT) group, should improve clarity in future trial reports.[36]

Compliance with CONSORT reporting guidelines should also clarify the proportion of carers who were in receipt of intervention. It seems that it is not uncommon for a large proportion of carers who have consented to engage in the research, and are available for follow-up over time, are not necessarily engaging in the intervention offered. For example, in a quasi-experimental trial of a caregiver education programme, 22 of 55 carers allocated to intervention dropped out after atten-

dance at one or no sessions.[37] In a review of respite interventions for caregivers of people with dementia, it was noted that between 22% and 91% of carers do not make use of the services offered to them within trials.[38] The assessment of patient preference even in a randomised controlled trial (RCT) may be useful, and has been recommended as it allows for the assessment of the interaction between preference, allocation and outcome.[39]

Recruitment and selection

Recruitment into trials has generally been through specialist diagnostic and treatment centres, or through contact with specialist voluntary organisations such as Alzheimer's societies with selection of carers made on the basis of characteristics of the care recipient rather than the carer, for example specifying the diagnosis of the care recipient (e.g. 'Alzheimer's disease'), or their living location (e.g. 'community' or 'non-institutional' dwelling; co-habiting with carer). Participants in studies have predominantly been female, spouses and older than 60. Although this is likely to be representative of the demographic of the carer population for people with dementia, it is not known to what extent results are generalisable to carers with other kin relationships (e.g. spouse, adult offspring).

Although quality of relationship and level of behavioural disturbance have been found to be predictors of caregiver distress, they have not been used as selection criteria for studies. Neither have carer psychological and social characteristics such as coping style and social support. Indeed, the absence of selection criteria has led to concerns over floor and ceiling effects in samples, for example using interventions for depression in non-depressed populations leaves no room for improvement on depression measures.

Measurement sensitivity

Typically outcome measures employed have fallen into the following categories: knowledge, psychological wellbeing, caregiver burden and social support. Other measures include quality of life/life satisfaction.

Earlier research tended to focus on the reduction of caregiver burden, but it has generally been acknowledged that burden measures are relatively insensitive to change, most particularly those questionnaires that focus on the objective challenges presented by the illness. In the review by Brodaty and colleagues only 1 of 20 (5%) interventions showed a statistically significant effect on burden, and the weighted average effect size for burden was only 0.09 (95% confidence interval (CI): −0.09 to 0.26).[6] In contrast, 23 of the total 34 interventions (68%) met criteria for study success

(significant change in one of the main outcome measures or an effect size of 0.5 or greater), and the mean effect sizes for caregiver knowledge and psychological morbidity were 0.51 (95% CI 0.05 to 0.98) and 0.31 (95% CI: 0.13 to 0.50) respectively.

Dosage

Brodaty et al classified interventions of one or two sessions as minimal, 3–5 sessions as moderate, 6–10 sessions as medium–high, and greater than 10 sessions as intensive.[6] Although Brodaty and colleagues found evidence for a linear relationship between intensity (dosage) and outcome, there was a suggestion that this relationship broke down at the most intensive interventions. This suggests there may be a threshold of intensity for CG interventions.

Follow-up

The need for follow-up over time is acknowledged in many caregiver studies, with 40% of studies reviewed by Cooke et al incorporating some form of follow-up assessment in addition to the post-intervention assessment.[11] Of these, three had assessments between 3 and 5 months post-intervention, six had follow-ups between 6 and 11 months, and three had follow-ups at 12 months or more. Three further studies had assessments at different time points on ongoing interventions.

Study design and analysis

Various study designs have been reviewed, including pre–post evaluation, comparison of group differences post-intervention, mixed between, within designs with repeated measures. Even where RCT methods are used, they are often implemented incompletely. Intention-to-treat analyses are largely not performed,[6] and Schulz et al found that only one of their 43 studies reported its use.[6,30]

The issue of study design is complex. Pre–post designs are fraught with difficulties such as the impact of attention (Hawthorne effect) as well as other issues of repeat measurement. RCTs, on the other hand, raise an important additional problem of allocation to groups, which may be undesired by some participants. In addition there is the preference controlled design where patients have the option of selecting the arm of the study. This design raises a number of issues of analysis and allocation to groups.[39]

Recent and future directions

Recent studies, and suggestions for improving the quality of future research, encourage measures to improve treatment integrity, awareness of caregiver heterogeneity, and development of the theoretical underpinnings of caregiver interventions.

Improving treatment integrity

A limitation in many caregiver intervention trials has been poor documentation of intervention content. In order to investigate treatment process variables, and to ensure that psychosocial interventions can be replicated, it is necessary to have precise descriptions of treatment components, and to ensure that the treatment delivered was indeed the treatment intended. In psychotherapy literature, this has been referred to as 'treatment integrity', a concept that has been developed and expanded by Lichstein et al,[40] and applied to caregiver intervention research in the 5-year Resources for Enhancing Alzheimer's Caregivers Health (REACH) cooperative programme.[41] Lichstein et al's 'treatment implementation' (TI) model outlines three treatment processes, namely delivery, receipt and enactment.[40] 'Treatment delivery' focuses on the interventionist's ability to present the intervention as it was intended, including ensuring the absence of aspects of other treatments. 'Treatment receipt' focuses on the degree to which the participant has received the intended treatment, and 'treatment enactment' focuses on the extent to which the participant has made the expected changes in behaviours, for example using the skills or knowledge taught in the intervention.

Two types of treatment implementation strategies are recommended: induction methods that enhance the probability that proper TI occurs, such as the written manuals produced by Olshevski et al,[20] and for all the interventions studied by the REACH collaborative,[6] and assessment methods that measure occurrence of the intended TI strategies.

Descriptions of interventions are required not only for the experimental condition, but also the control, the component features of usual care have generally been ill-defined in caregiver research. Control conditions have rarely been 'no intervention' and more commonly have been 'usual care'. Lack of description of 'usual care' makes comparisons across studies problematic, as an intervention that otherwise would have shown differences to the control condition may fail to reach significance because of the intensity of usual care. It also prevents consumers of research from understanding the study in the context in which it was carried out, and prevents comparison with 'usual care' in their local area. Given the variation in health and social care systems, descriptions of usual care are welcome, and have been provided by some authors. For example, Marriott et al were able to cite their 'usual care' as being that described by Lennon and Jolley.[18,42]

Empirically grounded interventions

Many caregiver intervention studies have been atheoretical, arising from pragmatic or opportunistic evaluations rather than either social or psychological theory.[23] Only limited reference has been made to basic research. Greater attention to theory would facilitate intervention selection, clarify intervention aims and guide the choice of outcome measures.[43] The frequent mismatch between stated aims and the selected measurement tools,[8] has been a common criticism of caregiver research. Having a clear theoretical basis to interventions would also assist in exploration of the mechanisms of change.

The advantages of a translational research process, in which theory/basic research informs intervention design and in which intervention results influence future intervention design and theory, has also been propounded by Salkovskis for psychotherapy research in general.[44] In this process, outcome trials also provide useful information to feedback to basic science, irrespective of the success of the intervention.

Aims and outcomes

In order to appropriately assess the impact of an intervention, we must be clear about the aim of that intervention. It is all too common for researchers to select a number of outcomes without any consideration of the specific objectives of the intervention. The selection of multiple outcomes where some would not be expected to change, raises not only the issue of multiple comparisons but also presents an unfocused approach to this area of study. Selection issues are also an important consideration. For example it is unlikely that an intervention will be able to improve an already positive state, such as increase satisfaction, or lessen a negative state, such as reduce depression. Questions in the selection of outcomes need to be asked such as: 'Will the intervention have physical, attitudinal, behavioural or emotional consequences?'; 'Do we expect a CG intervention to impact on the CG only, or on the PwD or both?'. Overall there is a need to bring the outcomes being assessed into line with the aims of a study.

Identifying active ingredients

As in other fields of psychological research use of multi-component interventions has made it difficult to determine which mechanism or mechanisms have brought about the effects.[5,23] Where multi-modal packages have been used, there have been calls to identify the 'active ingredients' within a package and the 'therapeutic dose' of those ingredients. However, in the same way as it has been difficult to pinpoint the impact of specific techniques within psychotherapy research, where 'generic therapeutic' and 'relationship' factors have been the most successful predictors of good outcome, it may be that the maintenance of an ongoing relationship with a consistent group of workers over time has greater benefit for carers

and care recipients than does any one particular intervention. The importance of maintaining contact over time was emphasised over a decade ago in John Hall's work on the 'psychology of caring'.[45]

Clinical meaningfulness

Using effect sizes alone as a measure of the value of interventions has been criticised, given that it is possible to demonstrate statistical significance of a clinically unimportant or insignificant effect by having sufficiently large or homogeneous samples.[46] The concept of clinical meaningfulness or clinical importance goes beyond statistical significance by considering the 'real life' practical value of the effects of an intervention,[47] and has become a topic of interest for caregiver intervention researchers.[5]

The definition of what constitutes clinically important change is likely to vary from carer to carer, and also within each carer over the course of the caregiving career. Carers' responses and state of wellbeing are influenced by factors such as gender, ethnicity and resources. Evaluation of clinically important change may need to take those into consideration. For example, each basic educational intervention may need to be conducted and evaluated differently to appropriately address the ethnic differences in understanding of Alzheimer's disease.[48] Rolland[14] recommends consideration of the work of Combrinck-Graham in which the three-generational family system is seen as oscillating through periods of family closeness (centripetal) and family disengagement (centrifugal).[49] In general, chronic disease exerts a centripetal pull on the family system, which may be in keeping or at odds with the natural momentum of the family life cycle.

Studies of caregiving are incomplete without consideration of 'life beyond caregiving', and with this in mind Schulz and colleagues have drawn together the caregiving and bereavement research to compare and contrast the various theoretical perspectives used to explain caregiving and bereavement outcomes.[50]

Conclusions

The lessons of caregiver intervention research are that many models of intervention exist, but they have proved difficult to classify or categorise. Research quality has improved over the years, as has treatment efficacy. Movement has been towards the development of multi-component packages tested in multi-site trials such as REACH II.[34] Disaggregation studies may assist in identifying which interventions are most successful for the which types of carer and care recipient in the short and longer term and at different times in the caregiving career. To facilitate development, a 'taxonomy of interventions' would be an advantage, as would consensus on definitions of clinically meaningful change,[5] and analyses of the influence of moderating and mediating variables on outcome.

As caregiving research moves towards the comparison of differing interventions rather than the current predominant design of intervention versus 'usual care' control, the size of trials will need to increase in order to gain enough power to detect a difference between active treatments. Increasingly it will be necessary for researchers to become mindful of where the carer is on their career of caring, and design interventions appropriately.

Acknowledgements

Ideas in this chapter have been developed during work supported by the following grants: MRC UK award (52024); NHS UK Responsive Funding; HTA 99/34/07; Alzheimer's Society UK Fellowship. Support of these organisations is gratefully acknowledged. Hilary Clarke is thanked for her assistance in preparation of the manuscript.

References

1. Schulz R, O'Brien AT, Bookwala J, Fleissner K. Psychiatric and physical morbidity effects of dementia caregiving: prevalence, correlates, and causes. Gerontologist 1995; 35(6):771–91.

2. Schulz R, Beach SR. Caregiving as a risk factor for mortality. The Caregiver Health Effects Study. JAMA, 1999; 282:2215–9.

3. Dhooper SS. Identifying and mobilizing social supports for the cardiac patient's family. J Cardiovasc Nurs 1990; 5:65–73.

4. Livingston G, Manela M, Katona C. Depression and other psychiatric morbidity in carers of elderly people living at home. BMJ 1996; 312:153–6.

5. Schulz R, O'Brien A, Czaja S et al. Dementia caregiver intervention research: in search of clinical significance. Gerontologist 2002; 42:589–602.

6. Brodaty H, Green A, Koschera A. Meta-analysis of psychosocial interventions for caregivers of people with dementia. J Am Geriatr Soc 2003; 51:657–64.

7. Sorensen S, Pinquart M, Duberstain P. How effective are interventions with caregivers? An updated meta-analysis. Gerontologist 2002; 42:356–72.

8. Knight BG, Lutzky SM, Macofsky-Urban F. A meta-analytic review of interventions for caregiver distress: recommendations for future research. Gerontologist 1993; 33:240–8.

9. Acton GJ, Winter MA. Interventions for family members caring for an elder with dementia. Ann Rev Nurs Res 2002; 20:149–79.

10. Pusey H, Richards D. A systematic review of the effectiveness of psychosocial interventions for carers of people with dementia. Aging Ment Health 2001; 5: 107–19.

11. Cooke DD, McNally L, Mulligan KT, Harrison MJG, Newman SP. Psychosocial interventions for caregivers of people with dementia: a systematic review. Aging Ment Health 2001; 5:120–35.

12. Aneshensel CS, Pearlin LI, Mullan JT, Zarit SH, Whitlatch CJ. Profiles in Caregiving: the Unexpected Career. London: Academic Press; 1995.

13. Rolland JS. Toward a psychosocial typology of chronic and life-threatening illness. Fam Syst Med 1984; 2:245–62.

14. Rolland JS. Chronic illness and the family life cycle. In: Carter B, McGoldrick M (eds). The Changing Family Life Cycle: a Framework for Family Therapy (2nd edn). Boston MA: Allyn Bacon; 1989: 433–56.

15. Bourgeois MS, Schulz R, Burgio L. Intervention for caregivers of patients with Alzheimer's disease: A review and analysis of content, process, and outcomes. Int J Aging Hum Dev 1996; 43:35–92.

16. Keady J, Nolan M et al. (1995) Listen to the voice of experience. J Dement Care 1995; 3:15–17.

17. Graham C, Ballard C, Sham P. Carers' knowledge of dementia and their expressed concerns. Int J Geriatr Psychiatry 1997; 12:470–3.

18. Marriott A, Donaldson C, Tarrier N, Burns A. Effectiveness of cognitive behavoural family intervention in reducing the burden of care in carers fo patients with Alzheimer's Disease. Br J Psychiatry 2000; 176:557–62.

19. Coon DW, Thompson L, Steffen A, Sorocca K, Gallagher-Thompson D. Anger and depression management: psychoeducational skill training interventions for women caregivers of a relative with dementia. Gerontologist 2003; 43:678–89.

20. Olshevski JL, Katz AD, Knight BG. Stress Reduction for Caregivers. London: Brunner/Mazel; 1999.

21. Gendron C, Poitras L, Dastoor DP, Perodeau G. Cognitive-behavioural group intervention for spousal caregivers: findings and clinical considerations. Clin Gerontol 1996; 17:3–19.

22. McNally ST, Newman S. Objective and subjective conceptualizations of social support. J Psychosom Res 1999; 46:309–14.

23. Pillemer K, Suitor JJ, Wethington E. Integrating theory, basic research, and intervention: two case studies from caregiving research. Gerontologist 2003; 43(special issue 1):19–28.

24. Briggs K, Askham J. The needs of people with dementia and those who care for them: a review the literature. London: Alzheimer's Society; 1999.

25. McNally S, Ben-Shlomo Y, Newman S. The effects of respite care on informal carers' well-being: a systematic review. Disab Rehab 1999; 21(1): 1–14.

26. Colvez A, Joel ME, Ponton-Sanchez A, Royer AC. Health status and work burden of Alzheimer's informal caregivers: comparisons of five different care programe in the European Union. Health Policy 2002; 60:219–33.

27. Hinchliffe AC, Hyman IL, Blizard B et al. Behavioural complications of dementia – can they be treated? International J Geriatr Psychiatry 1995; 10:839–47.

28. Brodaty H, Gresham M. Effect of a training programme to reduce stress in carers of patients with dementia. BMJ 1989; 299:1375–9.

29. Mittleman MS, Ferris SH, Shulman E et al. A comprehensive support program: effects on depression in spouse-caregivers of AD patients. Gerontologist 1995; 35:792–802.

30. Mittleman MS, Ferris SH, Shulman E, Steinberg G, Levin B. A family intervention to delay nursing home placement of patients with Alzheimer disease: a randomized controlled trial. JAMA 1996; 276:1725–31.

31. Moniz-Cook E, Agar S, Gibson G, Win T, Wang M. A preliminary study of the effects of early intervention with people with dementia and their families in a memory clinic. Aging Ment Health, 1998; 2:199–211.

32. Brodaty H, Gresham M, Luscombe G. The Prince Henry hospital dementia caregivers' training programme. Int J Geriatr Psychiatry 1997; 12:183–92.

33. Hunter JE, Schmidt FL. Methods of Meta-analysis: Correcting Error and Bias in Research Findings. Newbury Park, CA: Sage; 1990.

34. Schulz R, Burgio L, Burns R et al. Resources for enhancing Alzheimer's Caregiver health (REACH): overview, site-specific outcomes, and future directions. Gerontologist 2003; 43:514–20.

35. Clarke M, Oxman AD (eds). Cochrane Reviewers Handbook 4.1.1 In: The Cochrane Library. Oxford: Update Software; 2000.

36. Moher D, Schulz KF, Altman DG, for the CONSORT group. The CONSORT statement: revised recommendations for improving the quality of report of parallel-group randomized trials. Lancet 2001; 357:1191–4.

37. Brodaty H, Roberts K, Peters K. Quasi-experimental evaluation of an educational model for dementia caregivers. Int J Geriatr Psychiatry 1994; 9:195–204.

38. Gottlieb BH, Johnson J. Respite programs for caregivers of persons with dementia: a review with practice implications. Aging Ment Health 2000; 4:119–29.

39. Torgerson D, Sibbald B. Understanding controlled trials: what is a patient preference trial? BMJ, 1998; 316:360.

40. Lichstein KL, Riedel BW, Grieve R. Fair tests of clinical trials: a treatment implementation model. Adv Behav Res Ther 1994; 16:1–29.

41. Burgio L, Corcoran M, Lichstein KL et al, for the REACH investigators. Judging outcomes in psychosocial interventions for dementia caregivers: the problem of treatment implementation. Gerontologist, 2001; 41:481–9.

42. Lennon S, Jolley D. An urban psychogeriatric service. In: Jacoby R, Oppenheimer C (eds). Psychiatry in the Elderly. Oxford: Oxford University Press; 1991:322–9.

43. Carradice A, Beail N, Shankland MC. Interventions with family caregivers for people with dementia: efficacy problems and solution. J Psychiatr Mental Health Nurs 2003; 10:307–15.

44. Salkovskis PM. Empirically grounded clinical interventions. Behav Cogn Psychother 2002; 30:1–2.

45. Hall J. The psychology of caring. Brit J Clin Psychol 1990; 29:129–44.

46. Jacobson NW, Truax P. Clinical significance: a statistical approach to defining meaningful change in psychotherapy research. J Consult Clin Psychol 1991; 59:12–19.

47. Rockwood K, MacKnight C. Assessing the clinical importance of statistically significant changes in anti-dementia drug trials. Neuroepidemiology 2001; 20:51–6.

48. Ayalan L, Arean PA. Knowledge of Alzheimer's disease in four ethnic groups of older adults. Int J Geriatr Psychiatry 2004; 19:51–7.

49. Combrinck-Graham L. A developmental model for family systems. Fam Process 1985; 24:139–50.

50. Schulz R, Newson JT, Fleissner K, Decamp AR, Neiboer AP. The effects of bereavement after family caregiving. Ageing Ment Health 1997; 1:269–82.

20

Pharmacoeconomic Outcomes

Anders Wimo and Bengt Winblad

Introduction

A complete health-economical evaluation includes an analysis of both costs and outcomes. Furthermore, a comparator is also essential, such as a placebo group (or a 'do-nothing' alternative), or a comparison between two treatments.[1] 'Do nothing' does not mean to actually do nothing, it just implies that the treatment of interest is added to some kind of standard or established therapy. Examples of complete health-economic evaluations are cost-minimisation analysis (CMA), cost-effectiveness analysis (CEA), cost-utility analysis (CUA), and cost-benefit analysis (CBA). It is also important to stress the fact that it is the incremental result that is of interest, which can be explained as follows (the incremental cost-effectiveness ratio, ICER):

$$\Delta C / \Delta E = (C_A - C_B) / ((E_{AT1} - E_{AT0}) - (E_{BT1} - E_{BT0}))$$

where C = costs, E = effects, A, B = treatments, T0, T1 = time for assessments).

In this chapter we focus on outcomes. Costing care is a complicated issue and we refer to standard textbooks on health economy.

The various methods focus on different outcomes. In a CMA it is assumed that the outcomes (irrespective of what they are) are equal, which

means that the analysis focuses on identifying the treatment alternative with the lowest cost. In a CBA, both costs and outcomes are expressed in the same (monetary) units, such as dollars or euros (making it possible to calculate both a ratio and a difference). In CEA and CUA, the ratio illustrated above is the most common way to present results. The practical possibilities for using CMA have been questioned.[2] Even if CBA is preferable from a theoretical point of view, since the same units (money) are compared, it has rarely been used in the evaluation of dementia care. 'Willingness to pay' approaches, such as contingent valuation could be an interesting CBA approach in dementia care,[3] but its value needs to be proven (we are currently conducting a project of this kind). Thus, this presentation will focus on outcomes that may be of interest in CEA or CUA. In a CEA, the outcome is classified in quantifiable units, such as years of survival, days of remaining in own home, or months in a less severe state of dementia. In a CUA, the outcome is expressed in terms of 'utilities', most often as QALYs (quality-adjusted life years) gained.[4]

However, it is important to stress the fact that most health-economical evaluations of dementia care so far are models, where clinical data from short-term trials (3–12 months) are often extrapolated to longer periods. Resource use data are seldom collected prospectively, and the models have several underlying assumptions that may be questioned. Furthermore, the few studies that have been published with prospectively collected resource use and cost data,[5,6] have the design of cost-consequence analysis (CCA) rather than CEA or CUA as described above. The use of CCA must be judged critically since the selection of outcomes may be biased.[7]

Outcomes

The perspectives of who is paying, and what is being paid, are essential when costs are analysed (e.g. public payer, a county council, an insurance company, etc.). However, it is important to define a perspective even when outcomes are being regarded. Even if most studies focus on the effects of an intervention on the patients, it may also be of great interest to study the effects on the closest family members (most often spouses or children). Some intervention effects may be regarded as positive for the patients (such as postponing of institutionalisation), while the same effect may be more complex for the family members (it may be good for a spouse that her husband can stay at home for a longer period of time, but it may also increase the burden, see below).[8]

In most clinical trials of dementia care, there is a comprehensive set of efficacy measurements, focusing on cognition, activities of daily living (ADL) capacity, behavioural and psychological symptoms in dementia (BPSDs), etc. Such measurements can also be used as outcomes in pharmacoeconomic evaluations, but there is no real consensus as to how this should be done. These issues have been highlighted in the work by the International Working Group on Harmonization of dementia drug guidelines (IWG),[9] and on the international pharmacoeconomic conferences on Alzheimer's disease.[10,11] In most pharmacoeconomic models that have been published, the avoidance of deterioration to severe dementia (in CEA) or QALYs (in CUA) have been used.[12] In most cases, there is only one specified outcome,

in order to calculate the cost-effectiveness ratio. If there are different outcomes used in different studies there will, of course, be problems in making comparisons. Furthermore, if there are several efficacy measurements in a trial, which, as mentioned above, is common, there may be difficulties in choosing one of them. A common logical choice is the primary outcome in a trial. However, it is not certain that the primary outcome can be described in terms that make it suitable to put in the ICER formula mentioned above, nor is it certain that this adequately captures the preferences of patients and their caregivers. It is, at least from the pure mathematical point of view, easier in a CUA: QALYs are expressed as a figure between 0 (death) and 1 (perfect health).

Instruments

In evaluating the clinical effects of interventions in dementia care, irrespective of whether these are pharmacological or focused on care (or combinations), there are some outcome measures that are considered as mandatory, such as cognitive functioning and ADL capacity. There are, however, other outcomes that may be more important as measurements of care quality for the patients and the caregivers (Table 20.1). Quality of life, mood-depression, behavioural and psychiatric symptoms are probably of more interest than cognition in the long run. ADL capacity, behaviour and psychiatric symptoms are also important for care planning.

The Mini-Mental State Examination

In health-economic evaluations, the outcome chosen must be the relevant outcome of interest. Surrogate or intermediate endpoints are inappropriate to use in CEA.[13] The Mini-Mental State Examination (MMSE) is often used in dementia intervention studies,[14] as well as other measures of cognitive capacity such as the ADAS-Cog (Alzheimer's Disease Assessment Scale—cognitive subscale).[15] However, it is doubtful

Table 20.1 Examples of pharmacoeconomic papers where different outcomes have been used

Outcome	Drug	Reference
MMSE	Donepezil	Jonsson et al 1999[18]
QALYs	Donepezil	Neumann et al 1999[55]
	Donepezil	Ikeda et al 2002[56]
	Rivastigmine	Hauber et al 2000[57]
	Galantamine	Getsios et al 2001[58]
	Galantamine	Caro et al 2002[59]
	Galantamine	Ward et al 2003[60]
Staging	Donepezil	Stewart et al 1998[61]
	Donepezil	O'Brien et al 1999[62]

whether these measures are relevant outcomes for health-economic evaluations, since, in our opinion, they should be regarded as surrogate endpoints. It is not the cognitive decline itself that causes most problems in dementia care. However, due to its wide use, the MMSE has a special position in any evaluation of dementia care. Since the MMSE is highly correlated to costs of dementia care,[16,17] and to other outcomes, such as ADL capacity, position in care organisation, and stage of dementia (the MMSE is sometimes used for staging, see below), it has also been widely used in economical evaluations of dementia (see below). This may be necessary in a situation when there are no other options and the available database is small. Furthermore, a great advantage of the MMSE is that it is available in most longitudinal population-based studies, and so may be useful as a link between clinical intervention studies and population studies in the discussions of efficacy versus clinical effectiveness.

An outcome measure can be defined as a dichotomous variable, for example whether or not the patient's cognitive ability is below a certain level on the MMSE. If a MMSE score lower than 10 is defined as severe dementia, then the corresponding evaluation measure is incremental cost per day of avoiding severe dementia. However, this measure is not free from weaknesses either. This measure implicitly means that all MMSE scores above 10 have the same value for the patient and all MMSE scores below 10 have the same value. This is obviously a simplification. Even if the staging is refined to four or five levels, the problem regarding simplification and cut-off values for the staging remains.

The MMSE has been used in different ways in pharmacoeconomical studies. It can be used as a method for staging of the cognitive decline (e.g. in CEA),[18] but also as a 'vehicle' for costs in cost studies with no predefined outcome.[19–22]

Postponing institutionalisation

Delaying nursing home placement (NHP) and other forms of institutionalisation has been suggested and used as a clinically relevant outcome measure in intervention studies in dementia.[23–25] The basic idea is that a prolonged period at home (i.e. outside institution) improves the quality of life for the patient.

Even if this may be relevant for many patients, the situation is more complex for other family members/informal caregivers. On one hand, it may be an advantage from an emotional point of view that a married couple can live together for a longer period of time, but on the other hand, a prolonged period of staying at home may also produce more stress, burden, and also morbidity and mortality for the informal caregiver (and, as a consequence, a higher cost for informal care).[8,26] This potential effect has, however, not been confirmed in the two published studies where the amount of informal care has been measured in prospective randomised controlled trials (RCTs). In these studies, there were significant reductions, or a trend of a reduction in caregiver time combined with no significant decrease (although trends) in institutionalisation.[6,27] The place of residence is also, to a large extent, determined by availability and the provision of other services. Delay in NHP therefore depends both on how dementia care is organised and the patient's health status, making it difficult to interpret as an outcome variable.

Even if postponing institutionalisation is regarded as a relevant outcome, it has not been used so far as an outcome in CEA. Since institutionalisation itself is the major cost driver, both the numerator and the denominator in a cost-effectiveness ratio will be heavily burdened by the same factor. However, more or less implicitly, it is used as a proxy for cost savings.

Another aspect of the NHP-factor is the position of long-term care institutions in a country's care system. Postponing NHP is not a realistic goal in most developing countries since there are hardly any nursing homes.

Quality of life

Effects on quality of life (QoL) of both patients and caregivers are regarded by most dementia researchers as highly relevant outcomes. Health-related QoL, which is the concept that is used in evaluations of healthcare, includes several dimensions, such as ADL, social interaction, perception, pain, anxiety, economic status, etc. In principle, two kinds of instruments can be used: diagnosis-specific or generic instruments,[28] and there are a great number of scales.[28–31] A specific problem in any quality of life assessments in dementia is that since the patient's cognitive ability is deteriorated,

he/she cannot value his/her health state him/herself.[32] In particular, the subjective feeling, which is an important dimension of quality of life, may be difficult to measure. Thus, proxies (mostly caregivers) are often used (see below).

Dementia-specific instruments such as DQoL (the Dementia Quality of Life instrument),[33] and QoLAD (Quality Of Life-Alzheimer's Disease),[34,35] and QoLAS (Quality of Life Assessment Schedule; an instrument from the UK that has been used in dementia and other neuropsychiatric conditions)[11,36,37] may be useful in international intervention trials with a cost-consequence analysis design. However, these instruments are less appropriate for health-economic evaluations such as CUA, since these instruments are not preference based (see below).[38,39] It may be difficult to interpret a result saying that 'the cost for an improvement on the QoLAD with one point is 2000 euros'.

There are a great number of generic QoL scales. Examples of such instruments are the Sickness Impact Profile,[40] the Short Form 36 (SF-36),[41] the Health Utilities Index (HUI),[38,42] the EuroQoL/EQ-5D,[43]or the Index of Well Being/Quality of Well Being Scale (QWBS).[44] The HUI, EQ-5D, and QWBS can be used to calculate QALYs.[45]

QALYs are frequently used as outcome measures in health-economic research. This concept reflects both quantity of life and quality of life.[4,38] The use of QALYs also gives opportunities to make comparisons with other diseases. QALYs are expressed as figures between 0 (death) and 1 (perfect health) and are widely used in CUA. As with all QoL instruments used in dementia, the main problem with QALYs is how the preference score for a specific health state should be measured. Due to cognitive impairment, somebody else than the patient has to value the health state; some kind of indirect (proxy) measurement has to be used, e.g. with the help of instruments like EuroQol/EQ-5D, QWBS, or HUI. Indirect measurement *per se* is not necessarily a problem – the US Panel for Cost-effectiveness actually recommends that the general public should value health states and in that sense, an indirect measurement is possible.[39]

The respondent fills in the questionnaire, which is built upon a list of attributes (such as self-care, social capacity, pain, etc.) and then in the next step, the individuals' answers are mapped into the system used (such as HUI, QWBS or EQ-5D) to reflect societal preferences. Some attributes may be difficult to use with proxies, such as 'pain' or 'emotion', whereas others, such as 'mobility' and 'self-care' are easier to use with proxies. However, the methodological problems are illustrated by projects in dementia where proxy-rated QoL and self-rated QoL have been used simultaneously, resulting in great differences between the two approaches.[46]

Thus, the use of QALYs is not uncontroversial, particularly regarding the elderly.[47] One advantage with QALYs is that comparisons with other disorders are possible. However, this may disfavour chronic, incurable, progressive disorders when these are compared with, e.g. curative surgical treatment, such as cataract or hip replacement surgery. Another problematic issue is that if the proxy is a family member, the answers may partly reflect the situation and interests of the proxy.

There are other approaches that can be used in CUA, such as DALYs (disability-adjusted life years),[48] and HYE (healthy years equivalents).[49] These concepts are connected with even more methodological problems (DALYs focus on productivity more than quality of life, and HYEs require a great number of health scenarios to analyse[4]). The first choice in CUA therefore, in our opinion, remains QALYs.

Since caregivers are more or less cognitively intact in most cases, the number of QoL instruments that can be used to analyse their situation is comprehensive. The Short Form 36 (SF-36),[41] (and shorter versions SF 20 and SF 12) have been used, and in some sense, burden scales, such as the Burden Interview,[50] can be used as indicators of caregiver QoL. However, the interplay between patient characteristics, caregiver QoL, and burden is complex.[51] The caregiver quality of life index (CQLI) is a QoL instrument focusing on caregivers, making it possible to calculate QALYs.[52]

Severity staging of dementia

Since 'severity of dementia' is a wider concept than cognition, it has been suggested as a more relevant outcome. In several pharmacoeconomical studies, severity has thus been used, particularly in models. Severity includes dimensions such as cognition, personal and instrumental ADL, social functioning, and problem solving. Clinical Dementia Rating (CDR), and Global Deterioration Scale are widely used for staging of dementia,[53,54] and both these scales can be used in pharmacoeconomical evaluations. The cost-effectiveness judgement is often expressed as the cost of avoiding decline from one stage to a worse stage. One problem may be that the intervention effect must be strong in order to detect a shift in stage. It may demand large samples and long study periods (hardly less than one year, probably more). Therefore, different modelling approaches are often used, but the problem remains: how sensitive are the severity scales? There may be clinically relevant changes within a severity stage that may be missed. In addition, it is clear that treatment often results in patients having differential effects on symptoms, such that after a few years on treatment they can have symptoms which, in the untreated state might individually correspond to the mild, moderate or severe stages, but which, on treatment, are now seen in combination.

Conclusion

There is no obvious first choice of outcome in pharmacoeconomical evaluations. If there is a research question about cost-effectiveness, it is, however, logical to consider a generic outcome, making the calculations of QALYs possible. Future studies should consider methodological work to develop quality of life instruments that are useful for pharmacoeconomic evaluations. Furthermore, empirical studies including QALY calculations are needed, with clearly defined research questions about cost-effectiveness. The impact of BPSDs in pharmacoeconomical terms is also unclear. The use of models is still controversial and in order to improve the comparativity between studies, there is also a need for consensus work in line with the pharmacoeconomic conferences mentioned earlier.

References

1. Drummond MF, O'Brien B, Stoddart GL, Torrance GW. Methods for the economic evaluation of health care programmes. Oxford: Oxford University Press; 1997.

2. Briggs AH, O'Brien BJ. The death of cost-minimization analysis? Health Econ 2001; 10(2):179–84.

3. Johannesson M, Jonsson B. Economic evaluation in health care: is there a role for cost-benefit analysis? Health Policy 1991; 17(1):1–23.

4. Torrance G. Designing and conducting cost-utility analysis. In: Spilker B (ed). Quality of life and pharmacoeconomics in clinical trials. Philadelphia: Lippincott-Raven Publishers; 1996: 1105–21.

5. Wimo A, Winblad B, Stöffler A, Wirth Y, Möbius HJ. Resource utilization and cost analysis of memantine in patients with moderate to severe Alzheimer's disease. Pharmacoeconomics 2003; 21(5):327–40.

6. Wimo A, Winblad B, Engedal K et al. An economic evaluation of donepezil in mild to moderate Alzheimer's disease: results of a 1-year, double-blind, randomized trial. Dement Geriatr Cogn Disord 2003; 15(1):44–54.

7. Winblad B, Hill S, Beermann B, Post SG, Wimo A. Issues in the economic evaluation of treatment for dementia. Position paper from the International Working Group on Harmonization of Dementia Drug Guidelines. Alzheimer Dis Assoc Disord 1997; 11(Suppl 3):39–45.

8. Max W. The cost of Alzheimer's disease. Will drug treatment ease the burden? Pharmacoeconomics 1996; 9(1):5–10.

9. Whitehouse PJ. Harmonization of dementia drug guidelines (United States and Europe): a report of the International Working Group for the Harmonization for Dementia Drug Guidelines. Alzheimer Dis Assoc Disord 2000; 14(Suppl 1):S119–22.

10. Whitehouse PJ, Winblad B, Shostak D et al. First International Pharmacoeconomic Conference on Alzheimer's Disease: report and summary. Alzheimer Dis Assoc Disord 1998; 12(4):266–80.

11. Jonsson L, Jonsson B, Wimo A, Whitehouse P, Winblad B. Second International Pharmacoeconomic Conference on Alzheimer's Disease. Alzheimer Dis Assoc Disord 2000; 14(3):137–40.

12. Wimo A, Winblad B. Economic aspects on drug therapy of dementia. Curr Pharm Des 2004; 10:295–301.

13. Johannesson M, Jonsson B, Karlsson G. Outcome measurement in economic evaluation. Health Econ 1996; 5(4):279–96.

14. Folstein MF, Folstein SE, McHugh PR. 'Mini-mental state'. A practical method for grading the cognitive state of patients for the clinician. J Psychiatr Res 1975; 12(3):189–98.

15. Rosen WG, Mohs RC, Davis KL. A new rating scale for Alzheimer's disease. Am J Psychiatry 1984; 141(11):1356–64.

16. Jonsson L, Lindgren P, Wimo A, Jonsson B, Winblad B. Costs of Mini Mental State Examination-related cognitive impairment. Pharmacoeconomics 1999; 16(4):409–16.

17. Hux MJ, O'Brien BJ, Iskedjian M et al. Relation between severity of Alzheimer's disease and costs of caring. CMAJ 1998; 159(5):457–65.

18. Jonsson L, Lindgren P, Wimo A, Jonsson B, Winblad B. The cost-effectiveness of donepezil therapy in Swedish patients with Alzheimer's disease: a Markov model. Clin Ther 1999; 21(7):1230–40.

19. Lubeck DP, Mazonson PD, Bowe T. Potential effect of tacrine on expenditures for Alzheimer's disease. Med Interface 1994; 7(10):130–8.

20. Henke CJ, Burchmore MJ. The economic impact of the tacrine in the treatment of Alzheimer's disease. Clin Ther 1997; 19(2):330–45.

21. Wimo A, Karlsson G, Nordberg A, Winblad B. Treatment of Alzheimer disease with tacrine: a cost-analysis model. Alzheimer Dis Assoc Disord 1997; 11(4):191–200.

22. Fenn P, Gray A. Estimating long-term cost savings from treatment of Alzheimer's disease. A modelling approach. Pharmacoeconomics 1999; 16(2):165–74.

23. Knopman D, Schneider L, Davis K et al. Long-term tacrine (Cognex) treatment: effects on nursing home placement and mortality, Tacrine Study Group. Neurology 1996; 47(1):166–77.

24. Sano M, Ernesto C, Thomas RG et al. A controlled trial of selegiline, alpha-tocopherol, or both as treatment for Alzheimer's disease. The Alzheimer's Disease Cooperative Study. N Engl J Med 1997; 336(17):1216–22.

25. Geldmacher DS, Provenzano G, McRae T, Mastey V, Ieni JR. Donepezil is associated with delayed nursing home placement in patients with Alzheimer's disease. J Am Geriatr Soc 2003; 51(7):937–44.

26. Schulz R, Beach SR. Caregiving as a risk factor for mortality: the Caregiver Health Effects Study. JAMA 1999; 282(23):2215–19.

27. Wimo A, Winblad B, Stoffler A, Wirth Y, Mobius HJ. Resource utilisation and cost analysis of memantine in patients with moderate to severe Alzheimer's disease. Pharmacoeconomics 2003; 21(5):327–40.

28. Walker MD, Salek SS, Bayer AJ. A review of quality of life in Alzheimer's disease. Part 1: Issues in assessing disease impact. Pharmacoeconomics 1998; 14(5):499–530.

29. Spilker B. Quality of Life and Pharmacoeconomics in Clinical Trials. Philadelphia: Lippincott-Raven Publishers; 1996.

30. Bowling A. Measuring Health. A Review of Quality of Life Measurement Scales. Philadelphia, Milton Keynes: Open University Press; 1995.

31. Salek SS, Walker MD, Bayer AJ. A review of quality of life in Alzheimer's disease. Part 2: Issues in assessing drug effects. Pharmacoeconomics 1998; 14(6):613–27.

32. Stewart A, Brod M. Measuring health related quality of life in older and demented people. In: Spilker B (ed). Quality of Life and Pharmacoeconomics in Clinical Trials. Philadelphia: Lippincott-Raven Publishers; 1996: 819–30.

33. Brod M, Stewart AL, Sands L, Walton P. Conceptualization and measurement of quality of life in dementia: the dementia quality of life instrument (DQoL). Gerontologist 1999; 39(1):25–35.

34. Logsdon RG, Gibbons LE, McCurry SM, Teri L. Assessing quality of life in older adults with cognitive impairment. Psychosom Med 2002; 64(3):510–19.

35. Selai C, Vaughan A, Harvey RJ, Logsdon R. Using the QOL-AD in the UK. Int J Geriatr Psychiatry 2001; 16(5):537–8.

36. Elstner K, Selai CE, Trimble MR, Robertson MM. Quality of Life (QOL) of patients with Gilles de la Tourette's syndrome. Acta Psychiatr Scand 2001; 103(1):52–9.

37. Selai CE, Elstner K, Trimble MR. Quality of life pre and post epilepsy surgery. Epilepsy Res 2000; 38(1):67–74.

38. Torrance GW, Feeny DH, Furlong WJ et al. Multiattribute utility function for a comprehensive health status classification system. Health Utilities Index Mark 2. Med Care 1996; 34(7):702–22.

39. Siegel JE, Torrance GW, Russell LB et al. Guidelines for pharmacoeconomic studies. Recommendations from the panel on cost effectiveness in health and medicine. Panel on Cost Effectiveness in Health and Medicine. Pharmacoeconomics 1997; 11(2):159–68.

40. Bergner M, Bobbitt RA, Carter WB, Gilson BS. The Sickness Impact Profile: development and final revision of a health status measure. Med Care 1981; 19(8):787–805.

41. Ware JE Jr, Sherbourne CD. The MOS 36-item short-form health survey (SF-36). I. Conceptual framework and item selection. Med Care 1992; 30(6):473–83.

42. Neumann PJ, Sandberg EA, Araki SS et al. A comparison of HUI2 and HUI3 utility scores in Alzheimer's disease. Med Decis Making 2000; 20(4):413–22.

43. Coucill W, Bryan S, Bentham P, Buckley A, Laight A. EQ-5D in patients with dementia: an investigation of inter-rater agreement. Med Care 2001; 39(8):760–71.

44. Kerner DN, Patterson TL, Grant I, Kaplan RM. Validity of the quality of well-being scale for patients with Alzheimer's disease. J Aging Health 1998; 10(1):44–61.

45. Torrance GW. Preferences for health outcomes and cost-utility analysis. Am J Manag Care 1997; 3(Suppl):S8–20.

46. Jönsson L. Economic evaluation of treatments for Alzheimer's disease. Stockholm, Sweden: Karolinska Institutet; 2003.

47. Tsuchiya A, Dolan P, Shaw R. Measuring people's preferences regarding ageism in health: some methodological issues and some fresh evidence. Soc Sci Med 2003; 57(4):687–96.

48. Allotey P, Reidpath D, Kouame A, Cummins R. The DALY, context and the determinants of the severity of disease: an exploratory comparison of paraplegia

in Australia and Cameroon. Soc Sci Med 2003; 57(5):949–58.

49. Dolan P. A note on QALYs versus HYEs. Health states versus health profiles. Int J Technol Assess Health Care 2000; 16(4):1220–4.

50. Zarit SH, Reever KE, Bach-Peterson J. Relatives of the impaired elderly: correlates of feelings of burden. Gerontologist 1980; 20(6):649–55.

51. Deeken JF, Taylor KL, Mangan P, Yabroff KR, Ingham JM. Care for the caregivers: a review of self-report instruments developed to measure the burden, needs, and quality of life of informal caregivers. J Pain Symptom Manage 2003; 26(4):922–53.

52. Mohide EA, Torrance GW, Streiner DL, Pringle DM, Gilbert R. Measuring the wellbeing of family caregivers using the time trade-off technique. J Clin Epidemiol 1988; 41(5):475–82.

53. Hughes CP, Berg L, Danziger WL, Coben LA, Martin RL. A new clinical scale for the staging of dementia. Br J Psychiatry 1982; 140:566–72.

54. Reisberg B, Ferris SH, de Leon MJ, Crook T. Global Deterioration Scale (GDS). Psychopharmacol Bull 1988; 24(4):661–3.

55. Neumann PJ, Hermann RC, Kuntz KM et al. Cost-effectiveness of donepezil in the treatment of mild or moderate Alzheimer's disease. Neurology 1999; 52(6):1138–45.

56. Ikeda S, Yamada Y, Ikegami N. Economic evaluation of donepezil treatment for Alzheimer's disease in Japan. Dement Geriatr Cogn Disord 2002; 13(1):33–9.

57. Hauber AB, Gnanasakthy A, Mauskopf JA. Savings in the cost of caring for patients with Alzheimer's disease in Canada: an analysis of treatment with rivastigmine. Clin Ther 2000; 22(4):439–51.

58. Getsios D, Caro JJ, Caro G, Ishak K. Assessment of health economics in Alzheimer's disease (AHEAD): galantamine treatment in Canada. Neurology 2001; 57(6):972–8.

59. Caro JJ, Salas M, Ward A, Getsios D, Mehnert A. Economic analysis of galantamine, a cholinesterase inhibitor, in the treatment of patients with mild to moderate Alzheimer's disease in the Netherlands. Dement Geriatr Cogn Disord 2002; 14(2):84–9.

60. Ward A, Caro JJ, Getsios D et al. Assessment of health economics in Alzheimer's disease (AHEAD): treatment with galantamine in the UK. Int J Geriatr Psychiatry 2003; 18(8):740–7.

61. Stewart A, Phillips R, Dempsey G. Pharmacotherapy for people with Alzheimer's disease: a Markov-cycle evaluation of five years' therapy using donepezil. Int J Geriatr Psychiatry 1998; 13(7):445–53.

62. O'Brien BJ, Goeree R, Hux M et al. Economic evaluation of donepezil for the treatment of Alzheimer's disease in Canada. J Am Geriatr Soc 1999; 47(5):570–8.

21

Executive Dysfunction in Dementia

Roger Bullock and Sarah Voss

The concept of executive function

The assessment of different cognitive domains in dementia has been fundamental to the diagnosis and monitoring of disease progression in clinical trials. Diagnostic criteria, and consequently the tools used to assess these domains, are heavily inclined toward assessing memory impairment as the central feature of dementia. Recent research suggests memory impairment is perhaps not the most prominent feature of cognition to be impaired across the dementia spectrum, and that executive function is particularly important. This has been seen more when identifying vascular dementia,[1,2] but is also a strong indicator of mild or incipient Alzheimer's disease (AD).[3] Accordingly, the widely accepted dementia classification systems focusing on memory as the core cognitive domain have been reviewed and criticised. In its recommendations for future research, the American Academy of Neurology (AAN) advises that 'memory should no longer be a required part of the diagnosis of dementia'.[4]

Defining executive function(s)

Executive function or executive control function (ECF) is viewed by many as the highest and most complex of cognitive abilities. It refers to the cog-nitive capacity to plan and perform goal-directed behaviour,[5] is widely believed to be governed by the frontal lobe and can be affected by disruption to the fronto-subcortical networks.[6,7] ECF can be viewed as comprising four principal components: volition, planning, purposive action and effective performance.[8]

Volition is the ability to formulate goals and intentions. Patients with disorders in volition usually show signs of apathy. Assessing volition using cognitive tests is not feasible, and assessment is usually achieved through direct observation or caregiver reports on a patient's behaviour. However, it can be argued that apathetic patients would be likely to show reduced measures of attention and concentration, thus tests accessing sustained attention do correlate with volition and can be used as a surrogate.[9]

Planning involves a number of capacities and entails the identification and organisation of steps and elements needed to achieve a goal. Many neuropsychological tests indirectly assess planning, because it is necessary at some level to perform any cognitively demanding task. This aspect of ECF may actually be a confounding factor when interpreting many of the tests currently used (including those purporting to measure memory). More direct assessment of planning can be achieved through use of tests that entail the sequential and hierarchical development of ideas.

Purposive action refers to the correct delivery of behaviour. It involves the initiation, maintenance, switching and halting of behaviours in an integrated fashion. Purposive action can be delineated into two components: productivity and flexibility. Productivity is usually assessed through fluency tasks, whereas tasks to assess flexibility span different cognitive domains including attention, concept formation and motor programming.[8] Flexibility is assessed through tests of attention and working memory, or sorting and categorisation.

'Effective performance' is the ability to monitor, self-correct and regulate behaviour; an individual needs to be able to notice errors in order to correct them. Consequently, assessment of a patient's awareness of their own memory deficit is one way of evaluating this in dementia. Alternatively, analysis of errors made on tasks may give practical information about a patient's effective performance.

Assessing ECF in current practice and clinical trials

The 'Alzheimerisation' of dementia has meant that assessment of cognitive function in clinical trials for dementia has been very memory dominated. The Alzheimer's Disease Assessment Scale (ADAS)[10] was developed as an expanded Mini-Mental State Examination (MMSE),[11] and the memory changes that it demonstrates appear to correlate well with the increasing cholinergic deficits in AD. This makes it a sensible instrument for measuring the effect of a treatment designed to boost cholinergic function – and the US Food and Drug Administration (FDA) has endorsed this by making it a requirement in regulatory studies. The scale has now been used in non-AD dementias, more recently with additional items to try and reflect some of the suspected ECF components. However, ECF is unlikely to be mediated solely by cholinergic activity, so the scale may continue to be deficient.

Because the other scale that is required for regulatory studies was a global scale, ECF was not measured routinely until recent years. Ten years ago, activities of daily living (ADL) became a required measurement in Europe and now are routinely measured in all studies. Not all ADL scales have taken ECF into account. Most have focused on the patient's ability to carry out routine ADL as opposed to the level of initiation given to each task by the patient. Perhaps the scale that most effectively accesses ECF impairment from the functional performance perspective is the Disability in Alzheimer's Disease scale (DAD),[12] which measures intention (planning) and the effectiveness of the activity (purposive action). This has been used successfully, with consistent results in studies with AD and other dementias, as reviewed in Chapter 9. It is often reported that these functional improvements are more relevant to the well-being of the patient and carer than other measures, including 'cognitive' outcomes. It should be argued that because of the strength of the correlation between executive dysfunction and functional impairment, the evaluation of ADL in terms of initiation, planning and sequencing is also a cognitive measure. As such, a scoring system on the DAD that emphasises the elements of initiation, planning and organisation could facilitate the delineation of this cognitive aspect of ADL from the physical performance. Such an opportunity is not so apparent with the other existing ADL scales.

Neuropsychological assessment of ECF tends to focus on evaluation of performance on cognitive tasks that are believed to be governed by executive control. There have been a number of putative tests used in the literature for assessment of ECF such as: the Trail Making Test,[13] the Stroop Test[14] and the Wisconsin Card Sorting Test.[15] These tests are not ideal for use in a battery designed for the assessment of dementia as they are relatively lengthy, patients with more severe impairments find them difficult to complete, and their relationship to everyday executive performance – particularly for elderly people – is not always clear. It has been suggested that formal assessment of executive control functions can be conducted using a simple clock drawing test (CLOX)[16] or interview (EXIT).[17] However, these tests are designed to 'screen' for executive control function deficits, rather than offer full cognitive neuropsychological assessment.

As with any construct, research intended to access a particular paradigm results in a narrowing of the original concept. Tools designed to assess that construct actually begin to define it, and the concept mutates into a measurable entity; which is a diminutive element of the totality. If we take a step back and re-acquaint ourselves with the controlling qualities and broader form of ECF, we can perhaps appreciate more about its role in cognitive and functional deficits in dementia.

Broadening the boundaries of ECF

Early theories of ECF were based on the assumption of a central concept underpinning a cognitive hierarchy. More recently, developments have moved towards the proposal of multiple executive functions that operate in parallel.[5] Patients rarely show impairment in just one aspect of ECF; there may be evidence of a cluster of deficits, some of which are more prominent than others.[8] This is by definition a broader concept that embraces ECF as having components of both cognitive control and performance outcomes. This has led to some disagreement about what exactly constitutes ECF and how it can be reliably measured.

From the psychological perspective, it may be helpful to view ECF as being to global cognition what the central executive system (CES) is to the model of working memory.[18] Where the CES allocates attention to input to visual and verbal memory systems, ECF may be viewed as controlling attentional processes for cognitive tasks. Of course memory is a cognitive process, thus the two executives can not be discrete. Moreover, they are diffuse aspects constructed from an understanding of related processes. This analogy, though, helps to illustrate how different methods of assessment can

access the same construct. The precise operation of the CES has been researched in laboratory-type investigations through tasks that require strategy and resistance to interference, such as dual task techniques and random number generation tasks.[19] However, deficits in other components of working memory that impact on an individual's day-to-day living can also be attributed to dysfunction in the CES. Similarly, ECF might be accurately assessed using refined tools. However, deficits in cognitive tasks, including memory, that entail volition, planning, purposive action and effective performance may also be attributed to executive dysfunction.

Broadening the measurement of ECF: Its future in clinical trials

Clinical trials to date have used specific tests of ECF, including the Stroop Test, CLOX and EXIT. These have usually been those involving the treatment of vascular dementia, but rarely AD. Nonetheless, in order to avoid the concept becoming too limited and to maximise the chances of assessing the impact of ECF impairment, it can be argued that this assessment and its interpretation needs to be broadened. In fact, factor analytic studies have found that individual ECF measures do not load onto a single replicable construct, but instead relate to several different components that have been given a variety of names such as working memory, attentional control, rule discovery, visuospatial, set shifting and response inhibition (see discussion by Royall et al[20]) and these can transpose back to the original concepts of volition, planning, purposive action and effective performance. Even minimal dysfunction in ECF may impact on other areas of cognition and on functional ability. ECF should thus be viewed as an umbrella term for numerous higher cognitive processes that are accessed by different tasks.

Clearly, it would not be plausible to assess all aspects of ECF in great detail to evaluate the efficacy of treatments in clinical trials. However, given the discussion above it might be equally unwise to seek a 'gold standard' of measurement during the evolutionary phase of the concept. The middle ground is to keep the assessment broad enough to encompass its diversity, yet brief enough to be of use in a practical setting. Lezak's conceptualisation of ECF into volition, planning, purposive action and effective performance is consistently referenced in literature pertaining to its neuropsychological assessment. Cognitive tests such as verbal fluency,[21] complex figure tasks,[22] random number generation,[23] digit symbol substitution,[24] and the trail-making test[13] are suitable for use with mild to moderate dementia patients, and used together will access volition (sustained attention), planning, purposive action (productivity and flexibility) and effective performance (particularly if some attention is given to analysis of errors and the patient's awareness of these errors). In addition to these tasks, ADL assessment should gravitate towards an emphasis on initiation of tasks, and this should be delineated from physical execution of tasks. The initiation aspect can then be used as a cognitive rather than functional measure. Although the habitual use of 'memory' tests will be hard to break, researchers should begin to become aware that ECF may ultimately govern performance on list learning and paragraph recall tasks, and that the most useful method to assess the type of memory deficit associated with Alzheimer-dominated dementia is to assess recall of recent personal events, with verification from an appropriate caregiver.

The importance of ECF in understanding the patient and the disease

Clearly, the interpretation of ECF here is very broad and encompasses most areas of cognition that are non-memory. Nonetheless, this is plausible; language and visuo-spatial function are undoubtedly incorporated into the tests that are purported to assess ECF, and visuo-spatial skill, like ECF, has been found to correlate highly with functional ability.[25]

When assessing therapies for dementia, the focus needs to be on aspects of cognition that are of consequence for the patient. In the case of memory, it is more meaningful to know whether an individual's performance has improved enough to impact on day-to-day tasks; perhaps through evaluation of memory for recent personal events and episodic memory. Similarly, with ECF, attention and sequencing are higher cognitive processes that have a direct impact on functional ability and independent living. Although it is unlikely that researchers will agree on any one framework for understanding ECF in the near future, every theoretical approach to understanding the construct is currently informative.[5] Thus assessment needs to be broad enough to encompass ECF as a holistic entity, while remaining detailed enough to access the areas researchers believe to be important – plus those areas that may not yet have received due consideration. Until better understanding has been developed in practical settings across different disciplines, it is perhaps unwise to create a construct that is measurable on a single dimension. Instead, the four components of volition, planning, purposive action and effective performance should provide a framework to propose varied tests that may or may not be discrete from memory, but are believed to be governed to a greater or lesser extent by ECF.

If ECF is to be incorporated as a central feature of dementia, then understanding of the concept needs to be at a level that lends itself to assessment in order for impairment to be identified. In simple terms, ECF can be considered a mechanism whereby the frontal lobe 'manages' other cognitive activities in the brain. This management feature becomes impaired with age and in all current dementia syndromes, as measured by current standardised tests.[26–30] Moreover, experienced clinicians identify changes in executive function as important markers of their evaluation of the response to treatment in Alzheimer's disease.[31] In consequence, it can be argued that ECF is the central cognitive feature of dementia and not memory – as is now generally claimed.[32] Memory loss may be an additional cognitive factor, particularly in AD, where it may be the presenting complaint. In this instance memory loss may occur before an actual dementia exists – perhaps explaining the dichotomy between mild cognitive impairment (MCI) and dementia.

In order to explain this, a concept of frontal disconnection needs to be considered, whereby vital pathways between the frontal lobe and other cortical structures become damaged, either by primary or secondary degeneration, for example, from tangle formation or cerebrovascular disease. This would fit with the observations that Braak staging of tangles shows dementia occurring as tangles enter the frontal lobe,[33,34] and in the Nun Study, correlation between the amount of cerebrovascular disease and the degree of dementia was reported.[35]

If dysfunctions in the various components of ECF are a measure of frontal decline and disconnection, then the multiple tests possible may have characteristic patterns that could predict the particular dementia syndrome that is most likely to occur. ECF dysfunction is definitely present in MCI, and attentional speed declines from middle age.[36] Studying such asymptomatic patients may help target those most likely to need treatment in the future, and help select and study patients for future clinical trials of putative disease-modifying drugs.

Conclusion

Memory has dominated clinical trials in dementia to date. This is not surprising as the current treatments are cholinesterase inhibitors, designed to improve failing cholinergic function – with its more apparent effect on memory. Scales have been designed to capture this effect. However, cholinergic transmission is also fundamental in ECF, which may explain some of the improvements seen in existing published studies. Future studies will use other therapeutic agents that do not have a cholinergic basis, so current scales may prove relatively insensitive to their needs. Added

to this is the increasing insensitivity of the global measure currently used in dementia studies – especially outside AD, where placebo groups can remain stable over the 6-month duration of studies. Intuitively, this means that longer studies or new instruments are needed for the future.

What is needed is an acceptance of the concept of ECF and its importance across the dementia syndrome – namely that, whatever the aetiology of neurodegeneration, failure of ECF is a common result.[31] This must then be tied in to relevant patient outcomes; starting with neuropsychological tests that capture the four principal components of ECF, and relating these to a more global paradigm – for example, improvement in planning and execution on the DAD. As all these tests are not usually dependent on any single transmitter mechanism, they may identify more change across varied putative agents. What is required is an understanding of the expected change with treatment using each of these scales and what an acceptable range of improvement across the scales would be. What is clear is that an endpoint based on function will make the results more clinically interpretable and relevant to clinicians. One benefit of this could be the use of simpler yet more reliable measures in primary care, that allow clinicians there to see the benefit for themselves.

References

1. Almkvist O. Neuropsychological deficits in Vascular dementia in relation to Alzheimer's-disease – reviewing evidence for functional similarity or divergence. Dementia 1994; 5:203–9.

2. Royall DR, Executive cognitive impairment: a novel perspective on dementia. Neuroepidemiology 2000; 19(6):293–9.

3. Chen P, Ratcliff G, Belle SH et al. Cognitive tests that best discriminate between presymptomatic AD and those who remain nondemented. Neurology 2000; 55(12):1847–53.

4. Knopman D, Dekosky ST, Cummings JL et al. Practice parameter: diagnosis of dementia (an evidence-based review) – report of the Quality Standards

Subcommittee of the American Academy of Neurology. Neurology 2001; 56(9):1143–53.

5. Duke LM, Kaszniak A. Executive control functions in degenerative dementias: a comparative review. Neuropsychol Rev 2000; 10(2):75–99.

6. Mega M, Cummings JL. Frontal-subcortical circuits and neuropsychiatric disorders. J Neuropsychiatry 2001; 6:358–70.

7. Stauss DT, Benson DF. The Frontal Lobes. New York: Raven Press; 1986.

8. Lezak MD. Neuropsychological Assessment (3rd edn). Oxford: Oxford University Press; 1995.

9. McPherson S, Fairbanks L, Tiken S, Cummings JL, Back-Madruga C. Apathy and executive function in Alzheimer's disease. J Int Neuropsychol Soc 2002; 8(3):373–81.

10. Rosen WG, Mohs RC, Davis KL. A new rating-scale for Alzheimer's-disease. Am J Psychiatry 1984; 141(11):1356–64.

11. Folstein M, Folstein SE, McHugh PR. 'Mini-mental state'. A practical method for grading the cognitive state of patients for the clinician. J Psychiatr Res 1975; 12(3):189–98.

12. Gelinas I, Gauthier L, McIntyre M, Gauthier S. Development of a functional measure for persons with Alzheimer's disease: The disability assessment for dementia. Am J Occup Ther 1999; 53(5):471–81.

13. Reiten RM, Wolfson D. The Halstead–Reiten Neuropsychological Test Battery: Theory and Clinical Interpretation. Tucson, AZ: Neuropsychology Press; 1993.

14. Golden CJ. The Stroop Color and Word Test. Chicago: Stoelting; 1978.

15. Berg EA. A simple objective treatment for measuring flexibility in thinking. J Gen Psychol 1948; 39:15–22.

16. Royall DR, Cordes JA, Polk M. CLOX: an executive clock drawing task. J Neurol Neurosurg Psychiatry 1998; 64(5):588–94.

17. Royall DR, Mahurin R, Gray K. Bedside assessment of executive cognitive impairment – the executive interview. J Am Geriatr Soc 1992; 40(12):1221–6.

18. Baddeley A, Hitch GJ. Working memory. In: Bower G (ed). The Psychology of Learning and Motivation: Advances in Research and Theory. New York: Academic Press; 1974: 47–90.

19. Cohen G, Kiss G, Le Voi M. Memory: Current Issues (2nd edn). Buckingham: Open University Press; 1993.

20. Royall DR, Lauterbach EC, Cummings JL et al. Executive control function: a review of its promise and challenges for clinical research. J Neuropsychiatry Clin Neurosci 2002; 14(4):377–405.

21. Spreen O, Strauss M. A Compendium of Neuropsychological Tests. New York: Oxford University Press; 1991.

22. Coughlan AK, Hollows SE. The Adult Memory and Information Processing Battery (AMIPB). Leeds: AK Coughlan; 1985.

23. Baddeley A. The capacity for generating information by randomisation. Q J Exp Psychol A 1966; 18:119–29.

24. Smith A. Symbol Digit Modalities Test. Los Angeles: Western Psychological Services; 1982.

25. Perry RJ, Hodges JR. Relationship between functional and neuropsychological performance in early Alzheimer disease. Alzheimer Dis Assoc Disord 2000; 14(1):1–10.

26. See ST, Ryan EB. Cognitive mediation of adult age differences in language performance. Psychol Aging 1995; 10:458–68.

27. Cepeda NJ, Kramer AF, Gonzalez de Sather JC. Changes in executive control across the life span: examination of task-switching performance. Dev Psychol 2001; 37:715–30.

28. Span MM, Ridderinkhof KR, van der Molen MW. Age-related changes in the efficiency of cognitive processing across the life span. Acta Psychol (Amst) 2004; 117:155–83.

29. Royall DR, Chiodo LK, Polk MJ. Misclassification is likely in the assessment of mild cognitive impairment. Neuroepidemiology 2004; 23:185–91.

30. Royall DR, Palmer R, Chiodo LK, Polk MJ. Declining executive control in normal aging predicts change in functional status: the Freedom House Study. J Am Geriatr Soc 2004; 52:346–52.

31. Rockwood K, Black SE, Robillard A, Lussier I. Potential treatment effects of donepezil not detected in Alzheimer's disease clinical trials: a physician survey. Int J Geriatr Psychiatry 2004; 19:954–60.

32. Voss S, Bullock RA. Executive function: The core feature of dementia? Dement Geriatr Cogn Disord 2004; 18(2):207–16.

33. Braak H, Braak E. Neuropathological staging of Alzheimer-related changes. Acta Neuropathol 1991; 82(4):239–59.

34. Braak H, Braak E, Bohl J, Bratzke H. Evolution of Alzheimer's disease related cortical lesions. J Neural Transm Suppl 1998; 54:97–106.

35. Snowdon DA, Greiner LH, Mortimer JA et al. Brain infarction and the clinical expression of Alzheimer disease – The Nun Study. J Am Med Assoc 1997; 277(10):813–17.

36. Wesnes K, Bullock R. The emerging importance of attention and the speed of cognitive function in ageing and dementia research. CPD Bull Old Age Psychiatry 2001; 3(1):11–15.

22

Ethical Considerations in the Conduct of Clinical Trials for Alzheimer's Disease

John D Fisk

Ethical dilemmas are common in Alzheimer's disease and other neurodegenerative disorders that result in cognitive impairment and dementia. These are not limited to the situation of clinical trials or to the clinical decisions facing physicians and other healthcare providers. Rather, the situation in clinical trials needs to be considered in the context that ethical dilemmas arise whenever there is uncertainty about the course of action to be taken because values are in conflict and/or when harm may be done whatever the course of action. The values and beliefs that influence one's attitudes, actions and decisions, guide ethical decisions. When one considers clinical trials of potential treatments for Alzheimer's disease, it is important that everyone involved recognises the values and beliefs on which their decision making is based.

Here we review general guidelines on clinical practice and research in Alzheimer's disease, international and national guidelines on the conduct of research and clinical trials in general, and discussions and studies that examine the participation of cognitively impaired persons in research.

In 1997, the Alzheimer Society of Canada (ASC) released the results of a two-year consultative process in its Ethical Guidelines document.[1,2] This document was based on a 'discourse ethics' approach that had previously been employed in developing the 'Fairhill Guidelines'.[3,4] It uses focus group discussions with caregivers, individuals with mild Alzheimer's disease, and professionals such as physicians, nurses, lawyers, ethicists and administrators. The ASC extended this consultation process to a national level.[5] Both the Fairhill Guidelines and the ASC Ethical Guidelines addressed a number of important issues in clinical practice and everyday life for persons with Alzheimer's disease, such as diagnostic disclosure, driving privileges, autonomy in decision making, maintaining quality of life, appropriate use of restraints, and end-of-life care, but the ASC Ethical Guidelines also addressed genetic testing and participation in research. The need for continuous re-evaluation and revision of these guidelines was recognised. Developments that included increased awareness, earlier diagnosis, the case conceptualisation of mild cognitive impairment (MCI),[6] and the availability of treatments, resulted in a revision to these guidelines through another consultative process in 2001–2002. The revised ASC Ethical Guidelines illustrate the differences that can emerge between general international guidelines on the ethical conduct of research, and guidelines that are developed through a process involving individuals who are affected by specific medical conditions and their representatives. In particular, individuals affected by specific medical conditions and their advocates, while continuing to point out the importance of protecting the rights of vulnerable individuals, place greater emphasis on the need to ensure that the potential to benefit from participation in research is not denied to specific groups of individuals. This inclusive stance towards research participation has been articulated by High and colleagues: 'To deny persons access to research participation out of fear of exploitation of specific groups of persons is to avoid rather than accept and practice ethical responsibility'.[7]

Most general ethical guidelines regarding research focus on the individual in terms of the analysis of risk/benefit ratios and the consent process. Guidelines developed for persons with Alzheimer's disease and their caregivers, however, also acknowledge the need to consider the consequences of research participation on families and the importance of involving family members in the consent process.[2,8] Even for individuals with Alzheimer's disease who are competent to provide informed consent to participate in clinical trials of new treatments, consent may be required of their family member(s) since it is they who will be expected to provide transportation for the subject, provide observations of the subject's behaviour and monitor the administration of medications. Despite the existence of specific ethical guidelines, evidence of the implementation in clinical practice is not encouraging. For example, most guidelines strongly promote the disclosure of a diagnosis of dementia to the affected individual,[2,4,9] even though this viewpoint is not universally held.[10] In contrast to the view expressed in these guidelines, however, those few small studies of patients and their caregivers that have been conducted suggest that caregivers are told of the diagnosis more often than the affected individual,[11] and that caregivers frequently indicate they do not want the affected individual to be told.[12] The existence of social and cultural barriers to the implementation of existing ethical guidelines for clinical practice must be acknowledged, and the extent to which similar barriers exist in the imple-

mentation of guidelines for the ethical conduct of research is itself an important topic that warrants research.[13]

It is important for clinicians to be aware of the ethical issues that are likely to arise in the conduct of clinical trials for Alzheimer's disease so that they too can make informed decisions about whether to participate as clinician-scientists in such research. While attempts have been made to draw distinctions between the ethics of clinical care and the ethics of the conduct of research,[14] this view is strongly opposed by many,[15] and is inconsistent with most national and international ethical guidelines for research.[16–18] For the clinician-scientist, provision of the best medical care must always remain the priority, and must not conflict with an individual's participation in research. Tensions between these roles do exist, however, and the distinctions between the clinician-scientist's role in the management of the individual's healthcare and his/her role in the conduct of the clinical trial must be clearly understood by everyone from the outset. Throughout the study, the individual's understanding of the difference between research and clinical care procedures must be reconfirmed.

The principles that best describe ethical decision making regarding research participation in clinical trials are those articulated in the 'Belmont Report', produced by the US National Commission for the Protection of Human subjects of Biomedical and Behavioural Research.[19] These include the principles of respect for persons, beneficence (i.e. the obligation to do no harm and to maximise the potential for benefit while minimising the potential for harm), and justice. These principles, in turn, are applied to clinical research through consideration of informed consent, assessment of risks and benefits, and the selection of subjects. While differences in culture and in legislation between geographical regions might influence these considerations, the principles are basic to ethical decision making in most cultures and in most international regulatory research guidelines such as the World Medical Association's

Declaration of Helsinki,[18] the Ethical Guidelines for Biomedical Research of the Council for International Organisations of Medical Sciences (CIOMS),[17] and the Council of Europe's Convention on Human Rights and Biomedicine.[20] As is always the case for guidelines, however, the 'trick' is to find the correct balance of the application of the general principles to specific research contexts.

An obvious ethical dilemma associated with clinical trials for Alzheimer's disease is balancing respect for the autonomy of the individual with the protection of vulnerable persons (i.e. beneficence). Balancing autonomy and protection is but the first such issue. The ethics of placebo controls and the use of genetic testing are others. As our understanding, treatments, and research methods in Alzheimer's disease develop further, other ethical issues become increasingly important.

It is important to recognise and acknowledge the values and beliefs that guide one's own individual ethical decision making. This holds true when one considers how specific guidelines for the ethical conduct of research can be applied to Alzheimer's disease. For example, the International Conference on Harmonisation of Technical Requirements for Registration of Pharmaceuticals for Human Use Harmonised Tripartite Guideline (ICH)[21] represents internationally accepted guidelines for the conduct of clinical trials. These guidelines, often referred to as the 'good clinical practice' (ICH-GCP) guidelines, were produced as a joint regulatory and industry project to 'to facilitate the adoption of new or improved technical research and development approaches which update or replace current practices, where these permit a more economical use of human, animal and material resources, without compromising safety'.[21] Given their origins and mandate it is perhaps not surprising that despite recognising the need to address issues specific to the conduct of clinical trials in elderly populations, and specific reference to research on Alzheimer's disease,[22] ICH-GCP guidelines provide no insight into the prominent ethical issues that arise in those contexts. Instead, they defer these issues to others by requiring that each study be reviewed and approved by an appropriately constituted and independent ethical review board.

Most guidelines for ethical conduct employed by research ethics boards promote the view that 'the principle of respect for human dignity entails high ethical obligations to vulnerable populations'.[23] However, most also agree with the concept that the 'protection should be proportionate to the risk involved, with the least protection required when research involves minimal risk'.[24] This consideration of the limits of the acceptable risk/benefit ratio is central to the evaluation of all clinical trials.[25] However, it becomes particularly difficult when evaluating clinical trials for patients with Alzheimer's disease, given the potential for uncertainty about the subject's capacity for autonomous and informed decision making. A decade ago, High and colleagues noted 'no clear consensus exists either in the literature or in regulatory guidelines as to what constitutes an acceptable degree of risk when cognitively impaired persons are involved in research'.[7] This remains the case today.

The concept of 'minimal risk' has been used in most international ethical guidelines for research although it has been described in various ways.[25] Typically, this refers to the requirement that 'the probability and magnitude of harm or discomfort anticipated in the research are not greater in and of themselves than those ordinarily encountered in daily life or during the performance of routine physical or psychological examinations or tests'.[26] Such is rarely the case in clinical trials of new treatments for Alzheimer's disease, however. While it may be argued that it is acceptable to conduct research with cognitively impaired individuals that does not involve 'risk of harm beyond a minor increment over minimal',[27] such descriptive terms can be criticised for being poorly defined and open to wide interpretation. The difficulties in coming to terms with the issue of risk, even for an organisation whose mandate is the promotion of research in Alzheimer's disease, is reflected in the statement by the Alzheimer Association in the USA:

There is considerable disagreement in the ethics literature and in regulatory and statutory language about definitions of risk. In addition to the two polar definitions, 'minimal risk' and 'greater than minimal risk,' there are also various gradations, such as 'minor increase over minimal risk.' The Alzheimer's Association has chosen not to draw these finer distinctions in its position statement. However, this should not be construed to mean that the Association necessarily opposes involvement of decidedly impaired individuals in all nontherapeutic research involving minor increments over minimal risk. The degree of risk is a judgment call frequently and appropriately made by the Institutional Review Boards (IRBs).[28]

Thus, despite the potential for differences in viewpoints on this issue, both international regulatory bodies and organisations with the mandate of patient advocacy have deferred the decisions of ethical acceptability of clinical trials to individual ethical review boards.

A discussion of the tribulations of research ethics review boards is beyond the scope of this chapter but suffice to say that the growth of clinical research coupled with insufficiencies in resources, expertise, information and communication among review boards, all contribute to the delays in the review process, inconsistencies in decisions, and oversights that are sources of concern for all.[29,30] Also contributing to the inconsistencies in decisions is the use of varied ethical guidelines between, and even within, jurisdictions. For example, in the USA, while most research is conducted in accordance with federal regulations known as the 'common rule', privately sponsored research that is outside the jurisdiction of the Food and Drug Administration (FDA) does not require adherence to these regulations.[29] Similarly, in Canada, research on human subjects conducted at institutions that receive support from federal funding bodies must be conducted in accordance with guidelines established jointly by the Canadian Institutes for Health Research, the Social Sciences and Humanities Research Council, and the National Science and Engineering Research Council.[23] Research conducted outside such institutions, however, is governed by a patchwork of laws and voluntary industry guidelines such as the International Committee on Harmonization – Good Clinical Practice (ICH-GCP) guidelines,[21] which have little to say about the ethical issues of Alzheimer's disease research.

It has been argued that research ethics review boards spend a disproportionate amount of time addressing the issue of informed consent at the expense of addressing the analysis of risk.[25] Nevertheless, free and fully informed consent is a central concept in the consideration of research ethics, and one that poses problems for ethics review boards when they consider clinical trials for Alzheimer's disease. Once again, it is important to recognise that different perspectives can exist on this topic. Most national and international ethical guidelines take a relatively protectionist stance with regard to avoiding the potential for exploitation of individuals who are not competent to provide informed consent, but variability does exist in the nature and implementation of safeguards among various guidelines.[31] A diagnosis of Alzheimer's disease does not preclude competence to provide informed consent for participation in clinical trials or in non-therapeutic research that contains minimal risk. Improved diagnostic techniques mean that Alzheimer's disease and other dementias are detected at earlier stages in which many cognitive and functional abilities are relatively preserved. Moreover, competence to consent to research is not simply a matter of the stage of dementia for an individual. Competence is not a unitary construct and one can be competent in some aspects of one's life (e.g. competent to consent to research) without being competent in others (e.g. competent to drive a motor vehicle).[32] As the risk/benefit ratio and other aspects of a given study vary, the requirements for competence may vary also. Some ethical guidelines explicitly acknowledge this 'sliding-scale' of competency,[33] and emphasise that it must be considered in every research situation. For example, the Canadian Tri-Council Policy Statement: Ethical Conduct for Research Involving Humans says:

> Competence to participate in research, then, is not an all-or-nothing decision. It requires that they be competent to make an informed decision about participation in particular research. Competence is neither a global condition nor a static one; it may be temporary or permanent.[23]

Research on the topic of competency itself has primarily focused on competency to consent to treatment. While the use of standardised methods to determine competency to provide informed consent has been promoted, there remains no clear consensus on the best standardised assessment methods to address this.[32,34] Moreover, given the evidence of inconsistencies in expert clinical judgements for mildly affected patients,[35] the question of a 'gold standard' for evaluating such methods can be raised. Research on competency for the provision of informed consent in a research context has been less common. While Kim and colleagues have reported that even mild Alzheimer's disease has significant effects on competency to consent to research,[36] this issue remains an important topic for further research. Even if standardised assessments of competency to participate in clinical or research decision making were to become available, significant social, cultural and legislative variations will still limit their application in large multi-centre clinical trials. Despite this, and since changes in competency are to be expected over time in studies of AD, there is a clear need for ongoing monitoring and re-affirmation of consent/assent. Thus, it is reasonable for ethics review boards to expect that clinical trial protocols will describe how the ability of the potential subject to understand the nature of the research, the consequences of participation (i.e. potential risks and benefits) and alternative choices will be determined,[2,8] even if this is specific to individual sites within a multi-centre trial.

The difficulties presented by the absence of clear and consistent legislation regarding competency to participate in research and the important role of proxy decision making have been recognised for some time,[33,37] but still remain.[8] For example, in North America, legislation regarding substitute decision making for the participation of incompetent individuals in research is often unclear and varies among Canadian provinces and territories,[38] and between American states.[28] From a practical standpoint, this has meant that in most situations, research ethics boards and researchers have relied upon a family member of the person with Alzheimer's disease to provide 'third party authorisation' for participation in research, even when there has been no legal determination that the person with Alzheimer's disease lacks competence to provide informed consent. In some cases, proxy consent is aided by the availability of advance directives for research participation that specifically identifies proxy decision makers. But despite the fact that such advance directives have been promoted by some,[27] others have expressed concerns about their use, given that they are only infrequently available.[39] Even when consent from a legally authorised third party is given, the ongoing assent of the individual subject involved in a clinical trial for Alzheimer's disease is almost invariably necessary.

Reservations about the surrogate decision-making process for the participation of cognitively impaired subjects in research have been expressed for some time.[40] More recently, the factors that influence caregivers' decisions regarding research participation have themselves become the source of study. From the perspective of the clinician-scientist, a number of issues should be considered. First, informed decision making by the caregiver requires that they be informed about Alzheimer's disease. Unfortunately, studies that have examined caregiver knowledge about Alzheimer's disease have been limited and have reported relatively poor understanding of issues such as causes, symptoms and drug treatments.[41] This highlights the need to ensure adequate education of patients and their families about Alzheimer's disease, before raising the possibility of research participation. Karlawish et al, in a study of caregiver's views about the acceptance of risk associated with potential treatments for Alzheimer's disease, noted that caregivers' tolerance for risk of death associated with a hypothetical treatment was influenced by demographic factors.[42] Working adult children who were caring for early-stage patients were more willing to accept that risk than spouses. For the clinician-scientist, this points out the importance of ensuring that decisions made by proxies regarding research participation reflect the prior attitudes and values of the research subject. A study by Connell et al of caregiver attitudes toward participation in longitudinal, non-therapeutic research pointed out the presence of expectations of improved access to clinical care through research participation, as well as the existence of ethnic/racial differences in attitudes toward research that present barriers to equitable research participation.[43] This latter issue is an important consideration with regard to the ethical principle of justice and the equal distribution of the risks and benefits. The former issue illustrates that drawing a clear distinction between research participation and clinical care is essential for both the study subject and their caregiver(s). While improved monitoring and access to healthcare professionals may be the case in many clinical trials, it is possible that participation in such studies may reduce, rather than improve access to care. Such is the case, for example, when subjects are enrolled into the placebo arm of a clinical trial of a new treatment rather than initiating treatment with an established therapy.

Research directly related to proxy decisions about enrolment in clinical trials has been rare. Those studies that have been conducted suggest that while proxies actively involve the research subjects in the decision-making process,[44,45] they do not always consider their decisions to be consistent with the research subject's.[8] The clinician-scientist must recognise that, as with other demands of providing care to persons with Alzheimer's disease, the process of providing proxy consent can itself be burdensome for the caregivers who may not appreciate fully the risks of clinical trial participation.[45] As more is learned about Alzheimer's disease, and as new potential treatments become available for evaluation, decision making on the part of affected individuals and their caregivers will become increasingly complex and will be even more important topics for systematic evaluation. For the clinician-scientist involved in a clinical trial it is important to understand that the interests of the person with Alzheimer's disease and those of the substitute decision maker may be different. Thus, clinician-scientists have an obligation to determine, to the best of their ability, that the decision to participate in a clinical trial has been guided by the individual's wishes and/or that it has been made with the individual's best interests in mind.[7,8,23] Again, while this determination may be aided by the availability of an advance directive, their use remains limited.

In addition to the issues discussed above, a variety of new ethical issues regarding clinical trials for Alzheimer's disease have emerged in recent years. Among them, the rapid advances in our knowledge of genetics are having a significant impact on the ethics of clinical trials for Alzheimer's disease. In discussing the genetics of Alzheimer's disease, it is important to distinguish between predictive genetic testing and genetic risk assessment. Both can be relevant to the issues of clinical trials. However, it is risk factors which can potentially increase an unaffected individual's chance of developing Alzheimer's disease, such as the apolipoprotein E gene, that have become a source of interest as markers of potential treatment efficacy.[46] The presence or absence of genetic risk factors does not, however, enable us to predict who will get Alzheimer's disease and who will not, since Alzheimer's disease can exist in the absence of a given risk factor and can fail to develop in the presence of a risk factor. Nevertheless, since proxy consent by family members may be necessary, and since genetic testing can reveal information of direct relevance to those same family members, there is a potential for conflicts of interest that can pose ethical complications for the conduct of the trial. The results of genetic tests can have social, psychological and legal implications for the individual being tested as well as for their family, not limited to the potential for negative effects on employment and the ability to obtain insurance. Because of this, counselling by trained professionals has been recommended as a required part of any genetic testing,[38] including that which may take place in the setting of a clinical trial. The consistent availability of such counselling in clinical trials is not yet common practice.

Another more recent ethical dilemma in clinical trials for Alzheimer's disease is the use of placebo arms in randomised controlled trials. As medications to treat the symptoms of Alzheimer's disease have become available, debate has begun about whether it is still appropriate to use placebos in studies of new drugs for Alzheimer disease. The use of placebos in medicine and in medical research has a long and complex history.[47] The legal, ethical and scientific debate around the general question of placebo use is equally complex and well beyond the scope of this chapter (see for discussion Guess et al[48]). There is no one simple answer to this question that is appropriate for all situations in which this question arises, nor is there consistency among guidelines for the ethical conduct of research as to how this issue should be handled.[49] A recent national initiative to harmonise existing ethical and regulatory guidelines on the use of placebos in Canada, while not explicitly precluding the use of placebos in any clinical trials for Alzheimer's disease, has recommended that '. . . research subjects in the control group of a trial of a diagnostic, therapeutic or preventative intervention should receive an established effective therapy'.[49] Even for treatments approved by a regulatory agency, one could reasonably debate whether the totality of accumulated evidence makes them 'established effective therapy'. Moreover, the existence of established effective therapy still does not preclude the possibility of designing an ethically acceptable placebo-controlled trial.[49] However, the current availability of treatments for Alzheimer's disease make it more difficult to argue that withholding these treatments from persons enrolled in the placebo arm of a clinical trial is ethically acceptable and that it fulfils the clinician-scientist's duty to provide the best care to the study subject.[50,51] The design of new trials for Alzheimer's disease will continue to become increasingly complex and as with most ethical debates, this one will continue to evolve.

Finally, an issue that also deserves some consideration as new treatments for Alzheimer's disease are developed, is the fact that reimbursement criteria for these generally costly new medications by government and private insurers can be based on the inclusion/exclusion criteria used in clinical trials. While this is an issue for those designing the trials, it is also an issue that the clinician-scientist will eventually face in his/her clinical role. As described above, organisations with a patient advocacy mandate tend to favour a more inclusive approach to research participation, while most other ethical guidelines exclude enrolment of patients not competent to consent, unless it is necessary.[31] One could correctly argue that this is not a debate about ethics but rather a debate about the appropriateness of third-party reimbursement criteria. Nevertheless, it is important that the clinician-scientist ensures that the ethical concerns about exploitation of vulnerable populations, or the inconvenience of having to design ethical clinical trials that will accommodate more severely affected individuals, does not result in inappropriate restrictions for eventual access to treatment. The clinician-scientist must ensure that an absence of evidence of efficacy in a particular group of patients will not be misconstrued as evidence of an absence of efficacy.

In summary, it is important for the clinician-scientist involved in clinical trials for Alzheimer's disease to be aware of their primary duty to care for the individual and to clearly distinguish their roles as a clinician and researcher. They must also be aware of their own values and beliefs that influence their ethical decision making. The ethical principles of respect for persons, beneficence and justice represent a balance. In clinical trials for Alzheimer's disease, it is most often the principles of autonomy (respect for persons) and beneficence (do no harm/do good) that are in conflict. In such situations, different ethical guidelines may provide different perspectives on the balance between these principles, because of the values and beliefs that they reflect. Finding the balance of autonomy and beneficence requires the ethical analysis of risk and the process of informed consent. However, considerations of informed consent in clinical trials for Alzheimer's disease invariably come up against the issue of competency to consent on the part of the individual subject. Competency is not a unitary or static construct and the variability of existing legislation and current ethical guidelines on this issue may be of limited help to the individual clinician-scientist charged with enrolling subjects into a clinical trial. Proxy decision making for persons with Alzheimer's disease has its own set of challenges that also require careful consideration. As our knowledge of Alzheimer's disease and of treatments for Alzheimer's disease continues to expand, we will see the emergence of more ethical issues. Current examples include the debates surrounding the inclusion of genetic testing as a potential predictor of treatment responsiveness, and the use of placebo arms in controlled clinical trials. These are both exciting and challenging times for those involved in the design and implementation of clinical trials for Alzheimer's disease, particularly as we see the potential for treatments aimed at the disease process rather than symptom management. To face these challenges, clinician-scientists must be knowledgeable of ethical issues and must stay abreast of this developing field.

References

1. Alzheimer Society of Canada. Ethical Guidelines, 2003. Retrieved June *www.alzheimer.ca/english/care/ethics-intro.htm* (accessed February 15 2005).

2. Fisk JD, Sadovnik AD, Cohen CA et al. Ethical Guidelines of the Alzheimer Society of Canada. Can J Neurol Sci 1998; 25:242–8.

3. Habermas J. Justification and Application: Remarks on Discourse Ethics. Cambridge, MA: MIT Press; 1993.

4. Post SG, Whitehouse PJ. Fairhill guidelines on ethics of the care of people with Alzheimer's disease: A clinical summary. J Am Geriatr Soc 1995; 43:1423–9.

5. Cohen CA, Whitehouse PJ, Post SG et al. Ethical issues in Alzheimer's disease: The experience of a national Alzheimer Society Task Force. Alzheimer Dis Assoc Disord 1999; 13:66–70.

6. Petersen RC, Stevens JC, Ganguli M et al. Practice parameter: Early detection of dementia: mild cognitive impairment (an evidence-based review). Neurology 2001; 56:1133–42.

7. High DM, Whitehouse PJ, Post SJ, Berg L. Guidelines for addressing ethical and legal issues in Alzheimer disease research: A position paper. Alzheimer Dis Assoc Disord 1994; 8:66–74.

8. Karlawish JHT. Research involving cognitively impaired adults. N Engl J Med 2003; 348:1389–92.

9. Drickamer MA, Lachs MS. Should patients with Alzheimer's disease be told their diagnosis? N Engl J Med 1992; 326:947–51.

10. Carpenter B, Dave J. Disclosing a dementia diagnosis: A review of opinion and practice, and a proposed research agenda. Gerontologist 2004; 44:149–58.

11. Holroyd S, Turnbull Q, Wolf AM. What are patients and their families told about the diagnosis of dementia? Results of a family survey. Int J Geriatr Psychiatry 2002; 17, 218–221.

12. Fahy M, Wald C, Walker Z, Livingston G. Secrets and lies: the dilemma of disclosing the diagnosis to an adult with dementia. Age Ageing 2003; 32:439–41.

13. Graham C, Holmes C, Lindesay J. Clinical involvement in anti-dementia drug trials – why bother? Int J Geriatr Psychiatry 1999; 14:257–60.

14. Miller FG, Brody H. What makes placebo-controlled trials unethical? Am J Bioeth 2002; 2:3–9.

15. Lemmens T, Miller PB. (2002) Avoiding a Jekyll-and-Hyde approach to the ethics of clinical research and practice. Am J Bioeth 2002; 2:14–17.

16. American Medical Association. Code of Medical Ethics. *www.ama-assn.org/ama/pub/category/8292. html* (accessed February 15 2005).

17. Council for International Organizations of Medical Sciences. International Ethical Guidelines for Biomedical Research Involving Human Subjects, 2002. *www.cioms.ch/frame_guidelines_nov_2002.htm* (accessed February 15 2005).

18. World Medical Association, adopted 1964, last amended 2002, World Medical Association Declaration of Helsinki, Ethical Principles for Medical Research Involving Human Subjects. *www.wma.net/ e/policy/b3.htm* (accessed February 15 2005).

19. National Commission for the Protection of Human Subjects of Biomedical and Behavioural Research, The Belmont Report, Ethical Principles and Guidelines for the Protection of Human Subjects of Research, 1979, *http://www.hhs.gov/humansubjects/ guidance/belmont.htm*

20. Council of Europe, Convention for the Protection of Human Rights and Dignity of the Human Being with Regard to the Application of Biology and Medicine: Convention on Human Rights and Biomedicine, Oviedo, 4.IV.1997. *http://conventions.coe.int /treaty/en/Treaties/Html/164.htm* (accessed February 15 2005).

21. International Conference on Harmonization of Technical requirements for Registration of Pharmaceuticals for Human Use, 2004. Terms of reference of ICH. *www.ich.org/UrlGrpServer.jser?@_ID= 276&@_TEMPLATE=254* (accessed February 15 2005).

22. International Conference on Harmonization of Technical requirements for Registration of Pharmaceuticals for Human Use, 1993. ICHT, ripartite Guideline, Studies in Support of Special Populations: Geriatrics E7. *www.ich.org/ UrlGrpServer.jser?@_ID=276&@_TEMPLATE=254* (accessed February 15 2005).

23. Tri-Council Policy Statement: Ethical Conduct for Research Involving Humans, 1998 (with 2000, 2002 updates) *http://pre.ethics.gc.ca/english/policystatement/ policystatement.cfm* (accessed February 15 2005).

24. Office of Human Subjects Research, Information Sheet 7, National Institutes of Health. *http://ohsr.od.nih. gov/info/sheet3.html* (accessed February 15 2005).

25. Weijer C. The ethical analysis of risk. J Law Med Ethics 2000; 28:344–61.

26. Department of Health and Human Services. Code of Federal Regulations Title 45 Part 46 Regulations for the Protection of Human Subjects, Section 46.102.i, 2001.*http://www.hhs.gov/ohrp/human subjects/guidance/45cfr46.htm*

27. Keyserlingk EW, Glass K, Kogan S, Gauthier S. Proposed guidelines for the participation of persons with dementia as research subjects. Perspect Biol Med 1995; 38:319–62.

28. Alzheimer Association. Ethical Issues in Dementia Research, 2004. *www.alz.org/AboutUs/PositionState ments/overview.asp#ethical* (accessed February 15 2005).

29. Steinbrook R. Improving protection for research subjects. N Engl J Med 2002; 346:1425–30.

30. Slater EE. IRB reform. N Engl J Med 2002; 346:1402–4.

31. Wendler D, Prasad K, (2001). Core safe guards for clinical research with adults who are unable to consent. Ann J Intern Med 2001; 135:514–23.

32. Marson DC. Loss of competency in Alzheimer's disease: Conceptual and psychometric approaches. Int J Law Ethics, 2001; 24:267–83.

33. High DM. Research with Alzheimer's disease subjects: Informed consent and proxy decision-making. J Am Geriatr Soc 1992; 40:950–7.

34. Moye J, Karel M, Azar AR, Gurrera RJ. Capacity to consent to treatment: Empirical comparison of the instruments in older adults with and without dementia. Gerontologist 2004; 44:166–75.

35. Marson DC, McInturff B, Hawkins L, Bartolucci A, Harrell LE. Consistency of physician judgements of capacity to consent in Alzheimer's disease. J Am Geriatr Soc 1997; 45:453–7.

36. Kim SY, Caine ED, Currier GW, Leibovici A, Ryan JM. Assessing the competence of persons with Alzheimer's disease in providing informed consent for participation in research. Am J Psychiatry 2001; 158:712–17.

37. The Ethics and Humanities Subcommittees of the American Academy of Neurology. Ethical issues in clinical research in neurology: Advancing knowledge and protecting human research subjects. Neurology 1998; 50:592–95.

38. Alzheimer Society of Canada. Tough Issues: Ethical Guidelines of the Alzheimer Society of Canada. Tororonto: Alzheimer Society of Canada; 1997.

39. Sachs GA, Stocking CB, Stern R et al. Ethical aspects of dementia research: Informed consent and proxy consent. Clin Res 1994; 42:403–12.

40. Candilis PJ, Wesley RW, Wichman A. A survey of researchers using a consent policy for cognitively impaired human research subjects. IRB 1993; 15:1–4.

41. Werner P. Correlates of family caregivers knowledge about Alzheimer's disease. Int J Geriatr Psychiatry 2001; 16:32–8.

42. Karlawish JHT, Klocinski JL, Merz J, Clark CM, Asch DA. Caregivers' preferences for the treatment of patients with Alzheimer's disease. Neurology 2000; 55:1008–14.

43. Connell CM, Shaw BA, Holmes SB, Foster N. Caregiver's attitudes toward their family members' participation in Alzheimer disease research: implications for recruitment and retention. Alzheimer Dis Assoc Disord 2001; 15:137–45.

44. Karlawish JHT, Casarett D, Klocinski J, Sankar P. How do AD patients and their caregivers decide whether to enroll in a clinical trial? Neurology 2001; 56:789–92.

45. Sugarman J, Cain C, Wallace R, Welsh-Bohmer KA. How Proxies make decisions about research for patients with Alzheimer's disease. J Am Geriatr Soc 2001; 49:1110–19.

46. Issa AM, Keyserlingk EW. Apolipoprotein E genotyping for pharmacogenetic purposes in Alzheimer's disease: Emerging ethical issues. Can J Psychiatry 2000; 45:917–22.

47. Shapiro E, Shapiro A. The Powerful Placebo: From Ancient Priest to Modern Physician. Baltimore: The John Hopkins University Press Ltd; 1997.

48. Kleinman A, Guess HA, Kusek JW, Engel LW. The Science of the Placebo: Toward an Interdisciplinary Research Agenda. London: BMJ Publishing Group; 2002.

49. Final Report of the National Placebo Working Committee on the Appropriate Use of Placebos in Clinical Trials in Canada, Health Canada and the Canadian Institutes for Health Research, July 2004, www.cihr-irsc.gc.ca/e/25139.html (accessed May 23 2005).

50. Huston P, Peterson R. Withholding proven treatment in clinical research. N Engl J Med 2001; 345:912–14.

51. Freedman B, Glass K, Weijer C. Placebo orthodoxy in clinical research. II: Ethical, legal, and regulatory myths. J Law Med Ethics 1996; 24:252–9.

23

Designs for Trials when there is a Standard Therapy: Superiority, Equivalence, Non-inferiority

David L Streiner

Introduction

Despite their limitations, cholinesterase inhibitors are now generally accepted as the standard of care for patients with mild to moderate Alzheimer's disease (AD). Consequently, new compounds that are being tested for AD usually would not ethically be permitted to have placebo control arms, as to do so would violate the standard that efficacious treatment should not be withheld from patients.[1] While there may be special circumstances under which a placebo control condition is possible (e.g. if the duration of the trial is short, if stability without a cholinesterase inhibitor has been demonstrated, or if the patient is intolerant of or unwilling to try cholinesterase inhibitor therapy), the question remains how, in the majority of cases, the efficacy or effectiveness of new compounds should be tested.

Clearly, to determine effectiveness and efficacy, these new drugs, like any other intervention, would be subjected to randomised, controlled trials (RCTs).[2,3] Specifically, patients who meet the inclusion criteria are assigned at random to receive either the new, investigational treatment or a comparison intervention, a procedure essential to understanding whether the new compound produces clinically meaningful benefits.[4] Comparison of a new compound with an active one has

special considerations, however, which are the focus of this chapter. To understand them, however, it is useful to first review why placebo-controlled trials are so common, and to look at variations of the usual design of a treatment versus a placebo.

Use of placebos in controlled trials

RCTs have many well-known advantages, but perhaps chief amongst them is that, in theory, they allow treatment effects to be isolated by creating groups that closely resemble each other in all relevant ways, except for their treatment assignment. Using the terminology of Cook and Campbell, randomisation increases the *internal validity* of the study; that is, were the results due to the intervention or to other, spurious, causes?[5] The threats to the validity of the results (for example, whether certain pre-existing cognitive, behavioural or functional variables are associated with which drug a person receives, and hence the outcome) can be summarised as confounding, bias and chance. By balancing these baseline variables across groups, randomisation reduces the effects of confounding, of both known factors (e.g. stage of disease, referral to a particular clinic) and – more importantly – to unknown ones.[3,4] Bias is mitigated through blinding of anyone whose knowledge of group assignment could influence the outcome; and statistical tests (most of which are based on the assumption of randomisation) allow the play of chance to be evaluated formally and with reasonable precision.[6] Consequently, placebo-controlled trials provide the advantages of efficiency, and clear evidence about effectiveness.

Efficiency is related to the *power* of a statistical test; that is, its ability to show a significant difference when one actually exists. Research designs and statistical tests that are efficient need fewer study participants to yield statistical significance than non-efficient ones, given the same magnitude of an effect. Importantly, however, while the benefits of randomisation also occur with other types of trials, their efficiency is often lower, meaning that the number of participants needed to show a statistically significant effect is very often much larger, as will be discussed later. Similarly, while the results of a placebo-controlled trial are usually unequivocal, the answer often is not as clear-cut when the comparison group is receiving an active agent.

Use of active comparators in controlled trials

When an active agent is used as the comparison condition, there are three alternative questions that can be posed: is the new treatment superior, equivalent to, or not inferior to standard therapy?[7] As discussed below, each of these questions has importance nuances in how they are answered, and in how the answers are understood.

Superiority trials

The question being asked in a superiority trial is, unsurprisingly, whether the new drug is better than the standard one. 'Better' can mean either a greater therapeutic effect, fewer adverse events, or even less cost. In the present context of dementia studies, it is reasonable, for a variety of reasons, to focus on the first, although the arguments apply equally well to all three benefits. A superiority trial is predicated on the belief that one treatment is better than the other. In theory, this might require only a one-sided test of significance to evaluate the differences in effects between the two groups. This results in a lower sample size being needed to demonstrate statistical significance, but that gain comes at a cost which experience suggests is unacceptable. Briefly, strictly applied, 'significant' results in the opposite direction would have to be dismissed as chance findings. Experience in other therapeutic areas suggests that this would be unwise; any attempt to 'explain away' the contrary finding is an admission that the effect is real, and thus the rationale for the one-tailed hypothesis has been violated (i.e. it is due to chance). For example, in both the Cardiac Arrhythmia Suppression Trial (CAST) and the Finnish Trial, drugs that were tested to reduce ventricular arrhythmias and lower cardiac risk factors, respectively, resulted in study patients dying at a rate between 1½ and 3½ times that of the control group.[8,9] In short, while it might seem logical to posit that an intervention can only be helpful, experience shows that it is wise to consider that new interventions might, in fact, be harmful. Given that any drug can cause unexpected harm – witness the recent withdrawals of rofecoxib and hormone replacement therapy – superiority trials are almost never seen in drug studies using an active comparison therapy.

Equivalence trials

The goal of an equivalence trial is to demonstrate that two drugs are about the same, *within a given margin*. The rationale for proposing that drugs are equivalent within a given interval rather than that they are exactly equivalent is predicated on two facts. First, given a sufficient sample size, any difference, no matter how small, can be shown to be statistically significant; and second, some differences, even though statistically significant, may be clinically trivial.[10] For example, if we had two groups with 390 subjects in each, a difference between them of two points on the Brief Psychiatric Rating Scale (BPRS) would be statistically significant at $P < 0.05$, even though a difference this small (i.e., about one-fifth of a standard deviation) would be considered to be meaningless. As the interval gets smaller (i.e. the two drugs must be more similar), the sample size increases quite rapidly. In general, an effect size no larger than 0.10 (and in any case, not as big as 0.15) would fit the requirements of being large enough for the sample size to be feasible, yet small enough for the difference to be clinically unimportant. Trials with properly estimated equivalence intervals can be very large (and thus costly) and for this reason are often not conducted.

Note that an essential aspect of understanding the design of an equivalence study is that the roles of type I and type II errors are reversed. Recall that, with a type I error, we reject the null hypothesis when it is true (i.e. we accept as significant a difference which is not); whereas with a type II error, we accept the null hypothesis when it is false (i.e. we accept as not significant a relationship which actually is). In an equivalence design, niceties of the null hypothesis notwithstanding,[10] what we really mean to do is show that the drugs are the same; the real error that we wish to avoid is concluding that the drugs have the same effects when they do not. Consequently, we usually set α = 0.20 and β = 0.05.

Non-inferiority trials

In a non-inferiority trial, the hypothesis to be tested is that the new treatment is no worse than the existing standard therapy. The null hypothesis in a non-inferiority trial can thus be framed such that the new drug is worse than the standard one by at least some given amount, again expressed as an interval. The alternative hypothesis is that the superiority of the standard drug does not exceed this. Again, the roles of type I and type II errors are reversed in non-inferiority trials, and a one-tailed test is used.[10]

Problems with equivalence and non-inferiority trials

Despite their popularity, there are a number of problems with both equivalence and non-inferiority trials. First, the size of the non-inferiority region, I,

is quite arbitrary; what may be a clinically unimportant difference to one person could be quite important to another. Compounding this difficulty, the sample size is highly dependent on I;[10,11] as it shrinks (that is, the new drug must be less and less inferior to the standard), the sample size increases quite rapidly. Thus, there is great incentive to make the interval as large as possible.

This in turn leads to the second problem. The new drug can be less effective than the standard, but still pass the test of non-inferiority. It is easy to imagine a series of three or four trials, in which the new drug of one becomes the standard of the next. If, in each case, the new drug is worse than the standard, but not significantly so, the effectiveness of the interventions will decline from one study to the next.

The third problem is related to the previous two, and can be much more serious. In a placebo-controlled trial, there are two alternative outcomes: the new drug is better than the placebo, or it is not. That is, the results are unequivocal (ignoring type I and type II errors). This is not the case when a new drug is compared against an active comparison. Again there are two alternatives: the new drug is significantly worse (because, as has been said, it is highly unusual to test for superiority), or it is equivalent. If it is worse, there is no problem interpreting the findings (assuming they ever see the light of day); the new drug is worse than the old one. But, if the results show equivalence or non-inferiority, there are two possible reasons – both are equally effective, *or both were equally ineffective in this particular study*, and there is no way to determine which is the case.

We may like to assume that if the comparison drug has been shown to be effective, then there is no doubt – both drugs in the trial must have had some positive effect. However, this assumes that all trials were well conducted: that they have enrolled only participants who are likely to respond to the drug; that there are a sufficient number of people so that the study is not underpowered; that it was carried out competently (e.g. few people missing appointments, being erroneously given the wrong drug; not dropping out of the trial and so forth); that the outcome measures were both appropriate for the outcome of interest and administered in a reliable way; and on and on. Unfortunately, this is not always the case. For example, despite the fact that the efficacy of mono-amine oxidase inhibitors (MAOIs) for depression has been shown repeatedly, a large trial sponsored by the British Medical Research Council, and where one of the principal investigators (A Bradford Hill) was the person regarded as the father of the RCT, failed to find any effect of phenelzine, due to the enrolment of the wrong types of patients, inadequate levels of the drug, too short a duration of treatment, and an inappropriate outcome measure.[12] Similarly, Peet et al were unable to demonstrate any effect of either propranolol or chlorpromazine with schizophrenic patients, most likely because their sample size was woefully inadequate.[13,14] As mentioned above, the roles of type I and type II errors are reversed in equivalence and non-inferiority trials. Consequently, it is possible to show equivalence merely by running a poorly designed, badly executed, and low-powered study.[10,11]

The lesson is that even trials that are led by an experienced researcher, which use a proven drug, and are published in prestigious journals may be faulted on one or more grounds. Consequently, we cannot assume that a study that demonstrates no difference between a new drug and a standard has shown the equal efficacy of both. It is also possible that the drugs were equally ineffective in this particular trial, and in the absence of a placebo group, we often cannot determine which situation obtains.

Other considerations

In recent years, there has been a move to replace placebo-controlled trials with add-on designs; that is, comparing treatment as usual (TAU) with one drug versus TAU plus a second drug.[15] This leads to problems in sample size, but as we shall see shortly, makes the design of studies more straightforward. Because the TAU group is expected to improve more (or decline less) than a placebo group, it may be much more difficult to demonstrate the superiority of an additional intervention, because the patients may be closer to any physiological limit in terms of how much they can improve (or slow in their decline). In order to demonstrate statistical significance with smaller differences, larger sample sizes are required. For example, to show a statistically significant difference with an effect size (ES) of 0.5, 63 patients per group are required. If the ES is only 0.25, though, the required sample size is nearly four times as large – 252 in each group. Moreover, because the potential for drug interactions and an increased incidence of adverse events may make the combination worse than TAU, two-tailed tests should be mandatory, obviating the advantage of superiority trials.

Equivalence and non-inferiority trials similarly become meaningless with add-on trials. There is no sense in adding another drug, with all of the resultant costs and opportunities for adverse events and interactions, if the only consequence were to show that the combination is the same as or not worse than taking a single medication. Thus, with add-on trials, the only meaningful approach is a traditional two-group parallel study.

Comparative studies are common in cancer chemotherapy, where experience sometimes has been gained the hard way. A particularly salutary lesson came from the early experience with trying to improve upon the, to then, remarkable success of curative combination chemotherapy for patients with advanced stages of aggressive non-Hodgkin's lymphoma.[16] The so-called first generation of combination chemotherapy was with cyclophosphamide, doxorubicin, vincristine and prednisone, known as CHOP. This success led others to attempt to improve on it with newer combinations that were usually more expensive, often more toxic, and with improvements that were difficult to measure. Initial results seemed to demonstrate superiority, as survival with the newer combinations was usually in the range of 55 to 65%, compared with about 50% with CHOP, with the latter estimates usually coming from historical controls. However, when the Southwest Oncology Group carried out a sufficiently large and sufficiently unselected trial that compared the regimens to each other, and to CHOP, a different picture emerged. Overall, the three-year survival rates for the various regimens were about the same, at 50 to 54%. The highest survival was found with CHOP, in part because it had the fewest fatal toxic reactions (1%, compared with a range of 3–6% in the other groups). This led to CHOP again being established as the best available treatment for patients with advanced-stage intermediate-grade or high-grade non-Hodgkin's lymphoma, with trials from then on using it as the usual standard.[17] Still, not all lessons take, however, and even though CHOP chemotherapy became re-established as the best available treatment, five years later, the practice of using CHOP historical controls had recrudesced.[18] This was the case even though other studies with concurrent CHOP controls were showing no advantage of newer modalities.

Comparison trials and clinical meaningfulness

Clearly, the introduction of new compounds raises the question of whether they confer clinically meaningful benefits compared with standard treatment. In general, each of the above considerations apply: is the benefit meant to be in comparison to the new treatment, or in addition to it? As we have argued, if the comparison is to standard therapy, the only meaningful question is superiority, because there is no advantage in adding a new drug simply to get comparable results. That clinical meaningfulness in each of these contexts first requires that the studies be methodologically sound, and second that the results be statistically significant are not arguable, but once those have been determined, what other considerations should obtain?

A difference is only likely to be clinically meaningful if it is first detectable.[6] Detectable differences are appreciable both quantitatively and qualitatively. If the difference is readily detectable without special expertise or instrumentation (for the sake of argument, if it has an effect size more than ~0.5, which Cohen calls 'moderate'[19]), it is likely to be clinically meaningful. It would also be meaningful if what is detected clinically conforms to a known pattern that has a plausible biological basis (for example, if the pattern is one of improved executive function, and if the drug is known to favourably affect frontal-subcortical circuitry). In either case, the requirement of being able to detect effects will be more readily met if there is some sense of which effects are being sought. The latter is perhaps more an instrumentation issue than a design one, but also suggests that comparison trials should allow for important perspectives (including patient preferences) to be addressed.

Conclusions

There is little question that, when effective treatments exist, there are major ethical issues when new drugs are tested against placebos. However, there are many methodological problems when the comparison group consists of an active treatment:

- poorly executed trials with low power can be mistaken for 'proving' equivalence
- when the two arms yield comparable results, there is no guarantee that either one was effective in that particular trial
- there may be a tendency (conscious or unconscious) to use wide equivalence intervals to decrease sample size
- successive non-inferiority trials may lead to a gradual reduction of effectiveness
- superiority trials and equivalence trials with narrow interval require large sample sizes.

These difficulties raise ethical concerns themselves; not only from possibly erroneous findings, but also, if larger sample sizes are needed, from having more patients on a possibly less effective treatment and a delay in getting the drug to market.

It would be fatuous to say that all of these problems would be eliminated if only excellent trials were conducted. Such pleas have been made ever since RCTs were first run, to little avail; poorly designed, underpowered studies appear to be as prevalent now as when Cohen first said that most studies did not have sufficient power to test their main hypotheses.[20,21] My recommendation would be that, when an existing therapy exists, and if certain conditions obtain:

- studies should consist of three arms: the new drug, the existing drug and a placebo group
- the study should be adequately powered to detect a clinically important difference in superiority trials, or to rule out a type II error in equivalence and non-inferiority trials between the two drug arms
- the placebo arm need only be large enough to determine that the study as a whole was successful (i.e. to detect a difference between it and the pooled effect of the two treatments).

The conditions that should apply would include:

- the placebo arm should be as brief as possible
- the patient's condition is not expected to deteriorate rapidly
- patients are withdrawn if their deterioration 'is greater than that expected for normal clinical fluctuation in a patient with that diagnosis who is on standard therapy'[22]
- patients are automatically withdrawn if they begin to exhibit behaviours that may be dangerous to themselves or others, 'even if there is not sufficient deterioration in the overall monitoring to trigger disenrolment'[22]
- there is full and informed consent from the patient and/or the substitute decision maker.

A third arm would increase sample size somewhat, and does not address the issues of the size of the equivalence interval, or the potential gradual reduction in effectiveness, but would solve the major problem of the ambiguity of no difference. While placebo-controlled trials may be unethical, it is even more unethical to conduct active control studies when they can be scientifically meaningless or misleading.

References

1. Medical Research Council of Canada. Tri-Council Policy Statement: Ethical Conduct for Research Involving Humans, 2003. *http://pre.ethics.gc.ca/english/policystatement/policystatement.cfm* (accessed February 16 2005).

2. Feinstein AR. Clinical biostatistics: XLVIII. Efficacy of different research structures in preventing bias in the analysis of causation. Clin Pharmacol Ther 1979; 26:129–41.

3. Streiner DL, Norman GR. PDQ Epidemiology (2nd edn). Toronto: BC Decker; 1996.

4. Rockwood K, MacKnight C. Interpreting the clinical meaningfulness of statistically significant differences in anti-dementia drug trials. Neuroepidemiology 2001; 20:51–6.

5. Cook TD, Campbell DT. Quasi-experimentation: design and analysis issues for field settings. Boston: Houghton Mifflin; 1979.

6. Norman GR, Streiner DL. Biostatistics: the Bare Essentials (2nd edn). Toronto: BC Decker; 2000.

7. Fleischhacker WW, Czobor P, Hummer M et al. Placebo or active control trials of antipsychotic drugs? Arch Gen Psychiatry 2003; 60:458–64.

8. Akiyama T, Pawitan Y, Greenberg H et al. Increased risk of death and cardiac arrest from encainide and flecainide in patients after non-Q-wave acute myocardial infarction in the Cardiac Arrhythmia Suppression Trial. CAST Investigators. Am J Cardiol 1991; 68:1551–5.

9. Strandberg TE, Salomaa VV, Naukkarinen VA et al. Long-term mortality after 5–year multifactorial primary prevention of cardiovascular diseases in middle-aged men. JAMA 1991; 266:1225–9.

10. Streiner, DL. Unicorns *do* exist: a tutorial on 'proving' the null hypothesis. Can J Psychiatry 2003; 48:756–61.

11. Steiger JH. Beyond the *F* test: effect size confidence intervals and tests of close fit in the analysis of variance and contrast analysis. Psychol Methods 2004; 9:164–82.

12. Clinical Psychiatry Committee. Clinical trial of the treatment of depressive illness: report to the Medical Research Council. BMJ 1965; 1:881–6.

13. Peet M, Bethell MS, Coates A et al. Propranolol in schizophrenia: I. Comparisons of propranolol, chlorpromazine and placebo. Br J Psychiatry 1981; 139:105–11.

14. Streiner DL. Propranolol in schizophrenia [letter to the editor]. Br J Psychiatry 1982; 141:212–13.

15. Tariot PN, Farlow MR, Grossberg GT et al. Memantine treatment in patients with moderate to severe Alzheimer Disease already receiving donepezil. JAMA 2004; 291:317–24.

16. Fisher RI, Gaynor ER, Dahlberg S et al. Comparison of a standard regimen (CHOP) with three intensive chemotherapy regimens for advanced non-Hodgkin's lymphoma. N Engl J Med 1993; 328:1002–6.

17. Cameron DA, White JM, Proctor SJ et al. CHOP-based chemotherapy is as effective as alternating PEEC/CHOP chemotherapy in a randomised trial in high-grade non-Hodgkin's lymphoma. Scotland and Newcastle Lymphoma Group. Eur J Cancer 1997; 33:1195–201.

18. Cabanillas F, Rodriguez-Diaz Pavon J, Hagemeister FB et al. Alternating triple therapy for the treatment of intermediate grade and immunoblastic lymphoma. Ann Oncol 1998; 9:511–18.

19. Cohen J. Statistical power analysis for the behavioral sciences. Hillsdale NJ: Lawrence Erlbaum Associates; 1988.

20. Cohen J. The statistical power of abnormal-social psychological research: a review. J Abnorm Soc Psychol 1962; 65:145–53.

21. Maxwell SE. The persistence of underpowered studies in psychological research: causes, consequences, and remedies. Psychol Methods 2004; 9:147–63.

22. Orr JD. Guidelines for the use of placebo controls in clinical trials of psychopharmacologic agents. Psychiatr Serv 1996; 47:1262–4.

24

Conclusion: Lessons from Clinical Trials in Dementia

Kenneth Rockwood and Serge Gauthier

Introduction

Conducting a drug trial in patients who have Alzheimer's disease is fundamentally an act of hope, although exactly what people hope for varies by their interests and expectations. What most patients really would like – as one once told me: 'I want my old life back' – is only very rarely achieved. How we conceptualise the gap between what people want, and what can be achieved has been a subtext of many chapters herein. Its explicit consideration – rather than a summing up of all possible lessons learned – is how it might be best to conclude.

The shared hope held by all participants in the drug trials process – that patients do, in fact, get their lives back – provides only a small basis for consensus about what we aim for in the meantime. It is possible to be entirely cynical: drug companies want only profits, scientists acclaim, physicians acclaim and money, regulators compliance, families and patients fantasies that they cannot have, and funders for it all somehow to go away. Such cynicism would not entirely be groundless, but even a cynic would acknowledge that just focusing on the drugs now or soon to be available, there are important obstacles to helping patients get their lives back. While we know that some patients respond, we do not know who does, or for how long, or why, or even what we mean by 'respond'. Credible long-term data are lacking, as are comparative data: for example, thus far the head-to-head trials of cholinesterase inhibitors have consistently favoured the companies that funded the trials. Whether other trials exist, but have not seen the light of day is not clear, although publication bias operates to our certain personal knowledge, in that we have taken part in negative trials of several compounds whose publication has not yet seen the light of day, despite efforts to the contrary.[1] In this context, the effort to register trials if they are to be published in the major medical journals is an important one.[2]

Interpreting the internal validity of dementia clinical trials

Scepticism is not pointless if it can give rise to advances in our understanding. In this context, an important challenge has been issued by the results of the AD2000 trial. This English, two-year placebo-controlled study of donepezil showed no difference between treatment assignment groups in the rates of institutionalisation, and was vigorously presented in the media by its principal investigators as a 'negative' study.[3] AD2000 was not a perfect study (e.g. it was much smaller and took much longer to complete than intended, it did not incorporate patient preferences, it is not clear what choice patients had to participate in the trial, drop-out from the trial was not at random).[4] Strikingly to me, for a disease that induces so much misery, the tone of the write-up – that the study had failed – seemed more gleeful than rueful. Interestingly, it dismissed as irrelevant the statistically significant differences detected by even the crude tests of cognition and function that the study deployed, arguing that the numerical differences were too small, and, curiously, not considering standardised effect sizes. Still, the study is very useful in clarifying some of the challenges to be faced in dementia drug trial development, execution and expectation.

An important finding from AD2000 was that the investigators were unable to identify a group of consistent responders. Instead, they found that patients who initially did well were at risk of doing badly later. The most parsimonious explanation for this was that response is largely illusory, and represents only random fluctuation, with subsequent decline reflecting regression to the mean. This is a provocative hypothesis, and one clearly within the ability of the companies who maintain the clinical trials databases to test: how often does an apparent initial response presage later worse decline? Relatedly, who maintains the databases must change, and will need to do so if new guidelines for reporting clinical trials are to be followed.[2] Authors of reports from pivotal trials need to have some ability to re-analyse data published under their name, so that important hypotheses, such as regression to the mean of initial responders, can be tested.

One strength claimed of the AD2000 paper was that patient recruitment was more representative than in the usual clinical trail, which commonly excludes many people who would be treated in practice – and who, it seems do worse.[5] That representativeness is important is clear, but how it might be achieved is not. Regulatory requirements to be sure that a given drug works in a given indication can be at odds with including representative patients who often have more illnesses and are on more medications than analytical parsimony would allow. Importantly too, it is not clear in the AD2000 paper how the so-called 'uncertainty principle' played out. The authors stated that 'The doctor had to be substantially uncertain that the individual would obtain a worthwhile benefit clinical benefit from donepezil, taking into account the available evidence and clinical circumstances'. But at what level did the uncertainty exist? Did some doctors believe that it held only for some patients, or for all patients? For example, did different doctors enrol patients in different ways? Would such differences have been washed out in the randomisation? Could potential drug trial patients have a reimbursement choice? Against the 'real world' generalisability of AD2000 enrolment was the practice of multiple drug washouts, which would be quite uncommon in many parts of the world, and so too would be the high rate of institutionalisation in that study. Still, the study has strengths in employing a retrieved drop-outs design, in following people for up to two years and in following them to a 'hard' endpoint (institutionalisation) although it is not clear why, in contrast to therapeutics in other areas, treatment for AD must be found to be cost saving.

In addition to the challenges posed by AD2000, what other lessons should we take from this initial experience with dementia trials? Recalling that, on first principles, the purpose of the trials is to add to the evidence base in making valid claims about treatment effects, we can consider lessons about confounding, bias, chance and external validity. Broadly speaking, confounding occurs when two factors that are compared are undetectably associated with a third. In this context, it is worth considering that patients in both the active treatment and comparison groups share the factor of receiving treatment in a clinical trial. Inferences back to the general population of people who might be treated must consider that the analogies of both what happens with active treatment (reversal, slowing) and what happens without it (relentless progression) will often be inexact. In this context, some consideration of the experience of the placebo response in anti-dementia trials is salutary. Blass and Rabins, in their chapter (Chapter 18), drew attention to studies of people who received placebos in controlled trials of neuroleptics for behavioural and psychological symptoms of dementia (BPSD), noting an approximate 40% response rate to placebo (cf. 60% to active treatment). While they note that the high placebo response might be explained in part by a Hawthorne effect related to the study personnel and procedures, resulting in, amongst other things, an interested, perhaps optimistic but clearly therapeutic milieu. They also draw attention to other factors, including the importance of the benign natural course of BPSD in many individuals. As raised elsewhere,[6] both clinical and population-based studies allow the possibility of a slowly progressive form of AD. Might these patients be preferentially selected in drug trials?

Both the chapter by Hogan (Chapter 6) and that by Blass and Rabins (Chapter 18), as well as recent editorials,[4,7] consider recruitment or enrolment bias in the way that drug trials include and exclude patients, which has implications for the external validity/generalisability of the data. Clinical drug trials attempt to limit concomitant medications as a matter of routine trial design, so as to simplify interpretation of results by eliminating possible confounds and by limiting group heterogeneity. However, this practice significantly deviates from usual clinical care, where the use of more than one pharmaceutical agent is often necessary.

Use of more than one medication is certain to be an emerging trend in cognitive trials for cognitive effects and has methodological implications. For example, as elaborated in the chapter by Michael Borrie and Matthew Smith (Chapter 17), memantine most recently has been tested in combination with donepezil.[8] (On the other hand, the Forest Laboratories website (*www.frx.com*) reports on a large trial in which memantine was used in combination with a cholinesterase inhibitor (or more than one; the details are not spelled out) with no additional benefit to the combination.) If combination therapy can be reliably shown to be better than single treatment, this seems a worthwhile advance and a necessary improvement on placebo-controlled trials of more than a very short duration. Comparison studies can increase the level of complexity of studies – both in design (as reviewed in Chapter 23 by David Streiner) and in preclinical study requirements. As noted in consideration of the placebo response, however, dementia trials have long had the co-intervention of attentive care directed towards dementia symptoms. In consequence, the impact of substituting an active comparator group for a placebo group might well be attenuated in the context of dementia trials. Still, the use of an active comparison group is likely to require an increase in sample size to account for diminution of the effect size.

The responsiveness, or sensitivity to change, of the measures is an aspect of effect size and compares the signal coming from the groups (e.g. the mean difference in the value of the primary outcome between the intervention and comparison groups) to the noise (e.g. the pooled standard deviation of the change scores). This is an area of inquiry that has been comparatively under-appreciated, perhaps in consequence of the widespread description of raw difference scores (e.g. a two-point difference between treatment and control groups in the Alzheimer's Disease Assessment Scale cognitive subscale (ADAS-Cog)) as 'effect sizes'. As argued elsewhere (Chapter 7),[9,10] sensitivity to change is a crucial issue in drug trials, which measure differences in change scores. Often, the change in the raw score is interpreted only in the context of what might have been inferred from the untreated natural history, which is clearly inappropriate, due to the non-comparability between all-comers to clinics in the era before treatment, and patients selected for clinical trails. Other times, sceptics point to the theoretical range of the scale, arguing that a 3-point change on a 70-point scale (in the case of the ADAS-Cog) is inherently meaningless because it is too small to detect. Clinical detection, however, is not just a function of the signal, but also of the noise, and an apparently small signal (e.g. a 3-point change) can be readily detected if the noise is also small (e.g. a 5-point standard deviation). In this case, the ratio of the two is 0.60, which would be within the range of clinical detection by reasonably knowledgeable persons without special instrumentation. (This is also an achievable effect size with higher doses of cholinesterase inhibitors, for example.) As noted in Chapter 19, effect sizes have been widely used in producing quantitative syntheses in the caregiver literature (which shows a scale of treatment effects of ~0.15, or just below the threshold of clinical detection, to ~0.60, or well within that range). These results are only slightly lower than what is seen in the cholinesterase inhibitor trials,[10] including AD2000, where the effect sizes were in the range of 0.28–0.35.[3]

The ratio of signal to noise is also a concern in understanding the impact of missing data in the interpretation of clinical trials. For example, as discussed in Chapter 1, the supposedly conservative method using an intention-to-treat analysis, of following the last observation carried forward, can introduce bias in trials of drugs used in neurodegenerative disorders. In general, apart from caveats about missing data, it would seem reasonable to claim that the majority of published clinical trials meet the usual methodological standards, there are two exceptions. The first, as noted, are the head-to-head studies comparing drug A to drug B. A second group of studies in which there are legitimate concerns about their internal validity are those of long-term outcomes. Typically, the comparison group consists of historical controls – either estimates from untreated natural history studies, or estimates from the placebo arm of an unpublished, two-year controlled trial of an earlier unsuccessful compound. Neither group is particularly satisfactory, for reasons that are not surprising. That they are still used reflects a perceived lack of alternatives, but more advanced computational techniques – or even the production of confidence intervals for the comparison group, perhaps derived through bootstrapping – might

offer new insights in forecasting the range of alternative outcomes that might be available in patients who had not been treated.

One additional substantive factor needs to be considered, apart from the usual methodological cavils in studying long-term data, and that is equality of outcomes. This is perhaps best appreciated in the context of studies of dementia prevention, as in the so-called 'conversion' of people with mild cognitive impairment (MCI) to becoming patients with dementia. Reflecting most current practice – the approach of Morris et al in having classes of apparently mild dementia is an exception[11] – dementia is diagnosed dichotomously, but is actually a continuum. Also, the dementia to which they convert after treatment with donepezil might not be the dementia to which they convert after treatment with placebo. A similar concern obtains in understanding the validity of findings from trials in patients with moderate–severe dementia. Are the measures of disease progression in patients who are drug naïve the same as those in patients who have been on treatment but still arrive at a more advanced stage of dementia?

External validity and clinical meaningfulness of dementia trials

Even if we take into account these caveats about confounding, bias and chance, as well as substantive concerns in understanding the validity of clinical trials in dementia, there is still the question of external validity. Generalisability is not just a matter of whether the patients in trials have enough in common with the patients in practice. It is also closely linked to the question of clinical meaningfulness. This is vital, because if we recognise that we will not soon have a cure, then what we must settle for is clinically meaningful benefit. But what is that? As we have seen, it is different from the achievement of statistically significant differences (favouring treatment) in aggregate endpoints of well-conducted, valid clinical trials. Elsewhere, we have argued that clinical meaningfulness requires valid trials which show internal and external consistency, a dose–response effect, biological plausibility, and, especially, achievement of meaningful endpoints.[12]

Achievement of meaningful endpoints is a special challenge to trials in dementia, unlike, for example, trials in acute myocardial infarction, where mortality differences (binary, verifiable and non-arbitrary) are taken to have self-evident meaningfulness. The challenge is not unique, however; 'quality-of-life' considerations are paramount in many therapeutic areas, and few people would warrant a clinical trial in dental analgesia, for example, that did not measure patients' symptoms in favour of more 'objective' measures, such as heart rate, palmar sweat and pupillary dilatation. But in dementia, we must be concerned about the effect of the treatment on the essential aspects of an individual's sense of self, and, to some non-trivial extent, the ability of other people to recognise that sense of self in that person, a point to be pursued further, below.

The concern about understanding the clinical meaningfulness of therapeutic interventions in dementia is not limited to drug treatment, however. In a systematic review of interventions for caregivers, Schulz and colleagues found that, despite many studies demonstrating small–moderate statistically significant effects, very few studies had achieved 'clinically meaningful outcomes' such as impacts on service use, institutionalisation and the mental health of caregivers.[13] Despite this, many interventions were highly valued by caregivers. Why might this be?

Clinical trials as probes of brain function

When families tell us that one effect of using a cholinesterase inhibitor is that patients are 'more like themselves' and when physicians tell us that these are useful reports which inform their practice, then how we to interpret that information?[14] At present, such reports are nowhere captured by current measures – even the Neuropsychiatric Inventory (NPI) – used in dementia drug trials, so often there is no alternative to dismissing the reports as anecdotal. 'Sense of self' is not a traditional area of inquiry in dementia, and even in other areas often is seen in a social or psychological perspective, and not as reflecting specific brain injury.[15,16] Still, there is a neuroscience literature on this, which often derives from head injury, stroke or volunteer studies.[17,18] In short, we can learn from dementia drug trials, both to inform our understanding of disease, but also to gain insights into human cholinergic neurobiology. Indeed, to the extent that many of the gains which patients and caregivers describe are not easily modelled in preclinical studies, this is an essential – if usually unexploited – opportunity.

The sorts of studies – careful, descriptive, patient-driven, with open-ended questions and exploratory, qualitative analyses – that might inform inquiries into human cholinergic neurobiology, also could be used to achieve other needed ends. For example, as Ballard has pointed out (Chapter 10) the results of studies of behavioural and psychological symptoms in dementia are broadly similar, but consensus still is required to establish what constitute the main features, and whether subsyndromal clusters exist. For example, he points out that apathy appears to be improved by cholinesterase inhibitor (ChEI) therapy. Similarly, Blass and Rabins (Chapter 18) have noted that apathy in dementia might be part of a depressive syndrome, or might be part of a distinct (dysexecutive) syndrome seen in dementia. In this regard, a recent trial of donepezil for the treatment of BPSD in patients with Alzheimer's disease is of interest.[19] Patients with an average NPI score of 25 (not as severe, on average, as earlier BPSD trials, but higher than the average scores of ChEI trials) were initially treated with donepezil, after which they were randomised to either withdrawal and treatment with a placebo, or continued treatment with donepezil. The study establishes the merit of using donepezil in some patients with BPSD, but also gives the potential of understanding symptom profiles that respond to cholinergic manipulation. Interestingly, improvements in anxiety, apathy, disinhibition and depression are also seen in other ChEI studies,[14] and suggest cholinergic modulation of prefrontal function that is clinically detectable. The use of ChEI studies as probes of brain function is thus an essential process to inform meaningful treatment studies, and much could be accomplished with secondary studies of existing datasets. As experienced investigators know, however, gaining access to existing datasets for investigator-driven secondary analyses is operationally very difficult.

The stakes are high, however, as there is a potential for important discovery. For example, Grady et al have proposed the existence of a prefrontal compensatory network in Alzheimer's disease.[20] Comparing mildly demented patients with healthy elderly people, they found that, whereas the latter preferentially employed left hemisphere circuits in semantic and episodic memory tasks, AD patients uniquely showed bihemispheric activation. This suggests that patients with Alzheimer's disease have compensatory ability, and provides a basis for understanding why a model of effective treatment might consider enhancement of compensation, rather than recovery to the present state, as an appropriate metaphor for undertaking the search for clinically detectable effects.[21]

Clinical trials and the human enterprise

The gap between what people want and what we are able to provide offers, too, some insights into the human enterprise. The gap is not always acknowledged, so that perhaps we have indeed been too satisfied with too little, and with this satisfaction opportunities have been lost. It is hard to argue that there is inadequate effort being expended in the search for effective treatment for Alzheimer's disease, but there are many salutary lessons to be had from what has happened up to now. For example, the propentofylline story appears to offer a cautionary tale about the regulatory process as much as it does about the potential for glial cell inhibition in Alzheimer's disease.

The experience with anti-amyloid vaccination is another cautionary tale, of, depending on one's perspective, hope, hype, hubris and hard luck. Briefly, the rationale for a vaccine against β-amyloid is the presumed toxicity of the peptide. The mechanism whereby the peptide is neurotoxic is not clear, with, as reviewed elsewhere, a range of theories having been put forward, including oxidative stress, intracellular calcium overload, induction of apoptosis, and a variety of immunological and inflammatory processes.[22,23] The proliferation of theories, and the relatively poor correlations between many amyloid markers and cognitive decline have meant that the evidence for the toxicity of β-amyloid is famously and sometimes bitterly disputed.[24]

Still, beginning with vaccination studies in a mouse model,[25] there was excitement that a specific, mechanistically inspired treatment for Alzheimer's disease would become available. Despite promising results in other species, and in very early trials, phase II trials were stopped early when 18 patents (of 298) who had been actively immunised developed a meningoencephalitis syndrome.[26]

The lessons to be drawn from the vaccine experience are still unfolding, but it is difficult not to be humbled by the complexity of putatively mechanistic therapeutics. For example, successful development of antibodies to the fibrillar form of β-amyloid would require additional steps for the β-amyloid to be cleared. Just how this clearance would take place is not yet well understood, but to the extent to which it would require glial cell activation, then an anti-inflammatory strategy, which might be undertaken on other grounds, would run counter to successful vaccination. Moreover, the extent to which even complete clearance of fibrillar amyloid would result in clinically detectable improvement, or what that improvement would look like, is not understood.[27] Notwithstanding these difficulties, attempts to proceed with a passive immunisation strategy by intravenous injection of immunoglobulins containing antibodies against β-amyloid are under way.[28]

If at times our attempts at therapy outstrip our understanding of how therapies work, it is ever more important that we make efforts to use the studies to improve our understanding. Here, there appears to be no remedy other than continued, determined effort – an effort that often requires belief as much as evidence. For example, while many will have turned away from an anti-inflammatory strategy given initial, disappointing results,[29] others are motivated to look for other anti-inflammatory ones – for example, those that might inhibit the activation of microglia or the complement system – or for combinations of drugs aimed at different inflammatory targets.[30] Similarly, although hydergine is now rarely used, there is an old literature, which, as the subject of a meta-analysis and Cochrane review, resulted in the conclusion that it showed 'significant treatment effects when assessed by either global ratings or comprehensive rating scales'.[31] Such an effect might make it a candidate for combination trials, undertaken by a group with a scientific and clinical interest, rather than an exclusively proprietary one.

Even so, academic groups do not have just scientific and clinical interests. In a remarkably frank conference funded by the New York-based Institute for the Study of Aging, academic drug discovery groups were noted to be underfunded, to lack infrastructure, to be more concerned with disciplines than with problems, and to be insufficiently collaborative.[32] One important issue with respect to the latter was that the peer review process was often regarded as inadequately interdisciplinary. In short, it is not just proprietary concerns, but sometimes also professional ones, that make it more likely for perverse effects (e.g. unexpected or unintended outcomes) to be denied than used as the opportunity to advance knowledge. For example, at the clinical level, a perverse outcome would be when patient and caregiver preferences do not coincide – for example, when apparent improvement in the patient (manifested as greater insight or greater activity) causes distress to the caregiver. Brodaty and Thompson considered this

in Chapter 12 as it gives rises to the question of who is the target of treatment. While it is appropriate to include outcome measures that relate to caregiver activities and burden, we must consider how we would use evidence that the prescription of a drug to a patient is associated with improvement in the caregiver. The authors of that chapter conclude that 'benefits to second parties alone cannot be grounds for prescribing medication'.

A recurring theme throughout the book has been the importance attached to individualised interventions, whether it is in flexible drug dosing regimens or in matching caregiver need to intervention strategy.[33] Indeed, individualisation has been raised in several contexts. For example, Charlesworth and Newman in Chapter 19 noted that 'the definition of what constitutes clinically significant change is likely to vary from carer to carer, and also within each carer over the course of the caregiving career'. Similarly, the chapter on function (Chapter 9) noted the difficulty in establishing different levels of baseline performance based on different lifelong abilities. Thus there is a pragmatic need to distinguish between patients who do not undertake a certain function because they no longer do it, and those who do not undertake it because they have never done it. But as important a conceptual and instrumental advance as it might be, just individualising assessments will not fulfil the requirement of clinical meaningfulness unless it also incorporates patient and caregiver preferences. Consider again the individualised measurement of function. For the patient who has been used to intellectual pursuits throughout a lifetime, the ability to again comb his own hair again may seem to be too diminished a level of performance to be acceptably meaningful to him, or to the people who love him, even if it is measured in a way that accounts for baseline and lifelong performance level. In short, there is a critical gap between the information gathered on the drugs as they are being studied and the information gathered on the drugs for them to be best used, which will not be resolved until patient preferences are taken into account.

To date we have focused largely on the experience of working with drugs that manipulate the cholinergic system, reflecting the collectively greater experience with those compounds. As noted at various points throughout the book, however, a range of other compounds have been studied, so that lessons also have been learned from oestrogen studies, anti-inflammatory studies, complementary medicines, cholesterol-lowering drugs, antibiotics from drugs that manipulate the renin–angiotensin system. Although no brief summary can do justice to the full experience in these areas, a few points can be made.

As a general rule, we count on a reasonable understanding of disease mechanisms to provide us with a working hypothesis, following which the strongest test is replication of a given trial. An essential step, of course, is to test the correct compound from a given class. This is a tricky area: when are given effects unique to specific drugs, and when are they more likely to represent class effects? A precise answer to this question might well come against two unknowns: we do not know enough details of disease mechanisms, and we often do not understand all relevant drug mechanisms. For example, although the initial trials with statins have shown no persuasive effect on dementia prevention,[34] it has recently been shown that some, but not all statins are also selective butyrylcholinesterase inhibitors.[35]

A debate is still ongoing about the merit in pursuing oestrogen treatment in dementia. Most physicians will have been quite discouraged by the results of the Women's Health Initiative and particularly the Women's Health Initiative Memory Study, which showed that one of the compounds studied increased the risk of dementia.[36–38] Some investigators, however, while conceding that the data are definitive about oral conjugated oestrogen, note some persuasive evidence from other oestrogen trials, so that there might still be room for other compounds, and for alternative strategies in the timing of their administration.[39]

Of course, it is not just success, but failure of studies that requires replication. For example,

while the results of the rofecoxib trial were disappointing,[29] epidemiological studies still suggest a role for other anti-inflammatories.[40,41] In addition, a number of methodological objections have been raised,[42] as well as mechanistic ones – for example, that the dose required to achieve a putatively mechanistic effect (β-amyloid-lowering) was insufficient.[43]

Even if the right drug is chosen, it must be tested in the right patients. For example, it initially appeared that patients treated for hypertension with candesartin might have been too well to capture important impacts on cognition, in a study in which this was a primary outcome.[44] Another important problem faced in that trial was co-intervention – that many patents wound up being on other antihypertensives in addition to candesartin. But subsequent assessment only of patients who did not receive add-on anti-hypertensive therapy after randomisation, i.e. patients that best reflected the original intention of the placebo-controlled trial, still showed no impact on cognition.[45] Whether this is because there was no specific effect of angiotensin receptor blockers on cognition, or whether another agent (e.g. losartan) might do better remains unclear.[46,47] Moreover, although it was noted that lowering blood pressure did not adversely affect cognition in the candesartin trial,[44] it remains unclear how to optimise blood pressure in patients with or at risk for dementia.[48] Still, as discussed above, the emphasis on studying the 'right' patients, can be a moving target, as the propentofylline experience demonstrates.

It is also important to remember that, as in other areas of medicine, the role of knowing disease mechanisms can indeed be overstated, so that drug trials can be an important resource for understanding pathophysiology. Recent studies with the Chinese herbal medicine 'Ba Wei Di Huang Wan',[49] and with the antibiotics doxycycline and rifampin,[50] if replicated, offer the possibility for new insights into disease mechanisms.

Conclusions

Despite the reservations that we have raised in this book, we have been witness to remarkable progress in the last 20 years. We have moved from a time when patients could not realistically be offered treatment, to the present time, when many patients show profound benefit, including years of effective stabilisation. If we are sometimes harsh critics it is because of our felt need, as clinicians, to have most patients benefit in the way that only many do now. The era of cholinesterase inhibition is not over, but it is not too soon to take stock of what we have learned in the process of now having close to a decade's experience with their widespread use. Colleagues in academia, clinical practice, basic science, drug regulation, public policy and industry have reflected carefully on how treatment for dementia might proceed. We all need to take as broad a perspective as possible in considering how we might do better in the future.

References

1. Rockwood K, Beattie BL, Eastwood MR et al. A randomized, controlled trial of linopirdine in the treatment of Alzheimer's disease. Can J Neurol Sci 1997; 24:140–5.

2. De Angelis C, Drazen JM, Frizelle FA et al. Clinical trial registration: a statement from the International Committee of Medical Journal Editors. CMAJ 2004; 171:606–7.

3. AD2000. Long-term donepezil treatment in 565 patients with Alzheimer's disease (AD2000): randomised double-blind trial. Lancet 2004; 363:2105–15.

4. Schneider LS. AD2000: Donepezil in Alzheimer's disease. Lancet 2004; 363:2100–2101.

5. Gill SS, Bronskill SE, Mamdani M et al. Representation of patients with dementia in clinical trials of donepezil. Can J Clin Pharmacol 2004; 11:e274–85.

6. Fisk J, Rockwood K. Outcomes of incident mild cognitive impairment in relation to case definition. J Neurol Neurosurg Psychiatry 2005; 76(8).

7. Herrman N, Knopman D. Donepezil therapy for neuropsychiatric symptoms in AD: Methods make the message. Neurology 2004; 63:200–201.

8. Tariot PN, Farlow MR, Grossberg GT et al. Memantime treatment in patients with moderate to secure Alzheimer disease already receiving donepezil; a randomised controlled trial. JAMA 2004; 291:317–24.

9. Rockwood K, Stolee P. Responsiveness of outcome measures used in an antidementia drug trial. Alzheimer Dis Assoc Disord 2000; 14(3):182–5.

10. Rockwood K. Size of the treatment effect on cognition of cholinesterase inhibition in Alzheimer's disease. J Neurol Neurosurg Psychiatry 2004; 75(5):677–85.

11. Morris JC, Storandt M, Miller JP et al. Mild cognitive impairment represents early-stage Alzheimer disease. Arch Neurol 2001; 58:397–405.

12. Rockwood K, MacKnight C. Assessing the clinical importance of statistically significant improvement in anti-dementia drug trials. Neuroepidemiology 2001; 20(2):51–6.

13. Schulz R, O'Brien A, Czaja S et al. Dementia caregiver intervention research: in search of clinical significance. Gerontologist 2002; 42:589–602.

14. Rockwood K, Black SE, Robillard A, Lussier I. Potential treatment effects of donepezil not detected in Alzheimer's disease clinical trials: a physician survey. Int J Geriatr Psychiatry 2004; 19:954–9.

15. Nochi M. 'Loss of self' in the narratives of people with traumatic brain injuries: a qualitative analysis. Soc Sci Med 1998; 46:869–78.

16. Nochi M. Reconstructing self-narratives in coping with traumatic brain injury. Soc Sci Med 2000; 51:1795–804.

17. Stuss DT, Anderson V. The frontal lobes and theory of mind: developmental concepts from adult focal lesion research. Brain Cogn 2004; 55:69–83.

18. Levine B. Autobiographical memory and the self in time: brain lesion effects, functional neuroanatomy, and lifespan development. Brain Cogn 2004; 55(1):54–68.

19. Holmes C, Wilkinson D, Dean C et al. The efficacy of donepezil in the treatment of neuropsychiatric symptoms in Alzheimer disease. Neurology 2004; 63:214–19.

20. Grady CL, McIntosh AR, Beig S et al. Evidence from functional neuroimaging of a compensatory prefrontal network in Alzheimer's disease. J Neurosci 2003; 23:986–93.

21. Rockwood K, Wallack M, Tallis R. The treatment of Alzheimer's disease: success short of cure. Lancet Neurol 2003; 2(10):630–3.

22. Gelinas DS, DaSilva K, Fenili D et al. Immunotherapy for Alzheimer's disease. Proc Natl Acad Sci USA 2004; 101 Suppl 2:14657–62.

23. Dodel R, Hampel H, Depboylu C et al. Human antibodies against amyloid beta peptide: a potential treatment for Alzheimer's disease. Ann Neurol 2002; 52:253–6.

24. Koudinov AR, Berezov TT. Alzheimer's amyloid-beta (A beta) is an essential synaptic protein, not neurotoxic junk. Acta Neurobiol Exp (Wars) 2004; 64:71–9.

25. Schenk D, Barbour R, Dunn W et al. Immunization with amyloid-beta attenuates Alzheimer-disease-like pathology in the PDAPP mouse. Nature 1999; 400:173–7.

26. Orgogozo JM, Gilman S, Dartigues JF et al. Subacute meningoencephalitis in a subset of patients with AD after Abeta42 immunization. Neurology 2003;61:46–54.

27. Hock C, Konietzko U, Papassotiropoulos A et al. Generation of antibodies specific for beta-amyloid by vaccination of patients with Alzheimer disease. Nat Med 2002; 8:1270–5.

28. Dodel RC, Du Y, Depboylu C et al. Intravenous immunoglobulins containing antibodies against {beta}-amyloid for the treatment of Alzheimer's disease. J Neurol Neurosurg Psychiatry 2004; 75:1472–74.

29. Aisen PS, Schafer KA, Grundman M et al. Effects of rofecoxib or naproxen vs placebo on Alzheimer disease progression: a randomized controlled trial. JAMA 2003; 289:2819–26.

30. McGeer PL, McGeer EG. Local neuroinflammation and the progression of Alzheimer's disease. J Neurovirol 2002; 8:529–38.

31. Olin J, Schneider L, Novit A, Luczak S. Hydergine for dementia. Cochrane Database Syst Rev 2001; (2):CD000359.

32. Fillit HM, O'Connell AW, Brown WM et al. Barriers to drug discovery and development for Alzheimer disease. Alzheimer Dis Assoc Disord 2002; 16(Suppl 1):S1–8.

33. Acton GJ, Winter MA. Interventions for family members caring for an elder with dementia. Annu Rev Nurs Res 2002; 20:149–79.

34. Rockwood K, Darvesh S. The risk of dementia in relation to statins and other lipid lowering agents. Neurol Res 2003; 25:601–4.

35. Darvesh S, Martin E, Walsh R, Rockwood K. Differential effects of lipid-lowering agents on human cholinesterases. Clin Biochem 2004; 37:42–9.

36. Shumaker SA, Legault C, Rapp SR et al. Estrogen plus progestin and the incidence of dementia and mild cognitive impairment in postmenopausal women: the Women's Health Initiative Memory Study: a randomized controlled trial. JAMA 2003; 289:2651–62.

37. Rapp SR, Espeland MA, Shumaker SA et al. Effect of estrogen plus progestin on global cognitive function in postmenopausal women: the Women's Health Initiative Memory Study: a randomized controlled trial. JAMA 2003; 289:2663–72.

38. Espeland MA, Rapp SR, Shumaker SA et al. Conjugated equine estrogens and global cognitive function in postmenopausal women: Women's Health Initiative Memory Study. JAMA 2004; 291:2959–68.

39. Asthna S. Estrogen and cognition: a true relationship? J Am Geriatr Soc 2004; 52:316–8.

40. Etminan M, Gill S, Samii A. Effect of non-steroidal anti-inflammatory drugs on risk of Alzheimer's disease: systematic review and meta-analysis of observational studies. BMJ 2003; 327:128.

41. Szekely CA, Thorne JE, Zandi PP et al. Nonsteroidal anti-inflammatory drugs for the prevention of Alzheimer's disease: a systematic review. Neuroepidemiology 2004; 23:159–69.

42. Breitner JC. NSAIDs and Alzheimer's disease: how far to generalise from trials? Lancet Neurol 2003; 2:527.

43. Gasparini L, Ongini E, Wenk G. Non-steroidal anti-inflammatory drugs (NSAIDs) in Alzheimer's disease: old and new mechanisms of action. J Neurochem 2004; 91:521–36.

44. Lithell H, Hansson L, Skoog I et al. The Study on Cognition and Prognosis in the Elderly (SCOPE): principal results of a randomized double-blind intervention trial. J Hypertens 2003; 21:875–86.

45. Lithell H, Hansson L, Skoog I et al. The Study on Cognition and Prognosis in the Elderly (SCOPE): outcomes in patients not receiving add-on therapy after randomization. J Hypertens 2004; 22: 1605–12.

46. Tedesco MA, Ratti G, Di Salvo G, Natale F. Does the angiotensin II receptor antagonist losartan improve cognitive function? Drugs Aging 2002; 19:723–32.

47. Gard PR. Angiotensin as a target for the treatment of Alzheimer's disease, anxiety and depression. Expert Opin Ther Targets 2004; 8:7–14.

48. Majeski EI, Widener CE, Basile J. Hypertension and dementia: does blood pressure control favorably affect cognition? Curr Hypertens Rep 2004; 6:357–62.

49. Iwasaki K, Kobayashi S, Chimura Y et al. A randomized, double-blind, placebo-controlled clinical trial of the Chinese herbal medicine 'ba wei di huang wan' in the treatment of dementia. J Am Geriatr Soc 2004; 52:1518–21.

50. Loeb MB, Molloy DW, Smieja M et al. A randomized, controlled trial of doxycycline and rifampin for patients with Alzheimer's disease. J Am Geriatr Soc 2004; 52:381–7.

Index